Protein Tyrosine Phosphatase Control of Metabolism

Kendra K. Bence
Editor

Protein Tyrosine Phosphatase Control of Metabolism

Springer

Editor
Kendra K. Bence
Department of Animal Biology
University of Pennsylvania
Philadelphia, PA, USA

ISBN 978-1-4614-7854-6 ISBN 978-1-4614-7855-3 (eBook)
DOI 10.1007/978-1-4614-7855-3
Springer New York Heidelberg Dordrecht London

Library of Congress Control Number: 2013945401

Printed on acid-free paper

Springer is part of Springer Science+Business Media (www.springer.com)

Foreword

Tyrosine phosphorylation is a rapid and reversible protein modification catalyzed by the activities of protein tyrosine kinases (PTKs) and their cellular counterparts, protein tyrosine phosphatases (PTPs). Although phosphorylation of proteins on tyrosine is relatively rare compared to phosphorylation on serine or threonine residues, the past 2 decades of research into PTP function have led to a great appreciation of the critical role PTPs have in regulating basic cellular processes. Among these important roles is the regulation of cellular signaling pathways related to metabolism. This volume contains chapters which highlight many aspects of PTP function in the context of metabolism. Given the growing obesity and diabetes epidemics in the United States and throughout the world, the desire to identify possible therapeutic targets for treatment of these diseases is a high priority. In many ways, PTPs may be attractive drug targets since they are amenable to targeting with small molecules; however many challenges abound in making PTP inhibitors.

PTPs are encoded by a large family of 107 genes, the majority of which can be broadly classified into classical phosphotyrosine-specific phosphatases or dual-specificity phosphatases (which display serine, threonine, and tyrosine phosphatase activity). More than half of identified PTPs have been implicated in human disease to date, with a growing number of PTPs now known to play major roles in metabolic disease. Many metabolic signaling pathways invoke a feed-forward cascade of tyrosine-phosphorylated proteins; thus, PTPs have emerged as critical regulators of these pathways, including the insulin and leptin pathways.

The activity of PTPs is regulated in many ways within the cell, most notably by reversible oxidation of the catalytic cysteine residue by reactive oxygen species (ROS). In Chap. 1, assays to quantify redox regulation of PTPs are discussed in the context of metabolic signaling. Subsequent chapters in this volume discuss quantitative modeling approaches that may be effective in modeling PTP behavior, and the importance of identifying novel substrates of PTPs. Several chapters highlight the role of PTPs known to regulate metabolic signaling, including PTP1B, SHP2, TC-PTP, and RPTP epsilon. Over the past decade, mouse models of PTP-deficiency have provided important insight into the precise tissue-specific functions of many PTPs including PTP1B, TC-PTP, SHP2, and PTEN (which also functions as an inositol phospholipid phosphatase). Perhaps the most well-characterized PTP with a known metabolic role is the prototypical classical, non-receptor PTP, PTP1B. As such, several chapters in this volume are dedicated to the specific metabolic functions of PTP1B. More recently the MAPK phosphatases, or MKPs, have also emerged as important regulators of metabolic homeostasis. Finally, two chapters in this volume discuss the role of the low molecular weight class of PTPs (LMPTP) and the glycogen phosphatase laforin in human metabolic disease pathogenesis.

Overall, recent studies into PTP function in the context of metabolism highlight the importance of understanding the regulation/modifications of PTPs that affect activity, the subcellular localization of PTPs and how that affects their function, and the cell-type specificity of PTP functions. Going forward, it will be important to understand how PTPs function at the intersection of metabolic signaling and other pathways regulated by PTPs, including growth factor signaling and oncogenic signaling pathways, in order to sustain the growing interest in targeting PTPs for treating metabolic syndromes.

Philadelphia, PA Kendra K. Bence

Contents

Contributors

Kendra K. Bence, Ph.D. Department of Animal Biology, School of Veterinary Medicine, University of Pennsylvania, Philadelphia, PA, USA

Anton M. Bennett, Ph.D. Department of Pharmacology, Yale University School of Medicine, New Haven, CT, USA

Massimo Bottini, Ph.D. Infectious and Inflammatory Diseases Center, Sanford Burnham Medical Research Institute, La Jolla, CA, USA

Nunzio Bottini, M.D., Ph.D. Division of Cellular Biology, La Jolla Institute for Allergy and Immunology, La Jolla, CA, USA

Eric J. Belin de Chantemèle, Ph.D. Department of Physiology, Georgia Health Sciences University, Augusta, GA, USA

Mirela Delibegovic, Ph.D. Institute of Medical Sciences, University of Aberdeen, Aberdeen, UK

Anna A. DePaoli-Roach, Ph.D. Department of Biochemistry and Molecular Biology, Indiana University School of Medicine, Indianapolis, IN, USA

Ari Elson, Ph.D. Department of Molecular Genetics, The Weizmann Institute of Science, Rehovot, Israel

Gen-Sheng Feng, Ph.D. Department of Pathology, University of California, San Diego, CA, USA

Charalampia V. Geladari, M.D. Department of Medicine, Division of Cardiology, Beth Israel Deaconess Medical Center, Boston, MA, USA

Eleni V. Geladari, M.D. Department of Medicine, Division of Cardiology, Beth Israel Deaconess Medical Center, Boston, MA, USA

Águeda González-Rodríguez, Ph.D. Institute of Biomedicine Alberto Sols (CSIC/UAM), Consejo Superior de Investigaciones Científicas, Centro de Investigación Biomédica en Red de Diabetes y Enfermedades, Metabólicas Asociadas (CIBERDEM, ISCIII), Madrid, Spain

Fawaz G. Haj, M.Sc., D.Phil. Department of Nutrition, University of California Davis, Davis, CA, USA

Department of Internal Medicine, University of California Davis, Sacramento, CA, USA

Zhao He, Ph.D. Department of Pathology and Division of Biological Sciences, University of California, San Diego, La Jolla, CA, USA

School of Food Science and Technology, Jiangnan University, Wuxi, Jiangsu, China

Pimonrat Ketsawatsomkron, Ph.D. Department of Pharmacology, University of Iowa, Iowa City, IA, USA

Maria I. Kontaridis, Ph.D. Department of Medicine, Division of Cardiology, and Department of Biological Chemistry and Molecular Pharmacology, Beth Israel Deaconess Medical Center, Harvard Medical School, Boston

Ahmed Lawan, Ph.D. Department of Pharmacology, Yale University School of Medicine, New Haven, CT, USA

Matthew J. Lazzara, Ph.D. Department of Chemical and Biomolecular Engineering, University of Pennsylvania, Philadelphia, PA, USA

Nimesh Mody, Ph.D. Institute of Medical Sciences, University of Aberdeen, Aberdeen, UK

Benjamin G. Neel, M.D., Ph.D. Princess Margaret Cancer Center, University Health Network, Toronto, ON, Canada

Peter J. Roach, Ph.D. Department of Biochemistry and Molecular Biology, Indiana University School of Medicine, Indianapolis, IN, USA

Stephanie M. Stanford, Ph.D. Division of Cellular Biology, La Jolla Institute for Allergy and Immunology, La Jolla, CA, USA

David W. Stepp, Ph.D. Vascular Biology Center, Department of Physiology, Georgia Health Sciences University, Augusta, GA, USA

Tony Tiganis, Ph.D., B.Sc. Department of Biochemistry and Molecular Biology, Monash University, Clayton, VIC, Australia

Ángela M. Valverde, Ph.D. Institute of Biomedicine Alberto Sols (CSIC/UAM), Consejo Superior de Investigaciones Científicas, Centro de Investigación Biomédica en Red de Diabetes y Enfermedades Metabólicas Asociadas (CIBERDEM, ISCIII), Madrid, Spain

Yannan Xi, B.Sc. Department of Nutrition, University of California Davis, Davis, CA, USA

Yang Xu, B.S. Department of Medical Biophysics, Princess Margaret Cancer Center, University of Toronto/University Health Network, Toronto, ON, Canada

Jianxiu Yu, Ph.D. Department of Biochemistry and Molecular Cell Biology, Shanghai Jiao Tong University School of Medicine, Shanghai, China

Sharon S. Zhang, M.D., Ph.D. Department of Pathology, University of Chicago Medical Center, Chicago, IL, USA

Chapter 1
Redox Regulation of PTPs in Metabolism: Focus on Assays

Yang Xu and Benjamin G. Neel

Abstract Protein-tyrosine phosphatases (PTPs), along with protein-tyrosine kinases (PTKs), are the key regulators of phosphotyrosine signaling, and therefore are important contributors to normal metabolism and metabolic disease. Over the past 10 years, reactive oxygen species (ROS), which had long been viewed as toxic by-products of metabolism, have been recast as important second messengers, which act, at least in part, to regulate PTP activity by reversible oxidation. For example, ROS-catalyzed PTP oxidation can transiently inhibit PTP enzymatic activity and facilitate ligand-induced receptor tyrosine kinase (RTK) signaling. Identifying ROS-inactivated PTPs represents a key challenge to understanding the role of PTPs and redox regulation in physiology and pathology. Here, we briefly review ROS regulation of PTPs, focusing on existing assays and new approaches to identify and quantify PTP oxidation.

Abbreviations

ABP	Activity-based probe
AGE	Advanced glycation end product
Alk-β-KE	Alkyne β-ketoester
BBP-Biotin	α-Bromobenzylphosphonate biotin
BP1	Biotin-1,3-cyclopentanedione

Y. Xu, B.S. (✉)
Department of Medical Biophysics, Princess Margaret Cancer Center, University of Toronto/University Health Network, Toronto, ON, Canada M5G 1L7
e-mail: yangxu.xu@mail.utoronto.ca

B.G. Neel, M.D., Ph.D.
Princess Margaret Cancer Center, University Health Network, Toronto, ON, Canada M5G 2M9

K.K. Bence (ed.), *Protein Tyrosine Phosphatase Control of Metabolism*,
DOI 10.1007/978-1-4614-7855-3_1, © Springer Science+Business Media New York 2013

DSP	Dual-specificity PTP
DTT	Dithiothreitol
EGF	Epidermal growth factor
ER	Endoplasmic reticulum
FFA	Free fatty acid
GAPDH	Glyceraldehyde 3-phosphate dehydrogenase
GPX	Glutathione peroxidase
GRX	Glutaredoxin
GSH	Glutathione
H_2O_2	Hydrogen peroxide
HFD	High-fat diet
IAA	Iodoacetic acid
IAM	Iodoacetamide
IAP-Biotin	Iodoacetylpolyethylene oxide biotin
IB	Immunoblot
IF	Immunofluorescence
IKK	Inhibitor of κB kinase
IL	Interleukin
IP	Immunoprecipitation
IR	Insulin receptor
IRS	Insulin receptor substrate
JNK	c-Jun NH_2-terminal kinase
LA	α-Lipoic acid
LC-MS/MS	Liquid chromatography-tandem mass spectrometry
MAPK	Mitogen-activated protein kinase
MKP	MAPK phosphatase
MPB	3-(*N*-maleimido-propionyl)biocytin
MRM	Multiple reaction monitoring
NEM	*N*-ethylmaleimide
NF-κB	Nuclear factor-κB
NOX	NADPH oxidase
NRPTP	Non-receptor PTP
$O_2^{\bullet -}$	Superoxide anion
oxPTP Ab	Oxidized PTP active site antibody
PD	Pull down
PDGF	Platelet-derived growth factor
PI3K	Phosphatidylinositol 3-kinase
PKC	Protein kinase C
PROP	Purification of reversibly oxidized proteins
PRX	Peroxiredoxin
PTK	Protein-tyrosine kinase
PTP	Protein-tyrosine phosphatase
PTP1B-OX	Oxidized form of PTP1B
PV	Pervanadate
PVSN-N_3	4-(Azidomethyl)phenyl ethenesulfonate azide

ROS	Reactive oxygen species
RTK	Receptor tyrosine kinase
S^-	Thiolate
scFv	Single-chain variable fragment
SO_2H	Sulfinic acid
SO_3H	Sulfonic acid
SOD	Superoxide dismutase
SOH	Sulfenic acid
T2DM	Type 2 diabetes mellitus
TCA	Trichloroacetic acid
TCEP	Tris(2-carboxyethyl)phosphine
TNF-α	Tumor necrosis factor-α
TRX	Thioredoxin
VEGF	Vascular endothelial growth factor

Introduction

Tyrosine phosphorylation is one of the major regulatory mechanisms in signal transduction, and consequently, helps control many cellular processes, including cell growth, differentiation, migration, and metabolic homeostasis [1, 2]. The level of phosphotyrosine on any protein is regulated by the opposing actions of protein-tyrosine kinases (PTKs) and protein-tyrosine phosphatases (PTPs) [1–3]. Dysfunction of specific PTKs or PTPs is associated with several human diseases [4, 5], including metabolic disorders such as obesity, insulin resistance, and type 2 diabetes mellitus (T2DM) [6–9].

The PTP superfamily comprises 107 genes and can be subdivided into four families based on the amino acid sequence in their catalytic domains. Classes I, II, and III are cysteine-based PTPs defined by the consensus $HC(X)_5R$ motif, whereas class IV are aspartic acid-based PTPs [2, 3, 5]. The largest of these, the class I cysteine-based PTPs, containing 99 members, can be further divided into tyrosine-specific "classical PTPs" and dual-specificity PTPs (DSPs), which also can dephosphorylate Ser/Thr residues. The classical PTPs include 17 non-receptor PTPs (NRPTPs), and 21 receptor-like PTPs (RPTPs), each of which contains one, or for the RPTPs, often two, ~280 amino acid catalytic (PTP) domain(s), at least one of which contains a central, highly conserved signature motif [I/V]HCSXGXGR[S/T]G [10]. The invariant cysteinyl residue within the signature motif has a low *p*Ka (~4.5–5.5), enabling it to reside in the thiolate (S^-) state at physiological pH [11, 12]. This feature of the catalytic cysteine allows it to execute a nucleophilic attack on phosphotyrosine substrates [13], but also renders it highly susceptible to oxidation and inhibition by reactive oxygen species (ROS) [14, 15].

ROS have long been viewed as the toxic by-products of aerobic life and/or defense mechanisms used by phagocytic immune cells. However, studies over the past decade indicate that ROS, especially H_2O_2, also function as intracellular second

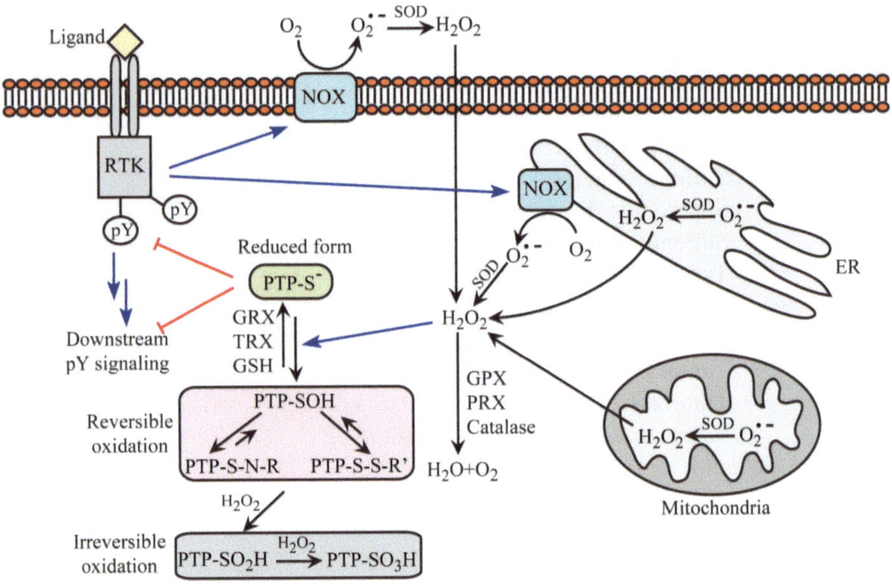

Fig. 1.1 Regulated PTP oxidation during RTK signaling and intracellular "ROS metabolism." Ligands (e.g., insulin, EGF, PDGF, and others) can bind to their receptors and trigger downstream phosphotyrosine signaling. RTK activation is accompanied by the stimulation of NOXs, which ultimately increase the intracellular H_2O_2 level and thus oxidize and transiently inactivate nearby PTPs. Higher levels of H_2O_2 (most likely seen only in pathological states) can result in irreversible PTP oxidation/inactivation. In addition to NOXs, the ER and mitochondria also can generate ROS during protein folding and electron transport, respectively, and all of these contribute to the basal ROS level in the cell. Superoxide can be converted to H_2O_2 by SOD, and H_2O_2 can be removed by antioxidant enzymes, including catalase, GPX, and PRX (*black arrow*: produce/convert to; *red arrow*: inhibit; *blue arrow*: promote)

messengers in multiple signaling pathways [16, 17]. In biological systems, most H_2O_2 arises from superoxide dismutase (SOD)-catalyzed breakdown of superoxide anions ($O_2^{\bullet-}$) [17, 18]. There are three main sources of regulated $O_2^{\bullet-}$ production inside cells (Fig. 1.1), each of which probably has important roles in cell signaling and disease. NADPH oxidases (NOXs) comprise a family of enzymes that use electrons from NADPH to reduce molecular oxygen to superoxide [19]. Mitochondria and the endoplasmic reticulum (ER) are additional major sources, using electrons that "leak" from complexes I/III of the electron transport chain [20] or arise from disulfide-bond formation during protein folding, respectively [21].

ROS have several targets, including kinases and transcription factors, but they also are well-documented regulators of PTPs (reviewed in [14, 22–24]). Growth factors, cytokines and hormones, can lead to transient and localized generation of ROS by stimulation of NOX(s) [19], although by mechanisms that remain poorly described (Fig. 1.1). The regulation of ROS production by mitochondria and ER also is not well understood and warrants more detailed investigation, particularly since major PTPs such as PTP1B and the 48 kD isoform of TC-PTP (TC48) reside on the

ER [25, 26], and several mitochondrial PTPs have been identified [27–30]. Ligand-induced ROS promote the reversible oxidation of PTPs from the thiolate (PTP-S^-) to the sulfenic acid (PTP-SOH) state, which is catalytically inactive, whereas supra-physiological ROS concentrations can cause irreversible oxidation to the sulfinic (PTP-SO_2H) or sulfonic (PTP-SO_3H) acid states [14, 31, 32]. The sulfenic acid (PTP-SOH) state of PTPs is labile and rapidly reacts with the adjacent main-chain nitrogen to form an intramolecular sulfenylamide (PTP-S-N-R) or with a proximal cysteinyl residue/glutathione (GSH) to form a disulfide bond (PTP-S-S-R′) [23, 33, 34]. The sulfenylamide and disulfide states probably serve to protect reversibly oxidized PTPs from hyper/irreversible oxidation, and thus preserve their ability to reduce back to the active (PTP-S^-) state upon reaction with antioxidants such as glutaredoxin (GRX) or thioredoxin (TRX) [35]. Whether different PTPs are differentially reduced by GRX or TRX remains unclear, although PTP1B appears to be reactivated more effectively by TRX [36].

The contemporaneous ligand-induced activation of RTKs and ROS-induced inactivation of PTPs probably have at least two related functions, both of which serve to optimize signal dynamics. First, "futile cycles" of phosphorylation and dephosphorylation are avoided, and second, RTK activation assumes "switch-like" characteristics [1, 14, 37]. Several studies have shown that specific PTPs undergo oxidation upon RTK activation: e.g., PTP1B oxidation during epidermal growth factor receptor (EGFR) signaling [36], SHP2 oxidation during platelet-derived growth factor receptor (PDGFR) signaling [38] and, of more immediate relevance to metabolic regulation, PTP1B, TCPTP, and PTEN oxidation during insulin receptor (IR) signaling [39–41].

Obviously, there are potential risks to using a diffusible toxic chemical as a second messenger. Consequently, H_2O_2 levels in normal cells are tightly regulated by antioxidants including catalase, peroxiredoxin (PRX), and glutathione peroxidase (GPX) [17, 37, 42, 43]. These systems can also be reversibly regulated during RTK signaling; for example, PRX1 becomes tyrosyl phosphorylated and transiently inactivated following growth factor stimulation [44]. The net result is likely to be exquisite local control of ROS production and disposition, although the precise mechanistic details of such control remain to be elucidated. Nevertheless, in pathological conditions, mitochondrial and/or ER-derived ROS, together with ROS from NOX, are elevated, and the increased basal oxidative stress leads to more global PTP oxidation/inactivation, which might contribute to the pathogenesis of several diseases, including cancer, obesity, and T2DM [45–48]. Moreover, release of ROS by infiltrating phagocytic cells probably contributes to the pathogenesis of acute and chronic inflammatory disorders [49–51].

As PTP oxidation is important in both physiological and pathological states, techniques that can detect and quantify PTP oxidation are of obvious value. Although for many years such techniques were largely elusive, there have been several recent improvements in the methodology for studying redox regulation of PTPs. Here, we summarize current assays that have been developed to study PTP oxidation and discuss the roles of ROS and redox-regulated PTPs in selected metabolic disorders.

Assays to Identify and Quantify PTP Oxidation

Basal or ligand-induced PTP oxidation generates two pools of PTPs, oxidized inactive PTPs (PTP-SOH, PTP-SO_2H, and PTP-SO_3H) and reduced active PTPs (PTP-S^-). Several techniques have been developed to identify and quantify reversibly oxidized PTPs (PTP-SOH) (summarized in Fig. 1.2; also reviewed in [52]).

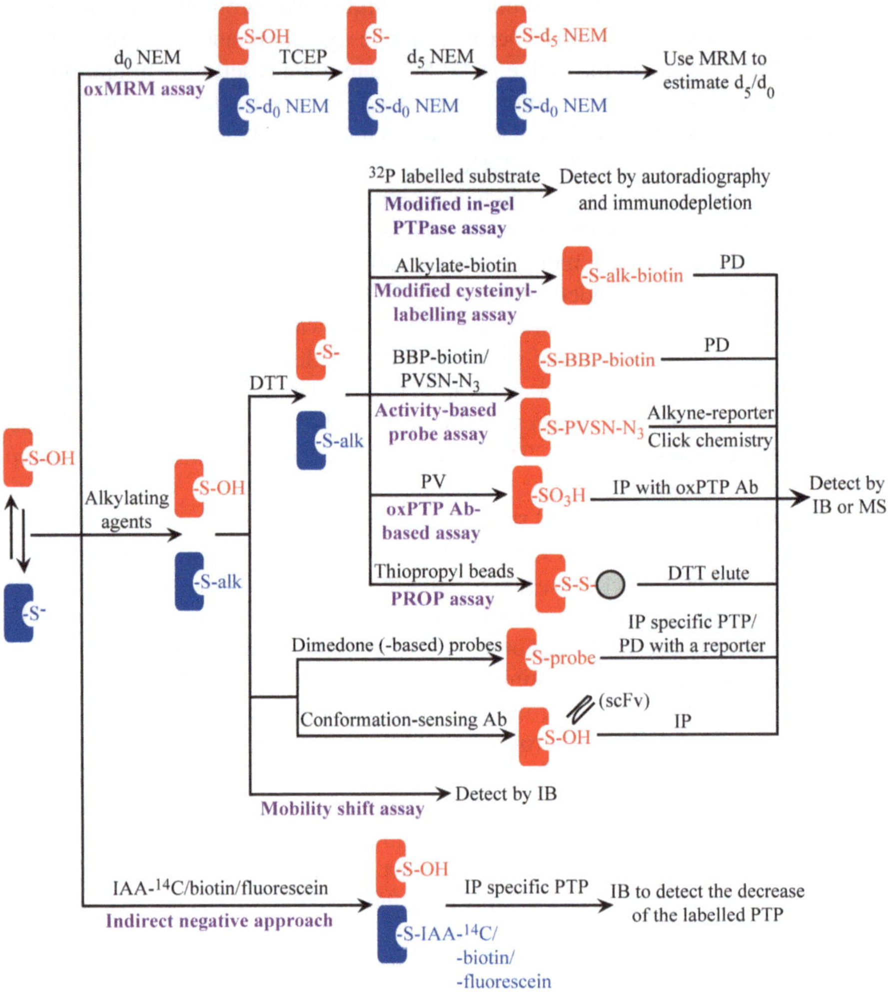

Fig. 1.2 Schematic of techniques to detect PTP oxidation. Techniques to detect protein oxidation generally can be divided into direct and indirect approaches, respectively. Most indirect approaches involve an alkylation step to inactivate reduced PTPs, followed by reduction of previously oxidized PTPs. The newly reduced PTPs can then be labelled and detected. By contrast, direct approaches directly detect oxidized PTPs without the reducing step, followed by immunoprecipitation (IP) or "pull-down" (PD) via an affinity tag. See text for details

These can be divided into two general categories: indirect approaches that assess PTP activity or the susceptibility of the reactive cysteinyl residue to various reagents, and direct approaches that rely on the distinct chemical reactivity of the S-OH group with specific probes or the structural changes induced in reversibly oxidized PTPs [52]. In addition, different techniques can be classified as "targeted" or "global" approaches. For the former, the PTP of interest must be immunoprecipitated (IP) and then its oxidation level is quantified, whereas in global approaches, the oxidation of multiple PTPs can be assessed simultaneously [52].

Because PTP thiolates are highly reactive, and even ambient oxygen levels can cause post-lysis PTP oxidation in the absence of reducing agents [53–55], many of the techniques that we describe involve common procedures to minimize artifactual oxidation: degassing of the lysis buffer to remove dissolved oxygen, performing labelling reactions under anaerobic conditions, and alkylation of reduced PTPs. The latter requires the stoichiometric and irreversible reaction of alkylation agents with reduced PTPs to render them inactive to probes that then either directly or indirectly label PTPs that have undergone oxidation. The commonly used alkylating agents contain either an α-haloacyl or maleimide functionality, such as iodoacetic acid (IAA), iodoacetamide (IAM), or *N*-ethylmaleimide (NEM).

Indirect Approaches

Indirect, Negative Approaches

Indirect approaches can be further divided into two general categories, based on whether the method detects a decrease in reduced PTPs ("negative" approaches) or an increase in oxidized PTPs ("positive" approaches) [52]. The negative approach of Lee et al. [36] was the first method developed to detect PTP oxidation. They used ^{14}C-labelled IAA at pH 6.5 (to enrich for labelling of highly reactive cysteines) to label the reduced form of PTPs, while leaving oxidized PTPs intact [36]. Following ^{14}C-IAA labelling, a PTP of interest could be immunoprecipitated and resolved on an SDS/PAGE gel to quantify the radioactivity of the specific PTP, with a decrease of radioactivity indicating increased PTP oxidation [36]. Using this method, Lee et al. reported that PTP1B was oxidized following EGF stimulation of A431 cells, with a maximum of 27 % of PTP1B oxidized after 10 min of stimulation and returned to baseline values after 40 min [36]. This result can be viewed as a landmark for ROS-regulated PTP oxidation as it provided the first direct evidence that PTPs are indeed physiological targets of ligand-evoked ROS, although the extent of EGF-evoked PTP1B oxidation might well have been overestimated [52]. In place of radioactive probes, several studies have successfully employed an analogous approach by using biotin- or fluorescein-tagged IAA to detect oxidation of PTPs (e.g., SHP2 oxidation during T-cell receptor activation) or other reactive cysteine-containing proteins [56–58].

A general problem with negative approaches, however, is that they measure decreases in labelling, which results in a relatively low signal to noise ratio compared with positive approaches [52]. Furthermore, this assay was introduced as a targeted method and is not easily adapted to a global approach. For this reason, negative approaches have largely been supplanted by positive methods.

Modified In-Gel Phosphatase (PTPase) Assay

The modified in-gel PTPase assay, developed by Meng et al. [38, 40, 59], provided the first positive approach and also the first global method for detecting oxidation of multiple PTPs simultaneously. As its name implies, the assay was derived from the earlier in-gel PTPase assay, developed to monitor PTP expression [60]. Cells are lysed in the presence of IAA at pH 5.5 to specifically alkylate reduced reactive cysteinyl residues, and lysates are then resolved on an SDS-PAGE gel containing a radioactive substrate [38, 59]. The gel is then denatured and renatured in the presence of reducing agents, leaving the originally reversibly oxidized PTPs in a reduced, active state capable of dephosphorylating the embedded substrate [38, 59]. PTPs that have been oxidized are detected by autoradiography as bands against an otherwise black (exposed) background, and their identity must be predicted by their molecular weights and confirmed by immunodepletion [38, 59]. Using this approach, Meng et al. first identified SHP2 oxidation in response to PDGF stimulation in Rat1 cells, consistent with a role for SHP2 in negative regulation of the PDGFR signaling pathway [38]. Using the same assay, this group showed subsequently that PTP1B and TCPTP (TC45) undergo reversible oxidation during insulin signaling [40].

The advent of the modified in-gel PTPase assay represented a substantial advance in our ability to study redox regulation of PTPs. Nevertheless, this approach has some significant limitations, including: (1) a detection bias towards non-transmembrane PTPs, because most RPTPs do not renature well; (2) the need for immunodepletion to identify specific PTPs, which of course, requires that an appropriate antibody is available; and (3) because it depends on (variable) PTP renaturation, it is not quantitative [52, 60].

Modified Cysteinyl-Labelling Assay

Another positive approach, the modified cysteinyl-labelling assay, employs an alkylating agent linked to a tag that facilitates its detection, such as 3-(*N-maleimido-propionyl*)biocytin (MPB), NEM-biotin, and iodoacetylpolyethylene oxide biotin (IAP-biotin). This assay was first used to assess the oxidation of non-PTP proteins [61], while later, Li and Whorton used it to measure the thiol redox status of PTP1B in response to NO donors in A431 or Jurkat cells [62]. Kwon et al. then used NEM-biotin to detect PTEN oxidation in response to EGF, PDGF, or insulin stimulation in different cell lines [41]. The general workflow for modified cysteinyl-labelling assays is to first alkylate reduced reactive cysteinyl residues, remove the free alkylating agent and reduce the former reversibly oxidized cysteinyl residues, and then label the newly reduced cysteinyl residues with the tagged alkylating agent. PTPs can be identified by immunoblotting (IB) or immunoprecipitation. In addition, the same approach can be adapted to a global, mass spectrometry (MS)-based approach to, in principle, detect all reversibly oxidized thiol-containing proteins. In practice, however, abundant labelled proteins tend to saturate the identification/quantification capacity of most current mass spectrometers, as illustrated by a recent study in which only three PTP-derived peptides were identified out of >1,000 cysteinyl-containing peptides, none of which contained the active site cysteine for that PTP [58].

To overcome these limitations, Boivin et al. developed an IAP-biotin-based method with some critical modifications, most importantly, lysing cells in non-denaturing buffers and carrying out probe labelling at pH 5.5, where most reactive thiols other than PTPs remain protonated [55, 63, 64]. Using this approach combined with immunoblotting, these investigators detected increased oxidation of LAR, SHP2, PTEN, and MAPK phosphatase (MKP-1) in PDGF-transformed angiomyolipoma cells compared with nontransformed cells [63]. It remains unclear if this method can be translated to a global MS-based assay, but particularly with the development of more rapid mass spectrometers (e.g., Q-Exactive), such an approach is certainly promising.

Activity-Based PTP Probes

Activity-based probes (ABPs) are designed to react with mechanistically related enzymes [65, 66]. PTPs contain two residues that are essential for their catalytic mechanism, the reactive cysteine in the signature motif and an aspartic acid residue in the conserved WPD loop [10], which function sequentially as a general acid and then a general base in the catalytic mechanism [13, 67]. Kumar et al. developed two ABPs, α-bromobenzylphosphonate (BBP)-biotin and 4-(azidomethyl)phenyl ethenesulfonate azide (PVSN-N_3), based on the PTP catalytic mechanism [68, 69]. BBP-biotin contains a biotin tag to facilitate its detection and has been used to visualize oxidation of LAR, SHP2, PTEN, and MKP-1 in PDGF-transformed angiomyolipoma cells [63]. By contrast, after labelling with PTPs, PVSN-N_3 adducts have to undergo click chemistry with alkyne-tagged reporter groups [69–71] to be detected. Although neither of these ABPs has been applied to MS, the PTP-specific property of ABPs makes them potential candidates for global approaches. In addition, since PVSN-N_3 does not have the bulky biotin tag compared with BBP-biotin, it is easier to synthesize and more cell permeable [52].

Oxidized PTP Antibody (oxPTP Ab)

Persson et al. developed an antibody (oxPTP Ab) that targets the signature motif VHCSAG peptide with its cysteinyl residue oxidized to the sulfonic acid (SO_3H) state [72, 73]. In their initial studies, PTPs of interest were immunoprecpitated from cell lysates prepared in the presence of IAA, reduced with DTT, and then hyperoxidized to the sulfonic acid state with pervanadate (PV) [72, 73]. Using this targeted approach, Persson et al. demonstrated that SHP2 was oxidized during PDGF stimulation of NIH 3T3 cells, and the oxPTP Ab also could successfully detect DEP-1, TCPTP, and rPTPα [72].

Our group modified the targeted oxPTP Ab-based method to a global MS-based approach able to detect and quantify classical PTP expression ("qPTPome") and oxidation ("q-oxPTPome") [32]. A monoclonal oxPTP Ab, rather than specific PTP antibodies, was used to immunoprecipitate PTPs, with efficient antibody binding achieved by protein-denaturation prior to immunoprecipitation [32]. Free alkylating agents and free DTT were removed through a gel filtration step [32]. The main difference between qPTPome and q-oxPTPome is the absence of the pre-alkylation step in qPTPome, thus permitting quantifying the expression of nearly all classical PTPs [32]. With the qPTPome assay, 17 classical PTPs could be identified and quantified in NIH 3T3 cells by liquid chromatography-tandem mass spectrometry (LC-MS/MS), including the D1 and D2 domains of several RPTPs [32]. Because the signature motif peptide is highly conserved across the classical PTP family, this assay was used for human, rat, and mouse PTPs, with 36 out 38 classical PTPs identified; the only exceptions were *PTPRN* and *PTPRN2*, which are thought to be catalytically inactive [32, 74]. By combining qPTPome and q-oxPTPome, the method can either directly or indirectly quantify all PTP oxidation states (i.e., S^-, SOH, SO_2H, SO_3H) and also can detect enhanced oxidation of specific PTPs in disease states (e.g., cancer). Another advantage is that it can be applied to tissues as well as cell lines, and quantification can be carried out "label-free." However, thus far, q-oxPTPome has not proven sensitive enough to detect ligand-induced PTP oxidation.

The Purification of Reversibly Oxidized Proteins (PROP) Assay

Templeton et al. recently reported a method that relies on thiopropyl agarose beads that contain dipyridyl disulfide groups, which can link to free thiol-containing proteins via spontaneous disulfide formation [75]. One of the advantages of these beads is that they can react with proteins in highly denaturing buffers, unlike streptavidin- or antibody-conjugated beads [75]. Cells are first fixed in situ with 10 % trichloroacetic acid (TCA) to rapidly inhibit post-lysis oxidation, and then alkylated with NEM [75]. Reversibly oxidized cysteinyl residues are then reduced by DTT and purified on thiopropyl agarose beads [75], with leftover NEM or DTT removed through methanol precipitation/resolubilization cycles [75]. Based on this approach, Templeton et al. successfully identified p38 mitogen-activated protein kinase (MAPK) oxidation in response to 10 mM H_2O_2 [75].

Subsequently, the same group applied PROP in a global MS assay to detect oxidized proteins and identify the site of oxidation in response to H_2O_2-treated HeLa cells [76]. About 300 proteins were identified, although these included only a few phosphatases [76]. There are several potential explanations for this low recovery: (1) the high concentration of H_2O_2 hyper-oxidized most PTPs, so they could not be reduced by DTT; (2) by performing the assay under denaturing conditions, most disulfide bond-forming cysteinyl residues would be accessible for labelling, which might increase the background and overwhelm the mass spectrometer; and/or (3) similar to the modified cysteinyl-labelling approach, PROP potentially labels all reversibly oxidized proteins, which increases the background against which (probably less abundant) PTPs can be detected. It would be interesting to see if the PROP approach could be employed with a low pH, non-denaturing lysis buffer to detect PTP oxidation in response to physiological stimuli.

Direct Approaches

A general advantage of the direct approaches is the ability to detect oxidized PTPs without the attendant losses of an intervening alkylation/reduction; therefore, direct approaches should, at least in theory, be more sensitive. Two types of direct approaches have been developed. The first relies on dimedone (5,5-dimethyl-1,3-cyclohexanedione) and its analogs, which are "soft nucleophiles" that employ a 1,3-diketone functional group to react selectively with protein-sulfenic acids. The second method uses conformation-sensing antibodies that specifically recognize structural changes in the PTP domains that occur during reversible oxidation.

Dimedone Antibodies

Dimedone is a cell-permeable small molecule that can undergo keto-enol tautomerism. Its enol state can react with the electrophilic sulfur atom of sulfenic acids, and hence, forms a thioether bond with oxidized proteins [52]. After dimedone labelling, dimedone-protein adducts can be detected by immunoblotting with anti-dimedone antibodies. Seo and Carroll generated such antibodies and showed that these could recognize oxidized glyceraldehyde 3-phosphate dehydrogenase (GAPDH), ACTIN, and PRX1 [77]. They also observed a global increase in antibody labelled, dimedone-reactive proteins following H_2O_2 stimulation of HeLa cells by immunofluorescence (IF) staining and profiled the sulfenic acid levels among different breast tumors and cell lines [77]. Maller et al. subsequently generated a similar reagent, although this appeared to label fewer proteins [78].

The dimedone antibody approach can be used to label reversibly oxidized proteins either pre- (in vivo) or post-cell lysis. However, in vivo labelling represents the accumulation of dimedone-reacted species, rather than a snapshot of the cell redox state, so it is likely to overestimate the oxidation levels in response to different stimuli. Furthermore, in vivo labelling itself probably alters redox balance. Also, existing dimedone antibodies are not useful for immunoprecipitations, and thus cannot be employed in global MS-based assays to identify unknown sulfenic acid-containing proteins.

Dimedone-Based Probes

Several studies have utilized a reporter/affinity tag linked to dimedone to facilitate the identification and recovery of dimedone-conjugated proteins. Poole et al. developed several such probes by conjugating dimedone with a fluorophore or biotin tag, including DCP-Bio1, which contains dimedone linked to a biotin tag through an ester bond [79–81]. The general procedure for using such probes includes cell lysis in the presence of alkylating agents and the probe, followed by purification/detection through the affinity/reporter tag. Using DCP-Bio1, Michalek et al. demonstrated reversible oxidation of SHP1, SHP2, and ACTIN following T-cell activation [82]. Later, Oshikawa et al. used DCP-Bio1 to show that extracellular SOD could induce PTP1B and DEP-1 oxidation during vascular endothelial growth factor (VEGF) signaling in endothelial cells [83], and Wani et al. used the same probe to detect AKT2 oxidation in response to PDGF in NIH 3T3 cells [84]. Despite these successes, the bulky tag impairs the cell-permeability of these tags, so they are not ideal for in vivo labelling. The biotin tag also might inhibit labelling of reversibly oxidized proteins due to steric hindrance. Therefore, Carroll et al. generated dimedone-based probes with a small azide/alkyne tag that can be linked subsequently to a biotin group by click chemistry [85, 86]. Using one such probe, DYn-2, they detected EGF-induced PTP1B, SHP2, and PTEN oxidation in A431 cells [86].

Although dimedone-based probes are a promising approach for detecting PTP oxidation, there have been limited applications of these reagents to global, MS-based assays. Charles et al. used a biotin-dimedone probe to label H_2O_2-oxidized proteins in rat ventricular mycocytes or rat hearts, then streptavidin-bound dimedonylated proteins were resolved on SDS/PAGE and analyzed by in-gel trypsinization and MS [87]. They identified 22 differentially oxidized proteins, but most of these were highly abundant, and none were PTPs [87]. Leonard et al., using a different probe called DAz-2, identified 189 sulfenic acid-containing proteins in HeLa cells; again, however, no PTP was detected [88]. Because most dimedone-based probes label sulfenic acid-containing proteins indiscriminately, and abundant oxidized proteins could prevent the identification of PTPs by MS, Leonard et al. recently developed "PTP-specific" dimedone-based probes containing the so-called PTP-binding module and showed that many of these had higher affinity for *Yersinia* phosphatase and PTP1B compared with DAz-1 (an analog of DAz-2) [89]. It will be interesting to see if such probes can improve the global detection of growth factor-induced PTP oxidation.

Probes with Dimedone-Like Functional Groups

Dimedone and dimedone-based probes have several disadvantages for use in global assays [90, 91]. Their synthesis often requires multiple synthetic steps, which decreases yield. Dimedone-conjugated peptides are not suitable for positive ion mode MS because they give rise to negatively charged species. Also, as indicated above, dimedone addition to cells can significantly increase basal ROS levels.

To circumvent these deficiencies, Qian et al. developed several probes that use 1,3-cyclopentanedione as the backbone [90]. Like dimedone, 1,3-cyclopentanedione can selectively react with sulfenic acids [90]. Moreover, derivatives of 1,3-cyclopentanedione, such as biotin-1,3-cyclopentanedione (BP1), can be generated conveniently by Michael addition [90]. Qian et al. observed a global increase in BP1-labelled proteins when treating the NIH 3T3 cell lysates with H_2O_2 [90]. The same authors subsequently used probes with a β-ketoester backbone, including alkyne β-ketoester (Alk-β-KE), to label sulfenic acid-containing proteins [91]. Compared with dimedone, Alk-β-KE has several advantages, including a higher reaction rate at neutral pH, lower cytotoxicity/ROS generation, and a cleavable biotin tag that yields a derivative amenable to MS [91]. These novel probes, especially Alk-β-KE, are promising candidates to detect ligand-induced PTP oxidation in vivo.

Conformation-Sensing oxPTP Antibodies

The most direct approach to detecting oxidized PTPs was developed by Haque et al., who developed antibodies that specifically detect the sulfenic acid form of PTP1B [92]. This approach was possible because oxidized PTP1B is quickly rearranged to a sulfenylamide, which exposes Tyr46 and causes other dramatic changes in the PTP1B active site [33, 34]. Interestingly, mutation of the catalytic cysteine and the adjacent serine residue to alanine (PTP1B-CASA) results in a similar conformation, which enabled the use of this mutant and phage display to screen for conformation-specific antibodies (in the form of single-chain variable fragments (scFvs)) [92]. The resultant "PTP1B-OX" antibodies specifically bind with high affinity to reversibly oxidized PTP1B and trap PTP1B in the oxidized, inactive form even in the presence of reducing agents [92]. Remarkably, PTP1B-OX antibodies do not bind to reversibly oxidized TCPTP, the closest relative of PTP1B in the PTP superfamily with ~75 % sequence identity in its catalytic domain [92]. The PTP1B-OX antibodies also were expressed inside cells as "intrabodies". Using this approach, Haque et al. provided the first (and only) direct evidence for the sulfenylamide form of PTP1B in vivo and also observed enhanced oxidation in response to insulin stimulation [92]. Interestingly, by "locking" PTP1B in the oxidized, sulfenylamide state, the intrabodies enhanced insulin signaling as well, raising the possibility that pharmacologic agents that have similar effects might act as highly selective PTP1B inhibitors [92] (see below).

Other Approaches

Held et al. introduced a method called "oxMRM" that utilizes multiple reaction monitoring [93, 94] to compare the relative amounts of reduced and oxidized proteins [95]. Cells are first lysed in TCA to protonate all thiol groups, and thus minimize post-lysis oxidation [95]. The cell pellet is resuspended in a denaturing buffer containing unlabelled NEM (d_0 NEM) to label all the free thiols [95]. The d_0 NEM is removed by TCA precipitation, previously oxidized cysteinyl residues are reduced by TCEP, and the lysate is then labelled with "heavy" NEM containing five deuteriums (d_5 NEM) [95]. Proteins of interest are immunoprecipitated and the relative amounts of d_5- and d_0-labelling are monitored by MS [95]. Based on this method, Held et al. were able to quantify site-specific cysteine oxidation of p53 and detected the oxidation of PTP1B following H_2O_2 stimulation [95]. Like all targeted methods, however, this requires immunoreagents for the protein of interest.

Lou et al. introduced a targeted MS-based method that could view all the PTP oxidation states (i.e., SH, SOH, SO_2H, and SO_3H) [31]. They lysed cells in the presence of IAA and the PTP of interest was then immunoprecipitated and resolved on an SDS/PAGE gel. The reversibly oxidized PTP was then reduced by the addition of sample buffer [31]. In-gel trypsinization-based MS could then be used to quantify the different PTP oxidation states (i.e., S-IAA, SH, SO_2H, and SO_3H) [31]. Using this method, Lou et al. reported that HepG2 and A431 cells had very high levels (>25 %) of reversible and irreversible oxidation of PTP1B [31]. Lou et al. also noted that the irreversibly oxidized and reduced forms of PTP1B migrated differently in SDS-PAGE [31]. Differential migration of reduced and reversibly oxidized forms has also observed for PTEN [96] and applied to the detection of PTEN oxidation in response to various stimuli [41]. However, such electrophoretic mobility shift assays are not applicable to a global approach; moreover, it seems unlikely that most PTPs would migrate differently in such assays [52].

PTP Oxidation in Type 2 Diabetes Mellitus and Obesity

Diabetes mellitus is one of the most common chronic diseases worldwide, with a prevalence of ~285 million people in 2010 [97]. T2DM is the more prevalent form, and its incidence continues to increase, primarily as a consequence of reduced physical activity and obesity [97, 98]. The incidence of obesity also is skyrocketing, with more than 1.1 billion adults worldwide considered overweight and 312 million frankly obese [99]. Indeed, for the first time in human history, disorders caused by food excess have outstripped starvation as a cause of human morbidity and mortality.

Obesity results from an imbalance between food intake and energy expenditure and is frequently associated with insulin resistance. Insulin resistance, in turn, is the first stage in and the major antecedent of T2DM pathogenesis, and ~90 % of T2DM is attributable to excess weight [99]. T2DM represents a late stage consequence of chronic insulin resistance, characterized by the failure of pancreatic β cells to secrete sufficient insulin to maintain euglycemia [100]. Therefore, the insulin signaling pathway and the mechanism of insulin resistance are keys to understanding these diseases and identifying novel therapeutic targets.

Insulin, of course, signals via the insulin receptor (IR), an RTK. As for other RTKs, ligand-binding (to the IR-α chain) promotes trans-phosphorylation of the IR β-subunits [101, 102]. The activated IR then tyrosine-phosphorylates several substrates, mainly insulin receptor substrate-1/2 (IRS-1/2) and SHC, triggering downstream signaling pathways, including the activation of phosphatidylinositol 3-kinase (PI3K)/AKT and RAS/ERK pathways. Of these, the former plays the major role in the metabolic effects of insulin. This leads to, for example, the translocation glucose transporter GLUT4 to the cell surface and promotes glucose uptake and glycogen synthesis and inhibits gluconeogenesis [101, 103].

ROS in Insulin Signaling and Resistance: A Double-Edged Sword

The role of ROS in insulin signaling, insulin resistance, and T2DM remains a topic of controversy [104]. This state of affairs is largely the result of "apples versus oranges" comparisons. For example, the effects of normal levels of ROS produced under physiological conditions are no doubt different from those of the excess ROS that accompany obesity-associated inflammation and/or mitochondrial dysfunction. Yet many investigations have extrapolated the effects of pathological ROS levels on cells and tissues to the normal, physiological state, and vice-versa. Moreover, the effects of ROS early in the pathogenesis of insulin-resistant states need not be the same as those in full-blown metabolic syndrome or T2DM. Nevertheless, the available data suggest that physiological ROS levels play an important role in normal insulin receptor signaling, as they do in other RTK pathways. Conversely, excessive, and possibly mislocalized, ROS production contributes to insulin resistance and its attendant complications.

Chronic ROS Production Promotes Insulin Resistance

Substantial evidence links oxidative stress and insulin resistance, although whether excessive ROS production is involved in the initial stages of insulin resistance (e.g., from high-fat feeding) or rather, contributes to progressive insulin insensitivity, remains unclear [46, 47, 105–107]. Several consequences of obesity and/or T2DM promote ROS generation (Fig. 1.3), including hyperglycemia, increased circulating free fatty acids (FFAs), and some pro-inflammatory cytokines (e.g., tumor necrosis factor-α (TNF-α) and interleukin (IL)-6) [108, 109]. Hyperglycemia and/or high FFAs can overwhelm mitochondria with surplus NADH and $FADH_2$, leading to electron leakage from electron transport chain complexes I and III [20, 104, 109] and superoxide production. Surplus cytosolic glucose can enter the polyol pathway, generating additional NADH and depleting NADPH [109–111]. Consequently, mitochondria are burdened further and replenishment of reduced GSH is inhibited, further altering redox tone. Furthermore, increased intracellular glucose can promote the generation of advanced glycation end products (AGEs), which can bind their receptor (RAGE) and stimulate ROS production through NOXs [109, 111].

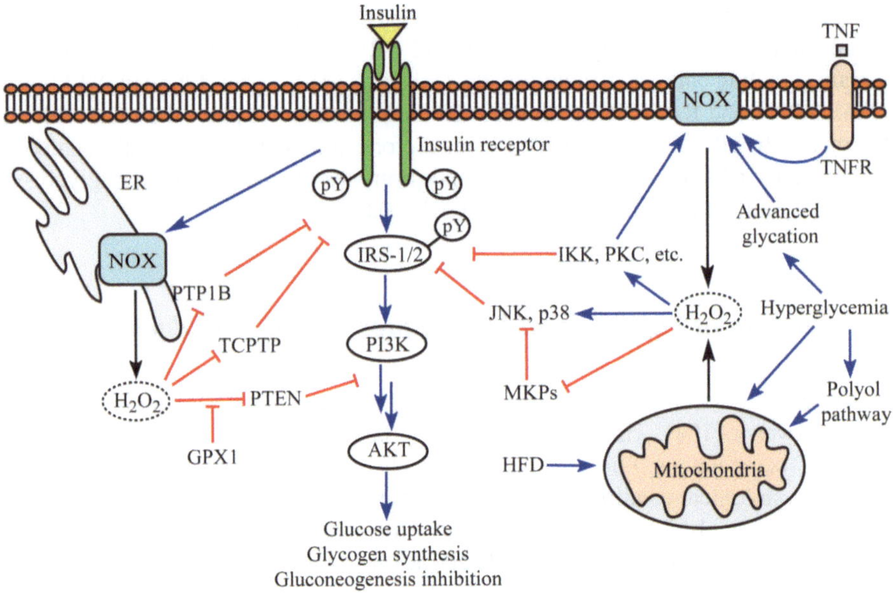

Fig. 1.3 Paradoxical roles of ROS and PTP inactivation in insulin signaling and resistance. Binding of insulin to its receptor triggers the tyrosine phosphorylation of IRS proteins and activation of the downstream PI3K/AKT pathway, which promotes glucose uptake and glycogen synthesis and inhibits gluconeogenesis. Hyperglycemia, high FFA, and/or elevated pro-inflammatory cytokines, e.g., in the context of obesity and T2DM, can increase basal ROS levels via mitochondria and/or NOX activity. In turn, elevated ROS can stimulate several stress-sensitive serine/threonine kinases, at least in part by oxidizing MKPs, which results in down-regulation of insulin signaling and further increases in ROS levels. On the other hand, several PTPs negatively regulate normal insulin signaling, and NOX-derived H_2O_2 can oxidize these PTPs and promote insulin-induced responses. Consistent with this notion, deletion of at least some antioxidant genes can increase PTP oxidation and protect cells from insulin resistance (*black arrow*: produce/convert to; *red arrow*: inhibit; *blue arrow*: promote)

Nutrient overload also is proposed to burden the ER protein-folding machinery and leads to elevated ROS [104]. In addition to excess nutrition, pro-inflammatory cytokines are elevated in obesity and T2DM patients [112]. TNF-α promotes ROS generation through NOXs [113, 114] and possibly mitochondria [115]. Together with elevated ROS levels, antioxidant deficiencies are reported in diabetes patients [116, 117], and both are likely to contribute to increased oxidative stress in those patients.

Much evidence suggests that oxidative stress contributes to the pathogenesis of insulin resistance, rather than being a mere consequence of chronic hyperglycemia [106, 118]. H_2O_2 treatment can impair insulin signaling and GLUT4 translocation in 3T3-L1 adipocytes [119, 120]. By contrast, the antioxidant α-lipoic acid (LA) can rescue the insulin-resistant phenotype induced by oxidative stress or high glucose in such cells [121, 122], and LA also has been reported to improve insulin sensitivity in T2DM patients [123, 124]. Moreover, Anderson et al. found that exposure to a high-fat diet (HFD) induces mitochondrial ROS production and insulin resistance in skeletal muscle of both rodents and humans, whereas mitochondrially targeted antioxidants can preserve insulin sensitivity even in the presence of HFD [47].

One proposed mechanism for ROS-induced insulin resistance involves the activation of several stress-sensitive protein kinases (Fig. 1.3, right): e.g., the MAPKs c-Jun NH_2-terminal kinase (JNK) and p38, as well as other serine/threonine kinases, such as protein kinase C (PKC) and inhibitor of κB kinase (IKK) [104, 108, 110, 118]. Once activated, these stress-sensitive kinases promote the serine phosphorylation of IR and IRS proteins. Serine phosphorylation of IRS-1/2 attenuates their ability to couple to the IR, and thereby reduces their tyrosine phosphorylation level coupling to downstream signaling components, most notably PI3K. Serine phosphorylation also promotes IRS-1 degradation [125, 126]. Under physiological conditions, these mechanisms might serve as a normal counterbalance to insulin stimulation; however, in the context of obesity or T2DM, chronic activation of these stress kinases by ROS contributes to, and thereby exacerbates, insulin resistance [108].

Although the precise mechanisms are still controversial, ROS-evoked activation of JNK and p38 MAPK is partially due to the oxidation and transient inactivation of MAPK phosphatases (MKPs), which belong to DSP family. For example, Kamata et al. found that oxidized MKPs migrate differently from reduced MKPs on SDS-PAGE under nonreducing conditions, and by combining this electrophoretic mobility shift assay with mutagenesis experiments, they provided evidence that the reactive cysteines of MKPs are oxidized by intracellular H_2O_2 [127]. However, ROS-induced activation of JNK also has been attributed to the activation of the upstream MAP kinase kinase ASK, via release of TRX [128]. Furthermore, in addition to phosphorylating IRS1, IKK can up-regulate the transcription of pro-inflammatory cytokines and RAGE via nuclear factor-κB (NF-κB), whereas PKC might activate NOXs; both of these events may contribute to "feed-forward" effects that can persistently increase cellular ROS levels [46, 110]. Excessive ROS production, in addition to its effects on insulin-responsive tissues, also is implicated in the eventual failure of pancreatic β cells seen in frank T2DM [46, 129, 130].

Overall, given the high basal oxidative level in insulin-responsive tissues in the context of obesity and T2DM, it is likely that *inappropriate* oxidation of other PTPs and other reactive cysteine-containing proteins also contribute to the pathology of insulin resistance. Thus far, however, the assays described above have yet to be applied to this problem. Indeed, such approaches would seem to represent a fertile area for future investigation.

Physiological ROS Levels Promote Insulin Signaling

Although chronic mitochondrial/NOX-induced ROS and accompanying MKP oxidation are related to insulin resistance, several PTPs are involved in the negative regulation of insulin signaling [8, 131] and the oxidation/inactivation of these PTPs through NOX-derived H_2O_2 is believed to promote insulin sensitivity [132].

Using several of the assays described above, PTEN, PTP1B, and TCPTP [39–41] have been shown to be transiently and reversibly oxidized by NOX-derived H_2O_2 [132, 133] (Fig. 1.3, left). Moreover, a recent study using both electrophoretic mobility shift and the IAP-biotin assays found that insulin-induced PTEN oxidation was elevated and prolonged in mouse embryo fibroblasts lacking the antioxidant enzyme GPX1, without any significant effect on PTP1B and TCPTP oxidation [134]. This observation suggests that oxidation of specific PTPs might be regulated by distinct cellular antioxidants. Although these events were not documented directly in insulin-responsive tissues, the same group showed that mice lacking GPX1 were protected from HFD-induced insulin resistance and showed increased insulin-induced AKT activation in muscle. Moreover, the enhanced insulin sensitivity in these mice could be reversed by administration of the antioxidant *N*-acetylcysteine [134]. These compelling genetic data certainly belie the longstanding view that ROS play a mainly pathogenic role in promoting insulin resistance.

In addition to their role in physiological regulation of RTK signaling, the properties of ROS-inactivated PTPs might provide a novel approach to drug development for metabolic disease. For example, PTP1B, which undergoes insulin-induced PTP oxidation, is a major regulator of IR and/or IRS1 [135–137]. As described in other chapters within this volume, PTP1B-deficient mice show tissue-specific increases in insulin responsiveness and insulin signaling events in skeletal muscle and liver and are protected from HFD-induced insulin resistance [138, 139]. These mice also are lean and resistant to HFD-induced obesity. For these reasons, PTP1B inhibitors might be valuable for treating these important metabolic disorders. Nevertheless, owing to the highly charged and reactive nature of active site of PTP1B (and the PTP family in general), it has been challenging to develop potent, specific and, most importantly, bioavailable PTP1B inhibitors [140, 141].

As discussed above, PTP1B (and most likely, other PTPs) undergoes a significant conformational change upon oxidation to the sulfenic acid/sulfenylamide state. In this conformation, the active site is no longer charged, and the geometry of the deep catalytic cleft is altered. The effects of the OX-PTP1B intrabodies (see above) strongly suggest that small molecule inhibitors that stabilize the sulfenylamide form of PTP1B could provide an alternative approach to inhibition of this enzyme.

Future Directions

The PTPs that have been found to undergo reversible oxidization in response to insulin are all non-transmembrane enzymes; however, RPTPs are also relevant to insulin signaling [8]. For example, LAR co-immunoprecipitates with IR and might dephosphorylate IR and IRS proteins [8, 142]. LAR knockdown in rat hepatoma cells leads to increased IR autophosphorylation with enhanced insulin-induced PI3K activity, while knockdown of LAR in human embryonic kidney (HEK 293) cells inhibits the insulin-induced activation of AKT and MAP kinases without affecting IR/IRS phosphorylation levels [143, 144]. Moreover, muscle-specific over-expression of LAR in mice results in insulin resistance, although glucose homeostasis remains normal [145]. RPTPα also has been suggested to target the IR [146], and rat adipose cells over-expressing this enzyme show significantly lower levels of cell surface GLUT4 following insulin stimulation [147].

Of course, RPTP activity also can be regulated by ROS. As discussed above, most RPTPs (including LAR and rPTPα) possess two tandem PTP domains, a membrane-proximal D1 domain and a membrane-distal D2 domain. Although most D2 domains are catalytically inactive, the D2 domain of RPTPα is more oxidation-sensitive than its D1 domain [72]. Reversible oxidation of the RPTPα D2 domain induces a conformational change that results in catalytically inactive dimers linked partly by an intermolecular disulfide bond in D2 [148, 149]. These observations suggest a model in which the D2 domain of RPTPs acts as a redox sensor to regulate RPTP activity [5]. However, the crystal structure of LAR suggests that this mode of inhibition might not be general [150]. The recent development of techniques that are not biased towards the detection of non-transmembrane PTPs should facilitate the exploration of RPTP oxidation under physiological and pathological conditions.

In addition to insulin, leptin is another key hormone involved in the regulation of energy balance, whose main function is to suppress food intake and promote energy expenditure [151, 152]. Similar to insulin, obesity also promotes leptin resistance, wherein elevated leptin levels produced by adipocytes fail to suppress feeding and promote weight loss [151]. Several PTPs are known to regulate leptin signaling and therefore are implicated in obesity and leptin resistance, with SHP2 generally promoting, and PTEN, PTP1B, TCPTP, and RPTPε inhibiting, leptin signaling [152]. Little is known about how these PTPs are regulated during leptin signaling, or whether and if so, how ROS-mediated oxidation of PTPs is involved.

Conclusion

The past 10 years have witnessed increasing realization that ROS are not merely toxic metabolic by-products that, when produced excessively, can cause disease, but also important second messengers. Evidence also is accumulating that ROS-induced PTP oxidation/inhibition represents an important mechanism for regulating PTP activity in response to various stimuli and, most likely, within different cellular compartments. Because so many pathways in metabolism utilize tyrosine-phosphorylation-based signaling mechanisms, and all such pathways are regulated by PTPs as well as PTKs, ROS might play a much more important role in normal signaling and disease than is appreciated currently.

Until recently, progress in understanding ROS-mediated signaling events has been impeded by a paucity of approaches that can detect reversible oxidation of PTPs and other proteins in response to physiological stimuli. As summarized herein, this limitation is rapidly being eliminated, and several assays that can detect at least some physiological oxidation events are now available. Although further improvements in assay sensitivity, as well as in the ability to detect localized PTP oxidation within the cellular and tissue context, are clearly needed, it seems likely that the next 10 years will see dramatic progress in understanding this new mode of signal transduction.

References

1. Lemmon MA, Schlessinger J (2010) Cell signaling by receptor tyrosine kinases. Cell 141(7):1117–1134
2. Julien SG, Dube N, Hardy S, Tremblay ML (2011) Inside the human cancer tyrosine phosphatome. Nat Rev Cancer 11(1):35–49
3. Alonso A, Sasin J, Bottini N, Friedberg I, Osterman A, Godzik A et al (2004) Protein tyrosine phosphatases in the human genome. Cell 117(6):699–711
4. Lahiry P, Torkamani A, Schork NJ, Hegele RA (2010) Kinase mutations in human disease: interpreting genotype-phenotype relationships. Nat Rev Genet 11(1):60–74
5. Tonks NK (2006) Protein tyrosine phosphatases: from genes, to function, to disease. Nat Rev Mol Cell Biol 7(11):833–846
6. Tsou RC, Bence KK (2012) Central regulation of metabolism by protein tyrosine phosphatases. Front Neurosci 6:192
7. Asante-Appiah E, Kennedy BP (2003) Protein tyrosine phosphatases: the quest for negative regulators of insulin action. Am J Physiol Endocrinol Metab 284(4):E663–E670
8. Cheng A, Dube N, Gu F, Tremblay ML (2002) Coordinated action of protein tyrosine phosphatases in insulin signal transduction. Eur J Biochem 269(4):1050–1059
9. Tiganis T (2013) PTP1B and TCPTP—nonredundant phosphatases in insulin signaling and glucose homeostasis. FEBS J 280(2):445–458
10. Andersen JN, Mortensen OH, Peters GH, Drake PG, Iversen LF, Olsen OH et al (2001) Structural and evolutionary relationships among protein tyrosine phosphatase domains. Mol Cell Biol 21(21):7117–7136
11. Zhang ZY, Dixon JE (1993) Active site labeling of the Yersinia protein tyrosine phosphatase: the determination of the pKa of the active site cysteine and the function of the conserved histidine 402. Biochemistry 32(36):9340–9345
12. Peters GH, Frimurer TM, Olsen OH (1998) Electrostatic evaluation of the signature motif (H/V)CX5R(S/T) in protein-tyrosine phosphatases. Biochemistry 37(16):5383–5393
13. Denu JM, Dixon JE (1998) Protein tyrosine phosphatases: mechanisms of catalysis and regulation. Curr Opin Chem Biol 2(5):633–641
14. Tonks NK (2005) Redox redux: revisiting PTPs and the control of cell signaling. Cell 121(5):667–670
15. Winterbourn CC, Hampton MB (2008) Thiol chemistry and specificity in redox signaling. Free Radic Biol Med 45(5):549–561
16. Rhee SG (2006) Cell signaling. H2O2, a necessary evil for cell signaling. Science 312(5782):1882–1883
17. Veal EA, Day AM, Morgan BA (2007) Hydrogen peroxide sensing and signaling. Mol Cell 26(1):1–14
18. Winterbourn CC (1993) Superoxide as an intracellular radical sink. Free Radic Biol Med 14(1):85–90
19. Lambeth JD (2004) NOX enzymes and the biology of reactive oxygen. Nat Rev Immunol 4(3):181–189
20. Murphy MP (2009) How mitochondria produce reactive oxygen species. Biochem J 417(1):1–13
21. Riemer J, Bulleid N, Herrmann JM (2009) Disulfide formation in the ER and mitochondria: two solutions to a common process. Science 324(5932):1284–1287
22. Salmeen A, Barford D (2005) Functions and mechanisms of redox regulation of cysteine-based phosphatases. Antioxid Redox Signal 7(5–6):560–577
23. Tanner JJ, Parsons ZD, Cummings AH, Zhou H, Gates KS (2011) Redox regulation of protein tyrosine phosphatases: structural and chemical aspects. Antioxid Redox Signal 15(1):77–97
24. den Hertog J, Groen A, van der Wijk T (2005) Redox regulation of protein-tyrosine phosphatases. Arch Biochem Biophys 434(1):11–15

25. Frangioni JV, Beahm PH, Shifrin V, Jost CA, Neel BG (1992) The nontransmembrane tyrosine phosphatase PTP-1B localizes to the endoplasmic reticulum via its 35 amino acid C-terminal sequence. Cell 68(3):545–560
26. Lorenzen JA, Dadabay CY, Fischer EH (1995) COOH-terminal sequence motifs target the T cell protein tyrosine phosphatase to the ER and nucleus. J Cell Biol 131(3):631–643
27. Pagliarini DJ, Wiley SE, Kimple ME, Dixon JR, Kelly P, Worby CA et al (2005) Involvement of a mitochondrial phosphatase in the regulation of ATP production and insulin secretion in pancreatic beta cells. Mol Cell 19(2):197–207
28. Pagliarini DJ, Dixon JE (2006) Mitochondrial modulation: reversible phosphorylation takes center stage? Trends Biochem Sci 31(1):26–34
29. Rardin MJ, Wiley SE, Murphy AN, Pagliarini DJ, Dixon JE (2008) Dual specificity phosphatases 18 and 21 target to opposing sides of the mitochondrial inner membrane. J Biol Chem 283(22):15440–15450
30. Zhang J, Guan Z, Murphy AN, Wiley SE, Perkins GA, Worby CA et al (2011) Mitochondrial phosphatase PTPMT1 is essential for cardiolipin biosynthesis. Cell Metab 13(6):690–700
31. Lou YW, Chen YY, Hsu SF, Chen RK, Lee CL, Khoo KH et al (2008) Redox regulation of the protein tyrosine phosphatase PTP1B in cancer cells. FEBS J 275(1):69–88
32. Karisch R, Fernandez M, Taylor P, Virtanen C, St-Germain JR, Jin LL et al (2011) Global proteomic assessment of the classical protein-tyrosine phosphatome and "Redoxome". Cell 146(5):826–840
33. Salmeen A, Andersen JN, Myers MP, Meng TC, Hinks JA, Tonks NK et al (2003) Redox regulation of protein tyrosine phosphatase 1B involves a sulphenyl-amide intermediate. Nature 423(6941):769–773
34. van Montfort RL, Congreve M, Tisi D, Carr R, Jhoti H (2003) Oxidation state of the active-site cysteine in protein tyrosine phosphatase 1B. Nature 423(6941):773–777
35. Holmgren A, Johansson C, Berndt C, Lonn ME, Hudemann C, Lillig CH (2005) Thiol redox control via thioredoxin and glutaredoxin systems. Biochem Soc Trans 33(Pt 6):1375–1377
36. Lee SR, Kwon KS, Kim SR, Rhee SG (1998) Reversible inactivation of protein-tyrosine phosphatase 1B in A431 cells stimulated with epidermal growth factor. J Biol Chem 273(25):15366–15372
37. Rhee SG, Kang SW, Jeong W, Chang TS, Yang KS, Woo HA (2005) Intracellular messenger function of hydrogen peroxide and its regulation by peroxiredoxins. Curr Opin Cell Biol 17(2):183–189
38. Meng TC, Fukada T, Tonks NK (2002) Reversible oxidation and inactivation of protein tyrosine phosphatases in vivo. Mol Cell 9(2):387–399
39. Mahadev K, Zilbering A, Zhu L, Goldstein BJ (2001) Insulin-stimulated hydrogen peroxide reversibly inhibits protein-tyrosine phosphatase 1b in vivo and enhances the early insulin action cascade. J Biol Chem 276(24):21938–21942
40. Meng TC, Buckley DA, Galic S, Tiganis T, Tonks NK (2004) Regulation of insulin signaling through reversible oxidation of the protein-tyrosine phosphatases TC45 and PTP1B. J Biol Chem 279(36):37716–37725
41. Kwon J, Lee SR, Yang KS, Ahn Y, Kim YJ, Stadtman ER et al (2004) Reversible oxidation and inactivation of the tumor suppressor PTEN in cells stimulated with peptide growth factors. Proc Natl Acad Sci U S A 101(47):16419–16424
42. Rhee SG, Chae HZ, Kim K (2005) Peroxiredoxins: a historical overview and speculative preview of novel mechanisms and emerging concepts in cell signaling. Free Radic Biol Med 38(12):1543–1552
43. Arthur JR (2000) The glutathione peroxidases. Cell Mol Life Sci 57(13–14):1825–1835
44. Woo HA, Yim SH, Shin DH, Kang D, Yu DY, Rhee SG (2010) Inactivation of peroxiredoxin I by phosphorylation allows localized H(2)O(2) accumulation for cell signaling. Cell 140(4):517–528
45. Ostman A, Hellberg C, Bohmer FD (2006) Protein-tyrosine phosphatases and cancer. Nat Rev Cancer 6(4):307–320

46. Newsholme P, Haber EP, Hirabara SM, Rebelato EL, Procopio J, Morgan D et al (2007) Diabetes associated cell stress and dysfunction: role of mitochondrial and non-mitochondrial ROS production and activity. J Physiol 583(pt 1):9–24
47. Anderson EJ, Lustig ME, Boyle KE, Woodlief TL, Kane DA, Lin CT et al (2009) Mitochondrial H2O2 emission and cellular redox state link excess fat intake to insulin resistance in both rodents and humans. J Clin Invest 119(3):573–581
48. Hotamisligil GS (2010) Endoplasmic reticulum stress and the inflammatory basis of metabolic disease. Cell 140(6):900–917
49. Fialkow L, Wang Y, Downey GP (2007) Reactive oxygen and nitrogen species as signaling molecules regulating neutrophil function. Free Radic Biol Med 42(2):153–164
50. Segel GB, Halterman MW, Lichtman MA (2011) The paradox of the neutrophil's role in tissue injury. J Leukoc Biol 89(3):359–372
51. Cannizzo ES, Clement CC, Sahu R, Follo C, Santambrogio L (2011) Oxidative stress, inflamm-aging and immunosenescence. J Proteomics 74(11):2313–2323
52. Karisch R, Neel BG (2013) Methods to monitor classical protein-tyrosine phosphatase oxidation. FEBS J 280(2):459–475
53. Tonks NK, Diltz CD, Fischer EH (1988) Characterization of the major protein-tyrosine-phosphatases of human placenta. J Biol Chem 263(14):6731–6737
54. Zhu L, Zilbering A, Wu X, Mahadev K, Joseph JI, Jabbour S et al (2001) Use of an anaerobic environment to preserve the endogenous activity of protein-tyrosine phosphatases isolated from intact cells. FASEB J 15(9):1637–1639
55. Boivin B, Tonks NK (2010) Analysis of the redox regulation of protein tyrosine phosphatase superfamily members utilizing a cysteinyl-labeling assay. Methods Enzymol 474:35–50
56. Kwon J, Qu CK, Maeng JS, Falahati R, Lee C, Williams MS (2005) Receptor-stimulated oxidation of SHP-2 promotes T-cell adhesion through SLP-76-ADAP. EMBO J 24(13): 2331–2341
57. Wu RF, Terada LS (2006) Oxidative modification of protein tyrosine phosphatases. Sci STKE 2006(332):pl2
58. Weerapana E, Wang C, Simon GM, Richter F, Khare S, Dillon MB et al (2010) Quantitative reactivity profiling predicts functional cysteines in proteomes. Nature 468(7325):790–795
59. Meng TC, Hsu SF, Tonks NK (2005) Development of a modified in-gel assay to identify protein tyrosine phosphatases that are oxidized and inactivated in vivo. Methods 35(1):28–36
60. Burridge K, Nelson A (1995) An in-gel assay for protein tyrosine phosphatase activity: detection of widespread distribution in cells and tissues. Anal Biochem 232(1):56–64
61. Wu HH, Momand J (1998) Pyrrolidine dithiocarbamate prevents p53 activation and promotes p53 cysteine residue oxidation. J Biol Chem 273(30):18898–18905
62. Li S, Whorton AR (2003) Regulation of protein tyrosine phosphatase 1B in intact cells by S-nitrosothiols. Arch Biochem Biophys 410(2):269–279
63. Boivin B, Zhang S, Arbiser JL, Zhang ZY, Tonks NK (2008) A modified cysteinyl-labeling assay reveals reversible oxidation of protein tyrosine phosphatases in angiomyolipoma cells. Proc Natl Acad Sci U S A 105(29):9959–9964
64. Boivin B, Yang M, Tonks NK (2010) Targeting the reversibly oxidized protein tyrosine phosphatase superfamily. Sci Signal 3(137):pl2
65. Cravatt BF, Wright AT, Kozarich JW (2008) Activity-based protein profiling: from enzyme chemistry to proteomic chemistry. Annu Rev Biochem 77:383–414
66. Jessani N, Cravatt BF (2004) The development and application of methods for activity-based protein profiling. Curr Opin Chem Biol 8(1):54–59
67. Zhang ZY (2003) Mechanistic studies on protein tyrosine phosphatases. Prog Nucleic Acid Res Mol Biol 73:171–220
68. Kumar S, Zhou B, Liang F, Wang WQ, Huang Z, Zhang ZY (2004) Activity-based probes for protein tyrosine phosphatases. Proc Natl Acad Sci U S A 101(21):7943–7948
69. Liu S, Zhou B, Yang H, He Y, Jiang ZX, Kumar S et al (2008) Aryl vinyl sulfonates and sulfones as active site-directed and mechanism-based probes for protein tyrosine phosphatases. J Am Chem Soc 130(26):8251–8260

70. Kolb HC, Finn MG, Sharpless KB (2001) Click chemistry: diverse chemical function from a few good reactions. Angew Chem Int Ed Engl 40(11):2004–2021
71. Sletten EM, Bertozzi CR (2009) Bioorthogonal chemistry: fishing for selectivity in a sea of functionality. Angew Chem Int Ed Engl 48(38):6974–6998
72. Persson C, Sjoblom T, Groen A, Kappert K, Engstrom U, Hellman U et al (2004) Preferential oxidation of the second phosphatase domain of receptor-like PTP-alpha revealed by an antibody against oxidized protein tyrosine phosphatases. Proc Natl Acad Sci U S A 101(7):1886–1891
73. Persson C, Kappert K, Engstrom U, Ostman A, Sjoblom T (2005) An antibody-based method for monitoring in vivo oxidation of protein tyrosine phosphatases. Methods 35(1):37–43
74. Gross S, Blanchetot C, Schepens J, Albet S, Lammers R, den Hertog J et al (2002) Multimerization of the protein-tyrosine phosphatase (PTP)-like insulin-dependent diabetes mellitus autoantigens IA-2 and IA-2beta with receptor PTPs (RPTPs). Inhibition of RPTPalpha enzymatic activity. J Biol Chem 277(50):48139–48145
75. Templeton DJ, Aye MS, Rady J, Xu F, Cross JV (2010) Purification of reversibly oxidized proteins (PROP) reveals a redox switch controlling p38 MAP kinase activity. PLoS One 5(11):e15012
76. Victor KG, Rady JM, Cross JV, Templeton DJ (2012) Proteomic profile of reversible protein oxidation using PROP, purification of reversibly oxidized proteins. PLoS One 7(2):e32527
77. Seo YH, Carroll KS (2009) Profiling protein thiol oxidation in tumor cells using sulfenic acid-specific antibodies. Proc Natl Acad Sci U S A 106(38):16163–16168
78. Maller C, Schroder E, Eaton P (2011) Glyceraldehyde 3-phosphate dehydrogenase is unlikely to mediate hydrogen peroxide signaling: studies with a novel anti-dimedone sulfenic acid antibody. Antioxid Redox Signal 14(1):49–60
79. Poole LB, Zeng BB, Knaggs SA, Yakubu M, King SB (2005) Synthesis of chemical probes to map sulfenic acid modifications on proteins. Bioconjug Chem 16(6):1624–1628
80. Poole LB, Klomsiri C, Knaggs SA, Furdui CM, Nelson KJ, Thomas MJ et al (2007) Fluorescent and affinity-based tools to detect cysteine sulfenic acid formation in proteins. Bioconjug Chem 18(6):2004–2017
81. Nelson KJ, Klomsiri C, Codreanu SG, Soito L, Liebler DC, Rogers LC et al (2010) Use of dimedone-based chemical probes for sulfenic acid detection methods to visualize and identify labeled proteins. Methods Enzymol 473:95–115
82. Michalek RD, Nelson KJ, Holbrook BC, Yi JS, Stridiron D, Daniel LW et al (2007) The requirement of reversible cysteine sulfenic acid formation for T cell activation and function. J Immunol 179(10):6456–6467
83. Oshikawa J, Urao N, Kim HW, Kaplan N, Razvi M, McKinney R et al (2010) Extracellular SOD-derived H2O2 promotes VEGF signaling in caveolae/lipid rafts and post-ischemic angiogenesis in mice. PLoS One 5(4):e10189
84. Wani R, Qian J, Yin L, Bechtold E, King SB, Poole LB et al (2011) Isoform-specific regulation of Akt by PDGF-induced reactive oxygen species. Proc Natl Acad Sci U S A 108(26):10550–10555
85. Reddie KG, Seo YH, Muse Iii WB, Leonard SE, Carroll KS (2008) A chemical approach for detecting sulfenic acid-modified proteins in living cells. Mol Biosyst 4(6):521–531
86. Paulsen CE, Truong TH, Garcia FJ, Homann A, Gupta V, Leonard SE et al (2012) Peroxide-dependent sulfenylation of the EGFR catalytic site enhances kinase activity. Nat Chem Biol 8(1):57–64
87. Charles RL, Schroder E, May G, Free P, Gaffney PR, Wait R et al (2007) Protein sulfenation as a redox sensor: proteomics studies using a novel biotinylated dimedone analogue. Mol Cell Proteomics 6(9):1473–1484
88. Leonard SE, Reddie KG, Carroll KS (2009) Mining the thiol proteome for sulfenic acid modifications reveals new targets for oxidation in cells. ACS Chem Biol 4(9):783–799
89. Leonard SE, Garcia FJ, Goodsell DS, Carroll KS (2011) Redox-based probes for protein tyrosine phosphatases. Angew Chem Int Ed Engl 50(19):4423–4427
90. Qian J, Klomsiri C, Wright MW, King SB, Tsang AW, Poole LB et al (2011) Simple synthesis of 1,3-cyclopentanedione derived probes for labeling sulfenic acid proteins. Chem Commun (Camb) 47(32):9203–9205

91. Qian J, Wani R, Klomsiri C, Poole LB, Tsang AW, Furdui CM (2012) A simple and effective strategy for labeling cysteine sulfenic acid in proteins by utilization of beta-ketoesters as cleavable probes. Chem Commun (Camb) 48(34):4091–4093
92. Haque A, Andersen JN, Salmeen A, Barford D, Tonks NK (2011) Conformation-sensing antibodies stabilize the oxidized form of PTP1B and inhibit its phosphatase activity. Cell 147(1):185–198
93. Elschenbroich S, Kislinger T (2011) Targeted proteomics by selected reaction monitoring mass spectrometry: applications to systems biology and biomarker discovery. Mol Biosyst 7(2):292–303
94. Picotti P, Bodenmiller B, Mueller LN, Domon B, Aebersold R (2009) Full dynamic range proteome analysis of S. cerevisiae by targeted proteomics. Cell 138(4):795–806
95. Held JM, Danielson SR, Behring JB, Atsriku C, Britton DJ, Puckett RL et al (2010) Targeted quantitation of site-specific cysteine oxidation in endogenous proteins using a differential alkylation and multiple reaction monitoring mass spectrometry approach. Mol Cell Proteomics 9(7):1400–1410
96. Lee SR, Yang KS, Kwon J, Lee C, Jeong W, Rhee SG (2002) Reversible inactivation of the tumor suppressor PTEN by H2O2. J Biol Chem 277(23):20336–20342
97. Shaw JE, Sicree RA, Zimmet PZ (2010) Global estimates of the prevalence of diabetes for 2010 and 2030. Diabetes Res Clin Pract 87(1):4–14
98. Wild S, Roglic G, Green A, Sicree R, King H (2004) Global prevalence of diabetes: estimates for the year 2000 and projections for 2030. Diabetes Care 27(5):1047–1053
99. Hossain P, Kawar B, El Nahas M (2007) Obesity and diabetes in the developing world—a growing challenge. N Engl J Med 356(3):213–215
100. Lin Y, Sun Z (2010) Current views on type 2 diabetes. J Endocrinol 204(1):1–11
101. Saltiel AR, Pessin JE (2002) Insulin signaling pathways in time and space. Trends Cell Biol 12(2):65–71
102. Taniguchi CM, Emanuelli B, Kahn CR (2006) Critical nodes in signalling pathways: insights into insulin action. Nat Rev Mol Cell Biol 7(2):85–96
103. Bryant NJ, Govers R, James DE (2002) Regulated transport of the glucose transporter GLUT4. Nat Rev Mol Cell Biol 3(4):267–277
104. Tiganis T (2011) Reactive oxygen species and insulin resistance: the good, the bad and the ugly. Trends Pharmacol Sci 32(2):82–89
105. Hoehn KL, Salmon AB, Hohnen-Behrens C, Turner N, Hoy AJ, Maghzal GJ et al (2009) Insulin resistance is a cellular antioxidant defense mechanism. Proc Natl Acad Sci U S A 106(42):17787–17792
106. Houstis N, Rosen ED, Lander ES (2006) Reactive oxygen species have a causal role in multiple forms of insulin resistance. Nature 440(7086):944–948
107. Koves TR, Ussher JR, Noland RC, Slentz D, Mosedale M, Ilkayeva O et al (2008) Mitochondrial overload and incomplete fatty acid oxidation contribute to skeletal muscle insulin resistance. Cell Metab 7(1):45–56
108. Bloch-Damti A, Bashan N (2005) Proposed mechanisms for the induction of insulin resistance by oxidative stress. Antioxid Redox Signal 7(11–12):1553–1567
109. Singh DK, Winocour P, Farrington K (2011) Oxidative stress in early diabetic nephropathy: fueling the fire. Nat Rev Endocrinol 7(3):176–184
110. Evans JL, Goldfine ID, Maddux BA, Grodsky GM (2002) Oxidative stress and stress-activated signaling pathways: a unifying hypothesis of type 2 diabetes. Endocr Rev 23(5):599–622
111. Forbes JM, Coughlan MT, Cooper ME (2008) Oxidative stress as a major culprit in kidney disease in diabetes. Diabetes 57(6):1446–1454
112. Hotamisligil GS (2006) Inflammation and metabolic disorders. Nature 444(7121):860–867
113. Kim YS, Morgan MJ, Choksi S, Liu ZG (2007) TNF-induced activation of the Nox1 NADPH oxidase and its role in the induction of necrotic cell death. Mol Cell 26(5):675–687
114. Dang PM, Stensballe A, Boussetta T, Raad H, Dewas C, Kroviarski Y et al (2006) A specific p47phox-serine phosphorylated by convergent MAPKs mediates neutrophil NADPH oxidase priming at inflammatory sites. J Clin Invest 116(7):2033–2043

115. Schulze-Osthoff K, Bakker AC, Vanhaesebroeck B, Beyaert R, Jacob WA, Fiers W (1992) Cytotoxic activity of tumor necrosis factor is mediated by early damage of mitochondrial functions. Evidence for the involvement of mitochondrial radical generation. J Biol Chem 267(8):5317–5323
116. Maxwell SR, Thomason H, Sandler D, Leguen C, Baxter MA, Thorpe GH et al (1997) Antioxidant status in patients with uncomplicated insulin-dependent and non-insulin-dependent diabetes mellitus. Eur J Clin Invest 27(6):484–490
117. Opara EC, Abdel-Rahman E, Soliman S, Kamel WA, Souka S, Lowe JE et al (1999) Depletion of total antioxidant capacity in type 2 diabetes. Metabolism 48(11):1414–1417
118. Evans JL, Maddux BA, Goldfine ID (2005) The molecular basis for oxidative stress-induced insulin resistance. Antioxid Redox Signal 7(7–8):1040–1052
119. Rudich A, Kozlovsky N, Potashnik R, Bashan N (1997) Oxidant stress reduces insulin responsiveness in 3T3-L1 adipocytes. Am J Physiol 272(5 pt 1):E935–E940
120. Tirosh A, Potashnik R, Bashan N, Rudich A (1999) Oxidative stress disrupts insulin-induced cellular redistribution of insulin receptor substrate-1 and phosphatidylinositol 3-kinase in 3T3-L1 adipocytes. A putative cellular mechanism for impaired protein kinase B activation and GLUT4 translocation. J Biol Chem 274(15):10595–10602
121. Greene EL, Nelson BA, Robinson KA, Buse MG (2001) alpha-Lipoic acid prevents the development of glucose-induced insulin resistance in 3T3-L1 adipocytes and accelerates the decline in immunoreactive insulin during cell incubation. Metabolism 50(9):1063–1069
122. Rudich A, Tirosh A, Potashnik R, Khamaisi M, Bashan N (1999) Lipoic acid protects against oxidative stress induced impairment in insulin stimulation of protein kinase B and glucose transport in 3T3-L1 adipocytes. Diabetologia 42(8):949–957
123. Jacob S, Henriksen EJ, Schiemann AL, Simon I, Clancy DE, Tritschler HJ et al (1995) Enhancement of glucose disposal in patients with type 2 diabetes by alpha-lipoic acid. Arzneimittelforschung 45(8):872–874
124. Jacob S, Ruus P, Hermann R, Tritschler HJ, Maerker E, Renn W et al (1999) Oral administration of RAC-alpha-lipoic acid modulates insulin sensitivity in patients with type-2 diabetes mellitus: a placebo-controlled pilot trial. Free Radic Biol Med 27(3–4):309–314
125. Gual P, Le Marchand-Brustel Y, Tanti JF (2005) Positive and negative regulation of insulin signaling through IRS-1 phosphorylation. Biochimie 87(1):99–109
126. Greene MW, Sakaue H, Wang L, Alessi DR, Roth RA (2003) Modulation of insulin-stimulated degradation of human insulin receptor substrate-1 by serine 312 phosphorylation. J Biol Chem 278(10):8199–8211
127. Kamata H, Honda S, Maeda S, Chang L, Hirata H, Karin M (2005) Reactive oxygen species promote TNFalpha-induced death and sustained JNK activation by inhibiting MAP kinase phosphatases. Cell 120(5):649–661
128. Liu H, Nishitoh H, Ichijo H, Kyriakis JM (2000) Activation of apoptosis signal-regulating kinase 1 (ASK1) by tumor necrosis factor receptor-associated factor 2 requires prior dissociation of the ASK1 inhibitor thioredoxin. Mol Cell Biol 20(6):2198–2208
129. Robertson RP, Harmon J, Tran PO, Poitout V (2004) Beta-cell glucose toxicity, lipotoxicity, and chronic oxidative stress in type 2 diabetes. Diabetes 53(suppl 1):S119–S124
130. Tang C, Han P, Oprescu AI, Lee SC, Gyulkhandanyan AV, Chan GN et al (2007) Evidence for a role of superoxide generation in glucose-induced beta-cell dysfunction in vivo. Diabetes 56(11):2722–2731
131. Vinciguerra M, Foti M (2006) PTEN and SHIP2 phosphoinositide phosphatases as negative regulators of insulin signalling. Arch Physiol Biochem 112(2):89–104
132. Mahadev K, Motoshima H, Wu X, Ruddy JM, Arnold RS, Cheng G et al (2004) The NAD(P)H oxidase homolog Nox4 modulates insulin-stimulated generation of H2O2 and plays an integral role in insulin signal transduction. Mol Cell Biol 24(5):1844–1854
133. Ostman A, Frijhoff J, Sandin A, Bohmer FD (2011) Regulation of protein tyrosine phosphatases by reversible oxidation. J Biochem 150(4):345–356
134. Loh K, Deng H, Fukushima A, Cai X, Boivin B, Galic S et al (2009) Reactive oxygen species enhance insulin sensitivity. Cell Metab 10(4):260–272

135. Bandyopadhyay D, Kusari A, Kenner KA, Liu F, Chernoff J, Gustafson TA et al (1997) Protein-tyrosine phosphatase 1B complexes with the insulin receptor in vivo and is tyrosine-phosphorylated in the presence of insulin. J Biol Chem 272(3):1639–1645
136. Salmeen A, Andersen JN, Myers MP, Tonks NK, Barford D (2000) Molecular basis for the dephosphorylation of the activation segment of the insulin receptor by protein tyrosine phosphatase 1B. Mol Cell 6(6):1401–1412
137. Goldstein BJ, Bittner-Kowalczyk A, White MF, Harbeck M (2000) Tyrosine dephosphorylation and deactivation of insulin receptor substrate-1 by protein-tyrosine phosphatase 1B. Possible facilitation by the formation of a ternary complex with the Grb2 adaptor protein. J Biol Chem 275(6):4283–4289
138. Elchebly M, Payette P, Michaliszyn E, Cromlish W, Collins S, Loy AL et al (1999) Increased insulin sensitivity and obesity resistance in mice lacking the protein tyrosine phosphatase-1B gene. Science 283(5407):1544–1548
139. Klaman LD, Boss O, Peroni OD, Kim JK, Martino JL, Zabolotny JM et al (2000) Increased energy expenditure, decreased adiposity, and tissue-specific insulin sensitivity in protein-tyrosine phosphatase 1B-deficient mice. Mol Cell Biol 20(15):5479–5489
140. Zhang S, Zhang ZY (2007) PTP1B as a drug target: recent developments in PTP1B inhibitor discovery. Drug Discov Today 12(9–10):373–381
141. Sobhia ME, Paul S, Shinde R, Potluri M, Gundam V, Kaur A et al (2012) Protein tyrosine phosphatase inhibitors: a patent review (2002–2011). Expert Opin Ther Pat 22(2):125–153
142. Ahmad F, Goldstein BJ (1997) Functional association between the insulin receptor and the transmembrane protein-tyrosine phosphatase LAR in intact cells. J Biol Chem 272(1): 448–457
143. Kulas DT, Zhang WR, Goldstein BJ, Furlanetto RW, Mooney RA (1995) Insulin receptor signaling is augmented by antisense inhibition of the protein tyrosine phosphatase LAR. J Biol Chem 270(6):2435–2438
144. Mander A, Hodgkinson CP, Sale GJ (2005) Knock-down of LAR protein tyrosine phosphatase induces insulin resistance. FEBS Lett 579(14):3024–3028
145. Zabolotny JM, Kim YB, Peroni OD, Kim JK, Pani MA, Boss O et al (2001) Overexpression of the LAR (leukocyte antigen-related) protein-tyrosine phosphatase in muscle causes insulin resistance. Proc Natl Acad Sci U S A 98(9):5187–5192
146. Lammers R, Moller NP, Ullrich A (1997) The transmembrane protein tyrosine phosphatase alpha dephosphorylates the insulin receptor in intact cells. FEBS Lett 404(1):37–40
147. Cong LN, Chen H, Li Y, Lin CH, Sap J, Quon MJ (1999) Overexpression of protein tyrosine phosphatase-alpha (PTP-alpha) but not PTP-kappa inhibits translocation of GLUT4 in rat adipose cells. Biochem Biophys Res Commun 255(2):200–207
148. Blanchetot C, Tertoolen LG, den Hertog J (2002) Regulation of receptor protein-tyrosine phosphatase alpha by oxidative stress. EMBO J 21(4):493–503
149. van der Wijk T, Overvoorde J, den Hertog J (2004) H2O2-induced intermolecular disulfide bond formation between receptor protein-tyrosine phosphatases. J Biol Chem 279(43): 44355–44361
150. Nam HJ, Poy F, Krueger NX, Saito H, Frederick CA (1999) Crystal structure of the tandem phosphatase domains of RPTP LAR. Cell 97(4):449–457
151. Myers MG, Cowley MA, Munzberg H (2008) Mechanisms of leptin action and leptin resistance. Annu Rev Physiol 70:537–556
152. St-Pierre J, Tremblay ML (2012) Modulation of leptin resistance by protein tyrosine phosphatases. Cell Metab 15(3):292–297

Chapter 2
Quantitative Modeling Approaches for Understanding the Role of Phosphatases in Cell Signaling Regulation: Applications in Metabolism

Matthew J. Lazzara

Abstract The regulation of cell signaling occurs through a complex set of coupled processes occurring over multiple length and time scales. Computational modeling approaches have been applied to dissect this complexity over these various time and length scales, which range from the molecular level to the cell and tissue level, but these approaches have not focused heavily on the regulation or roles of phosphatases. Because of the clear importance of phosphatases in cell signaling, significant opportunities exist to expand our understanding of the regulation of cell signaling in metabolism and other cell regulatory processes by focusing modern computational approaches on phosphatases and the processes they regulate. The aim of this chapter is to provide a brief review of some computational modeling approaches that have been usefully applied to study the regulation of signaling, mainly by kinases, over a range of length and time scales and to describe opportunities to apply similar approaches for understanding signaling regulation by phosphatases. Some specific examples of key relevance to metabolism are described.

Abbreviations

AKT	Protein kinase B
DEP1/RPTPJ	Density-enhanced phosphatase 1/receptor-like protein tyrosine phosphatase J
EGF	Epidermal growth factor
EGFR	Epidermal growth factor receptor
GAB1	GRB2-associated binder 1

M.J. Lazzara, Ph.D. (✉)
Department of Chemical and Biomolecular Engineering, University of Pennsylvania, Philadelphia, PA 19104-6393, USA
e-mail: mlazzara@seas.upenn.edu

K.K. Bence (ed.), *Protein Tyrosine Phosphatase Control of Metabolism*,
DOI 10.1007/978-1-4614-7855-3_2, © Springer Science+Business Media New York 2013

GLUT4	Glucose transporter 4
GRB2	Growth factor receptor-bound protein 2
HER2	Human epidermal growth factor receptor 2
HER3/ErbB3	Human epidermal growth factor receptor 3
IR	Insulin receptor
IRS	Insulin receptor substrate
MAP	Mitogen-activated protein
MAPK	Mitogen-activated protein kinase
MEK	Mitogen-activated protein kinase kinase
MKP3/DUSP6	MAP kinase phosphatase 3/dual specificity phosphatase 6
NSCLC	Non-small cell lung cancer
PAG/Cbp	Phosphoprotein associated with glycosphingolipid-enriched microdomains/Csk-binding protein
PDGFR	Platelet-derived growth factor receptor
PDK1	Phosphoinositide-dependent kinase-1
PI3K	Phosphoinositide 3-kinase
PLSR	Partial least squares regression
PTEN	Phosphatase and tensin homolog
PTP	Protein tyrosine phosphatase
PTP1B	Protein tyrosine phosphatase 1B
Raf	Mitogen-activated protein kinase kinase kinase for ERK pathway
ROS	Reactive oxygen species
RPTP	Receptor-like protein tyrosine phosphatase
RTK	Receptor tyrosine kinase
SH2	Src homology 2
SHP1	Src homology 2 domain containing phosphatase 1
SHP2	Src homology 2 domain containing phosphatase 2
Src	Proto-oncogene tyrosine kinase

Introduction

Over the last several decades, there has been a dramatic increase in the application of computational modeling approaches to the study of cell biological phenomena. In the study of mechanisms of cell signaling, a spectrum of modeling approaches has been brought to bear to understand how structural changes in molecules involved in signal transduction, interactions between receptors and ligands, receptor trafficking processes (e.g., endocytosis, recycling), and various posttranslational modifications are ultimately translated into cellular decisions through their impact on the regulation of multivariate intracellular signaling pathways [1]. These computational studies have addressed phenomena spanning multiple length and time scales, from molecular level studies with atomistic detail to approaches encompassing signaling at the network level and attempting to connect changes in signaling network regulation to altered system output at the cellular or tissue level (e.g., [2–4]). Of course, it is increasingly appreciated that the study of these phenomena with single-cell resolution provides a deeper and more realistic picture of the complexities which dictate phenotypes at the tissue level. Because of continuing experimental advances that enable the acquisition of large data sets at the single-cell level, it is increasingly possible, and indeed useful, to apply computational approaches to understand how variability in the regulation of cell signaling from cell to cell arises and influences the fates of populations of cells [5].

In the broad field of cell signaling research, the vast majority of computational work has focused on the function of kinases and the phosphorylation events they catalyze. By contrast, relatively little direct effort has focused on dissecting the role of protein phosphatases through computational modeling. However, the unique protein structures and functions of phosphatases [6–8], and their clear importance in the regulation of signal transduction in health and disease [9–11], make them ideal subjects for computational modeling studies. Indeed, application of computational approaches to study phosphatases and the processes they regulate across the same varied length and time scales as have been explored for kinases is particularly timely as we continue to uncover more about the critical and complex roles played by phosphatases in signal transduction. Of specific relevance to this book, an increasingly important role continues to emerge for phosphatases in the regulation of metabolism (e.g., [12–16]).

Since there is not a large body of work already dedicated to the application of computational approaches to study phosphatases, this chapter will mainly attempt to address ways in which various computational approaches that have been applied to well-studied kinases could be usefully implemented to aid our understanding of how protein phosphatases function across multiple length and time scales. Protein tyrosine phosphatases (PTPs) will be a focus of discussion, but the points presented are broadly applicable to other classes of phosphatases. Our discussion will begin at the atomistic and molecular length scales, then progress to the signaling network level, discuss specific multiscale approaches, and end with a brief discussion of specific potential applications of these approaches to the study of metabolism.

Models at the Molecular Level

For receptor and non-receptor kinases, a great deal of computational work has been undertaken to explore the structure/function relationships which exist in wild-type proteins and how these relationships can be perturbed in meaningful ways through mutations arising in disease. While models at the molecular level based on molecular docking simulations or molecular dynamics simulations have both been in use for some time, simulations based on molecular dynamics have generally been less common because of the challenges associated with these approaches and the computational resources required. As a result of advances in computational methods and increased availability of computational resources, we have more recently seen increased application of molecular dynamics. Such simulations may include detail at the atomic level, with calculation of atomistic trajectories based upon solutions of Newton's equations with consideration of interatomic potential force fields [17–19]. These approaches enable new kinetic understanding of the effects of specific polymorphisms in intramolecular and intermolecular interactions that may regulate protein activity and binding.

For example, the epidermal growth factor receptor (EGFR) has been the focus of recent effort to understand, based in part upon modern molecular dynamics simulations with full atomistic detail, how recently identified polymorphisms may alter the catalytic activity of the receptor kinase and downstream phosphorylation of receptor cytoplasmic tyrosines, as well as how such mutations might alter the binding of targeted kinase inhibitors and ATP itself [20, 21]. The same kinds of approaches have been used to understand how the structure of the HER3/ErbB3 pseudokinase determines the low, but non-zero, level of activity exhibited by this member of the ErbB family of receptor tyrosine kinases (RTKs) [22, 23]. The power of such approaches is that they can be utilized to predict, prior to experimental analysis, the molecular level consequences for specific mutations in terms of likely changes to activity or other functions (e.g., binding) or to understand the specific impact of mutations identified in disease (e.g., [24–26]). Such information can be used to prioritize the experimental analysis of mutations identified in genomic screens of tissues isolated from patients as part of the study of specific diseases. This is potentially quite useful since the functional consequences of many mutations identified through screens are completely unknown. In part, the most useful implementation of these approaches relies upon unique structural information which has been gleaned through crystallography about the importance of specific structural motifs in regulating the activity of receptor kinases. Recent advances in these crystallographic approaches have enabled more reliable predictions from these kinds of models [27].

Although these kinds of modern approaches have not yet been widely applied to the study of protein phosphatases, their application, with suitable experimental validation, could indeed yield a great deal of needed insight. The cytosolic PTPs SHP1 and SHP2, both of which contain N-terminal SH2 domains, provide useful examples for discussion [28]. A now well-characterized aspect of the regulation of these phosphatases is that the binding of the "back side" of their N-terminal SH2 domains (i.e., the face of the SH2 domain not involved in binding to phosphotyrosines) to the catalytic domain of the phosphatases is thermodynamically favored when the SH2 domains are not engaged by a phosphotyrosine on a cognate receptor or intracellular adapter protein such as GAB1 [28]. When the SH2 domains become engaged, however, the SH2 domain tends to disengage from the catalytic domain of the receptor, which exposes the catalytic domain and enables functional phosphatase activity. Of course, a number of SHP2 mutations have been characterized in diseases including Noonan and LEOPARD syndromes [29–31]. Interestingly, these diseases share some clinical features, but are associated with different families of *PTPN11* mutations [32]. Whereas Noonan Syndrome is characterized by gain-of-function SHP2 mutations [33], the most common mutations found in LEOPARD syndrome appear to lead to SHP2 deactivation [34]. Intriguingly, these SHP2-deactivating mutations involve an open SHP2 confirmation, which may result from a disruption of the structure of SHP2's catalytic pocket. While these findings were interpreted using fairly standard molecular modeling and visualization approaches, the basis for these differences could be further explored by comparing the effects of these mutations through molecular dynamics simulations and deciphering what structures may become preferentially stabilized as a result of each mutation. Of course, this same approach can be used for the purpose of predicting the effects of new SHP2 mutations as they are identified and may lead to new inferences on the exact structural consequences of already catalogued SHP2 mutations.

These types of molecular dynamics simulations may also be useful in the design and optimization of candidate therapeutics aimed at the specific inhibition of PTPs such as SHP2, PTP1B, and others. Some potent inhibitors of PTPs are indeed available (e.g., sodium orthovanadate), but they generally suffer from a lack of specificity, which has been found in trials to lead to undesirable clinical outcomes [35, 36]. The primary challenge in synthesizing specific PTP inhibitors has been attributed mainly to the shallow and highly conserved catalytic domain found in PTPs [35, 36]. It may be possible, however, to overcome this issue by capitalizing on allosteric mechanisms for regulating PTP activity that are controlled by protein epitopes located outside the catalytic domain, as may be possible for PTP1B for example [37]. Molecular dynamics simulations may be a useful tool for understanding the applicability of this approach for other PTPs.

The application of similar computational approaches would also be of substantial benefit in the elucidation of molecular mechanisms underpinning the activation of receptor-like PTPs. Unlike many RTKs, where dimerization is a critical step leading to receptor phosphorylation, prototypical receptor-like PTPs tend to become inhibited through dimerization [6, 38, 39]. Relatively little has been done at the level of molecular dynamics simulations to understand the basis for this effect.

Mechanistic Models with Dynamic Resolution at Various Levels of Network Complexity

A number of investigators have developed computational models with purely dynamic considerations for the regulation of intracellular signaling. These models often take the form of coupled systems of ordinary differential equations, with one equation for each distinct species considered in the model. Such models frequently include, and may be limited to, consideration of the activation of an upstream initiator of signaling such as the RTKs EGFR and PDGFR (e.g., [40–42]). Receptor-level models of this type may need to consider many receptor-containing species (e.g., ligand-bound, ATP-bound) in order to capture all of the experimentally observed dynamics intended to be recapitulated by the model. This combinatorial complexity can lead to relatively large models (e.g., tens to hundreds of coupled differential equations) even with consideration of receptors alone. In many cases, these models may also attempt to connect upstream receptor activation with the activation of one or more downstream pathways, such as those leading to phosphorylation and activation of the extracellular regulated kinase (ERK) and AKT [43–47]. Not surprisingly, the additional consideration of these downstream pathways greatly expands the size and complexities of these dynamic models. Generally, the goal of these studies is to gain predictive understanding of how system perturbations of different kinds (e.g., receptor mutations affecting kinase activity or ligand binding, changes in protein expression) and/or alterations in network topology ultimately lead to alterations in the dynamics of receptor or downstream pathway phosphorylation or the emergence of new system properties such as ultrasensitivity [2, 42, 43, 48–53].

Focusing first on receptor kinases which initiate downstream signaling, a substantial amount of prior work has focused on EGFR/ErbB1 and its sister ErbB receptors (HER2/ErbB2, ErbB3, and ErbB4). In many cases, these models have taken the form of compartmentalized mechanistic models built upon systems of coupled ordinary differential equations that describe the dynamics of the evolution of the molecular species under consideration (e.g., free EGFR, ligand-bound EGFR). Compartmentalization of these models into plasma membrane and internal compartments can be an effective way to understand the rates and roles of receptor trafficking processes and the influence of various cellular perturbations (e.g., HER2 over-expression) on receptor trafficking. Lauffenburger and colleagues, as well as other investigators, have implemented a number of such models (e.g., [43, 52, 53]), for example to quantitatively determine the role of HER2 in regulating EGFR endocytosis in cultured breast epithelial cells.

In terms of individual signaling pathways, perhaps none has received more direct attention than the mitogen-activated protein kinase (MAPK) cascade leading to ERK phosphorylation. This pathway has been thoroughly investigated in isolation to uncover the basis for signaling amplification from upstream initiators, to uncover the role of various feedback mechanisms, and to uncover the basis for ultrasensitivity of the pathway (e.g., [54–56]). In some recent models, investigators have attempted to integrate from the upstream level of receptor activation down to the ultimate effectors of intracellular pathways such as ERK and AKT [43, 45, 52, 53]. Despite the increased complexity and size of the models, some useful and testable hypotheses can be generated through such models. For example, the impact of perturbations to receptor endocytosis and kinase activity on ligand-mediated ERK and AKT activation has been investigated for EGFR to probe the functional consequences of a family of kinase-activating EGFR mutations that arise in non-small cell lung cancer (NSCLC) [43].

One issue which plagues these models, and indeed all mechanistic models of this general type, is the need for realistic parameters to characterize the various rate processes under consideration. This has led to the development of numerous new approaches to address the issue of model identifiability (e.g., [57, 58]). Where pathways including EGFR and ERK are involved, modelers have benefited from a relative wealth of available parameters in the literature. Of course, when a relatively small number of parameters are unknown, these can often be fit through regression or training of the model against appropriate kinetic data. Finally, it is also worth noting that in many instances the behavior of the model as a function of various topological variations can be quite informative, even without the availability of well-quantified rate constants. Computational methods to aid in topology inference and analysis have also been described in the literature (e.g., [59, 60]).

In some cases, published models have attempted to quantify the activities of phosphatases with respect to a particular phosphorylated site on a receptor or downstream signaling intermediate through model inference and parameter fitting. For example, some authors have assumed forcing functions for dephosphorylation kinetics coupled with model assumptions about the intracellular site of receptor dephosphorylation to fit rate constants for dephosphorylation; these are treated as unknowns in models which can be regressed against dynamic data for the transient induction of phosphorylation of proteins in response to an upstream stimulatory effect (e.g., [52]). An inherent difficulty with this approach, however, is that fitting such models against data for a phosphorylation response to ligand alone may not fully reveal the kinetics associated with the dephosphorylation process. For example, with respect to the EGFR, it is clear that receptor tyrosines cycle through phosphates relatively quickly compared to the time scale with which receptor phosphorylation is diminished in time as part of a response to ligand alone [41, 61, 62]. This rapid, or futile, phosphate cycling is most readily revealed by following the kinetics of EGFR dephosphorylation in response to cell treatment with inhibitors of the receptor kinase, which leads to nearly complete dephosphorylation approximately 1 min after inhibitor addition [41]. Thus, to fully reveal and quantify the rates of regulation of particular phosphosites by PTPs, it is necessary to consider multiple kinds of phosphorylation response data in model regression, including response to ligands, kinase inhibitors, and phosphatase inhibitors. Experiments with expression and knockdown of specific PTPs and their mutants are of course also useful.

In fact, this is the basic approach we recently adopted in a study aimed at quantifying the kinetics of dephosphorylation of EGFR at a specific tyrosine (1068), either at the cell surface (prior to endocytosis) or in the cell interior (after internalization) [41]. Our approach centered on the construction of a novel kinetic, compartmentalized model of EGFR dynamics, with consideration of the dynamics of ligand binding, receptor dimerization, phosphorylation, endocytosis, and dephosphorylation. Based on extensive prior quantitative characterization of EGFR-mediated cellular process reported in the literature, we were able to construct the model with relatively few unknown parameters. In addition to rate constants for the decoupling of receptor dimer species, we also left as unknowns the rate constants describing the dephosphorylation of EGFR at the cell surface and in the cell interior. To determine these unknowns, we fit our model against data for the dynamics of phosphorylation of EGFR in response to a diverse set of perturbations including the addition of: exogenous EGF, EGF chased by the EGFR kinase inhibitor gefitinib, and PTP inhibition using pervanadate. We repeated a subset of these experiments in cells expressing dominant negative (K44A) dynamin and also utilized quantitative measurements for the dynamics of internalization of radiolabeled EGF. By asking the model to optimally and simultaneously explain all of the diverse data we obtained with a single set of parameters (using a simulated annealing algorithm), the model arrived at best-fit results indicating that EGFR dephosphorylation at the plasma membrane and in the cell interior are equally rapid. Moreover, this rate of dephosphorylation is so rapid that EGFR is dephosphorylated and rephosphorylated several times after initial kinase activation and prior to internalization. Thus, EGFR tyrosines undergo a futile cycling process on a time scale which is small compared to receptor trafficking.

This central finding has a number of important potential implications for our understanding of the regulation of cell signaling by RTKs such as EGFR. For example, given that EGFR tyrosine phosphorylation at Y1068 is required for normal ligand-induced clathrin-mediated EGFR internalization regulated through the process of Grb2 binding [63–65], the finding that EGFR is dephosphorylated prior to its endocytosis suggests the possibility that phosphatases could regulate receptor endocytosis. In fact, one recent paper by Yarden and coworkers suggests that the RPTPs DEP1/RPTPJ and RPTPK may in fact regulate EGFR endocytosis in this way [66]. This possibility is particularly interesting to consider given that the classical view of RTK-mediated signaling has been that RTKs are activated and phosphorylated at the cell surface and dephosphorylated in a signal-attenuating mechanism ultimately after endocytosis. The new picture painted by our recent modeling study [41], and indeed by the results of others' experimental analysis, suggests a significantly more complex picture wherein EGFR tyrosine phosphorylation is highly transient.

Yet another potential barrier to the realistic implementation of models including effects of PTPs is that there is generally scarce information on how the activities of PTPs are regulated in response to upstream signaling activators such as RTKs. For the ERK cascade, for example, knowledge of the degree to which MKP3/DUSP6 activity is induced, as well as the activities of phosphatases for MEK and Raf, in response to upstream pathway activation would be useful in developing improved models. For most PTPs, however, the dynamics of activity regulation in response to upstream activators has not been well characterized. SHP2 is perhaps one notable exception [67]. A recently described experimental approach may also provide a means to quantitatively determine the activities of multiple phosphatases from cell extracts in high-throughput fashion [68]. Future studies which attempt to more realistically recapitulate regulation by phosphatases will need to include detailed cell biological measurements of induction of PTP activity in time. Of course, it should be noted that a number of studies have demonstrated the importance of the transcriptional regulation of phosphatases such as MKPs as an immediate early gene response to growth factor receptor activation [69, 70]. Consideration of these effects in kinetic models would also clearly aid in more realistically capturing signaling dynamics.

Mechanistic Models with Spatiotemporal Resolution of the Intracellular Distribution and Function of Phosphatases

The network level models of signaling described in the previous section deal with dynamics alone. That is, these models most often take the form of coupled ordinary differential equations describing the evolution in time of the various molecular species involved and the rate processes in which they participate, without directly addressing the possibility that spatial variations within the cell in the concentrations of these species may exist and influence the overall rate processes considered. Such spatial effects may arise for several reasons, including protein intracellular localization effects and diffusional limitations.

PTP1B provides a useful example of a phosphatase whose intracellular localization may lead to potentially important spatial regulation of signaling initiated by EGFR, which is an established substrate of PTP1B [71, 72]. PTP1B contains an N-terminal tag which tethers it to the ER [73]. This has led to speculation by some that PTP1B may act upon EGFR primarily in the cell interior after ligand-mediated endocytosis. To whatever extent PTP1B activity with respect to EGFR is spatially biased to the cell interior, however, this almost certainly does not mean that EGFR tyrosines are dephosphorylated only in the cell interior. Indeed, several studies provide unambiguous evidence that EGFR can be rapidly dephosphorylated at the plasma membrane [41, 61, 62], which may suggest that multiple PTPs are likely to be involved in the regulation of EGFR tyrosine phosphorylation/dephosphorylation. Interestingly, recent evidence has also been provided that PTP1B tethered to the endoplasmic reticulum may still be able to interact with plasma membrane-localized substrates, such as Src, at sites where the ER membrane comes into close proximity with the plasma membrane [74]. In settings where this effect is present, the length scale separating PTP1B from other well-known substrates including EGFR is greatly reduced.

Yet another aspect of the spatial regulation of PTP activity which deserves attention and has been ignored for the most part in the literature is the extent to which the intracellular localization of specific phosphatases may change in response to cellular perturbations such as growth factor receptor activation. It is clear that receptor-like PTPs such as DEP1/PTPRJ are capable of acting upon their substrates while those substrates remain in the plasma membrane, but whether or not membrane-bound PTPs such as DEP1 follow EGFR into endocytic pits or not has not been well explored, at least as far as we are aware. Moreover, downstream trafficking processes such as endocytic recycling or routing for degradation have also not been investigated in the literature. Quantitative measurements of these trafficking processes can be made, however, using well-established techniques, utilizing radiolabeling of ligand or antibodies and imaging approaches, which have been employed extensively for the analysis of receptor trafficking. More importantly for the purposes of this discussion, such quantitative measurements can be used directly in network level models of signaling. Specific RPTPs of interest can be treated as trafficking-competent species in the same way as receptors are often treated in such models. The extent to which these RPTPs participate in established trafficking processes in which RTKs participate may have extremely important implications for the dynamics of signaling, which can be elucidated in part through the implementation of such models.

Spatial regulation in PTP-mediated signaling may also involve potential diffusional limitations which may arise in the regulation of certain PTPs. An intriguing example arises through consideration of SHP2, whose activity is greatly augmented through engagement of N-terminal SH2 domains by cognate phosphotyrosines [28]. This aspect of SHP2 functionality leads to an interesting and important question regarding how long it takes SHP2 to return to the basal auto-inhibited state after it disengages from a phosphotyrosine-containing protein. The answer to this question is important because it will influence the length scale over which SHP2 diffuses after disengaging from a phosphotyrosine-containing protein and while remaining catalytically active. If that time scale is arbitrarily small, SHP2 activity effectively requires constant engagement of its SH2 domain. If, however, that time scale is sufficiently large (that is, large compared to the time scale for diffusion over the cellular length scale, which is $\sim R^2/D$ where R is the radius of a cell and D is the diffusivity of the SHP2), it may be possible for SHP2 to be active anywhere within the cell before returning to the basal auto-inhibited state. Since the time scale for the free diffusion of SHP2 substrates within the cell is likely to be relatively small given the sizes of typical signaling proteins, this potential aspect of SHP2's regulation in space would likely be most important in cases where SHP2 substrates are not freely diffusible within the cell interior. Given SHP2's role in regulating signaling initiated from focal adhesion complexes [75, 76], such scenarios may indeed arise. The determination of the time scale for SHP2's return to the basal auto-inhibited state could be estimated through the implementation of modern molecular dynamics simulations which consider the movement and momentum of individual water molecules and attractive and repulsive intramolecular forces existing between different protein domains which regulate conformation. This time scale would also be influenced to some extent by rotational entropic effects that partially dictate whether or not the SH2 domain is in proper alignment to bind to the catalytic domain when the two encounter one another.

Yet another interesting and illustrative example of the potential importance of understanding reaction-diffusion aspects of spatiotemporal regulation of PTPs comes from the effects of reactive oxygen species (ROS) on PTP activity. The conserved cysteine present in the highly conserved catalytic domain of PTPs is susceptible to oxidation, which leads to PTP inactivation [77, 78]. The consequences of this for the regulation of PTP activity are manifold, but an interesting example for our present discussion arises from consideration of the fact that the activation of RTKs, including EGFR, may generate ROS [79, 80]. Thus, the very event which catalyzes the nucleation of signaling complexes containing SHP2 by EGFR can also lead to the inactivation of SHP2. ROS are generally extremely short-lived species, however, and the extent to which ROS can diffuse away from the site of generation and react with SHP2 or other PTPs which may be RTK-activated prior to ROS decomposition is not clear. The relevant length scale over which this ROS diffusion must occur for PTP inactivation would also be affected by the extent to which SHP2-containing complexes are able to dissociate and diffuse away from the RTK which initiated protein complex formation and ROS generation. These considerations have not been previously explored, but mechanistic computational models with spatial resolution would be useful for addressing these issues.

From a computational modeling perspective, the relevant equations to address the aspects of spatiotemporal PTP activity arising from reaction-diffusion mechanisms are well described, as are the approaches to solve those equations [81]. In the simplest examples of problems where the effects of convection (i.e., bulk fluid motion in the cytosol) are not considered, the basic conservation equations for species diffusion within cells are coupled second-order partial differential equations for individual species. Subject to appropriate boundary and initial conditions, such as reaction or flux conditions at the plasma membrane, and specification of rate constants and parameters, these equations can be solved using readily available commercial software or through the implementation of ad hoc implicit solvers.

Data-Driven Models at the Signaling Network Level

A fairly recent approach for studying the quantitative relationships between multivariate signaling processes and cellular decision processes (e.g., complex phenotypic processes such as proliferation, differentiation, and apoptosis) is partial least squares regression (PLSR). This computational approach has been applied widely outside of biological research, for example to the analysis of combustion, which is also characterized by the presence of numerous chemical species, but has more recently been adopted to address the problem of quantitatively predicting cell fate determination from multivariate data for the dynamic regulation of numerous signaling pathway intermediates in response to cell stimuli [82–84].

The PLSR method is based in part upon linear algebraic approaches for identifying a series of so-called principal components from a matrix of signaling data, which is typically arranged with different cellular perturbations in rows (e.g., treatment of cells with a particular cytokine or inhibitor) and a variety of measures of different signaling analytes in columns (e.g., ERK phosphorylation 1 h after treatment of cells with a particular ligand). The principal components consist of linear combinations of signaling metrics which together maximally describe the most important variations in the signaling data observed as a function of the perturbations ordered in rows. Parallel cell phenotypic data (e.g., proliferation response to the perturbations) are then regressed to determine what linear combinations of the principal components best predict cell outcome decisions with the highest degree of confidence. The results of this calculation enable one to pick out signaling analytes which are maximally loaded into the highest ranking principal components, and thereby identify those pathways which are likely to exert the strongest control over a particular cell decision process. The essence of this approach is to embrace the fact that complex cell fate decision processes are generally determined not through the activity of an individual signaling pathway but instead through the integration of changes in the activities of many pathways. Without consideration of the mechanistic details of how these pathways are actually activated or interconnected with each other, this approach, in its most common implementation, assumes that fold-changes in different representations of the phosphorylation state of individual proteins (e.g., level, integral over time, or instantaneous rate of change) contribute to the determination of cell fate decisions. Interestingly, recent work demonstrates the potential to obtain mechanistic insights (e.g., the identification of key time scales or the identification of especially important reactions in a larger network of reactions) from the large data sets which typically accompany data-driven modeling studies through consideration of the dynamic changes in measures of covariance for analyte concentrations [85].

This approach depends upon the availability of multivariate sets of signaling and cell response data. A variety of approaches have been used individually or in combination to gather the necessary signaling data, including mass spectrometry phosphoproteomics and bead-based multiplexed sandwich assay approaches such as Luminex [82, 83, 86]. Many of these techniques tend to focus on the regulation of kinases, typically because of reagent availability. For those that do not have an inherent bias in gathering data with coverage of kinases, the analysis of data still tends to focus mainly on kinases, in large part because the functions of phosphorylated sites are generally more well characterized for kinases than those of phosphatases.

In the future, it would be of great interest to apply these kinds of data-driven modeling approaches to model the ability of phosphatases to control cell fate decision processes. A challenge for implementing this approach for phosphatases, however, is that the regulation of phosphatase activity cannot generally be as well monitored by measurement of proxies such as the phosphorylation of key residues, as is true with many kinases and transcription factors. One approach to circumvent this present difficulty could be to utilize specific PTP activity assays, as has been done for some previous PLSR studies for a variety of kinases. Yet another approach might be to attempt to develop phosphatase-centric PLSR models using PCR microarray data to look for changes in transcription of phosphatases as a function of time in response to cellular perturbations leading to phenotypes of interest. Indeed, a number of tools are already commercially available for the high-throughput analysis of phosphatase expression.

Models Integrating Multiple Scales

Increasingly, computational biologists are attempting to construct models that integrate phenomena occurring at multiple diverse scales in order to understand how changes at the molecular level ultimately translate into new system level responses. Again, EGFR provides a useful point of discussion. Recently, Radhakrishnan and coworkers developed a set of such models to elucidate the molecular mechanisms that may underlie the improper regulation of EGFR-mediated signaling, as often arises in cancer [21, 87]. Specifically, in recent papers, these authors have investigated how recently identified somatic mutations in EGFR ultimately translate into downstream signaling through models incorporating effects at the level of EGFR dimerization and kinase activation, membrane dynamics considerations governing the rate of EGFR endocytosis, and connection of phosphorylated EGFR to downstream signaling pathways. At the molecular level, these models consider the ability of differential formation of hydrogen bonds and salt bridges arising with or without EGFR oncogenic mutations to stabilize different EGFR kinase conformations. Receptor-mediated initiation of signaling is then modeled using deterministic frameworks, in some cases allowing for branching of network activities arising from preferential binding of intracellular adapters to different EGFR regulatory tyrosines. Such models may also include the effects of EGFR internalization, as in recent papers utilizing hybrid methods based on both discrete and continuum stochastic dynamics for the consideration of membrane deformation (field-based continuum models) and protein diffusion with respect to budding endocytic vesicles [88, 89].

The integration of multiscale phenomena in models of phosphatase regulation of signaling would also be possible, and indeed useful, and a number of analogies can be easily drawn based upon the work that has already been undertaken for the regulation of kinases. For example, based upon points already discussed in this chapter, multiscale models of how specific point mutations arising in phosphatases such as SHP2, or various RPTPs such as DEP1, influence the ability of these PTPs to regulate downstream signaling would require the integration of molecular dynamics simulations to understand how mutations alter PTP structure and ability to interact with relevant adapters and receptors. To the extent that these PTPs regulate receptor-level processes such as trafficking, as has been reported for at least some PTPs recently [66], these multiscale models might also require consideration of the same kinds of hybrid approaches that have been applied to study EGFR endocytosis and recruitment of signaling proteins to endocytic vesicles.

Rules-Based Approaches for Understanding Phosphatase Regulation

As mentioned previously, there is generally less information available on the roles of specific phosphorylation sites on protein phosphatases, as well as generally less information on whether phosphorylation of these sites regulates the structure and function of phosphatases in meaningful ways. In some cases, this incomplete understanding arises because the structure and phosphorylation sites of phosphatases, as well as the intrinsic catalytic activity of the phosphatase, allow for a large number of possible intra- and intermolecular interactions. SHP2 again provides a useful focus for discussion. SHP2's N-terminal SH2 domains are involved in the regulation of its activity, but are also likely to be involved in the intracellular localization of SHP2. Moreover, it is possible for these N-terminal SH2 domains to become engaged by SHP2 C-terminal tyrosines Y542 and Y580. Such intramolecular engagement would clearly alter SHP2's ability to engage binding partners through its SH2 domains and may also alter SHP2's activity. Moreover, SHP2 may also exhibit phosphatase activity for its own C-terminal phosphotyrosines, perhaps even via an intramolecular mechanism, and may also play a role in dephosphorylating the tyrosines on adapter proteins such as GAB1 (Y627) that lead to SHP2's activation downstream of RTK activation.

Rules-based models represent one computational approach for dealing with this kind of complexity. The rules-based approach allows a user to define an array of molecular species, each with defined characteristics such as biochemical activities, posttranslational modification sites, and interaction domains. Based on the rules ascribed to each molecular species, rules-based models generate a network of species that can arise due to the prescribed allowable interactions and enable calculation of dynamics for the network. This general approach was used by Haugh and coworkers, who developed a rules-based kinetic model of SHP2 function downstream of a generic dimerized RTK [90]. They reported that even for a relatively small number of interacting proteins considered, the network of achievable binding and phosphorylation states exceeds 1,000 different species. Within this framework, Haugh and coworkers identified regimes of binding where the deletion of either the N-terminal SH2 domain or C-terminal tyrosines could either promote or antagonize the dephosphorylation of the generic RTK considered. Conversely, they found that deletion of the C-terminal SH2 domain mitigated the ability of SHP2 to regulate RTK phosphorylation.

Of course, rules-based approaches for the study of cell signaling have been used more broadly than in the study of phosphatases. Indeed, this approach has been applied in numerous more general studies of cell signaling networks and in the analysis of transcriptional regulation (e.g., [91–93]). Software packages such as BioNetGen are available to provide a user-friendly computing environment for implementing these models [94, 95].

Specific Applications to Metabolism

Based on the preceding discussion and upon growing evidence for the roles of phosphatases in the regulation of metabolism, it is clear that opportunities exist to apply computational models to the study of cell signaling and metabolic regulation. The results of such computational modeling studies may offer new insights for how to rationally intervene in diseases where PTPs have been identified as playing important roles. Several specific examples of potential applications in metabolism are now discussed.

The same kinds of mechanistic modeling approaches which have been heavily applied to the study of signaling regulated by EGFR could be applied to the study of key signaling processes in metabolism, such as signaling mediated by the insulin receptor (IR) and the associated regulation of insulin receptor substrates (IRS), phosphoinositide 3-kinase (PI3K), phosphatase and tensin homolog (PTEN), and protein kinase B (Akt). There is substantial evidence in the literature that important spatiotemporal considerations must be taken into account for each of these elements of the critical pathway regulating the expression and trafficking of GLUT4 and uptake of glucose. Given the complexity of the mechanisms which have already been identified in this spatiotemporal regulation, however, it is clear that the implementation of mechanistic models will not only be useful, but in fact be required, in order to parse the importance of various processes in the overall function of this pathway. As in other examples already discussed in this chapter, useful models of insulin-mediated signaling would also require proper consideration of the action of phosphatases.

At the level of IR itself, relatively little is quantitatively understood about the dynamics with which the receptor is dephosphorylated at regulatory tyrosines. The importance of PTP1B in regulating IR phosphorylation is clear, as previously discussed, but where and how quickly this occurs during IR-mediated signaling is unclear. As with what has been the classical view of EGFR's regulation by PTPs, there is some sense in the IR literature that PTPs exert their control over IR only after the dissociation of ligand [96]. While the final dissociation of ligand may lead to final termination of signaling via the action of PTPs such as PTP1B, it is possible that IR, as with EGFR [41], is under the constant control of PTPs and that it undergoes numerous cycles of phosphorylation and dephosphorylation during the time with which the receptor is ligand-bound and prior to final dissociation of ligand or degradation of the receptor. To our knowledge, this has not yet been definitively demonstrated for IR. It is certainly conceivable, however, that differences in the kinetics of such cycling could underlie the issue discussed by some that growth factor receptors such as platelet-derived growth factor receptor (PDGFR), interleukin receptors, and adhesion molecules which appear to stimulate PI3K activity to a similar level as IR have only minimal effects on GLUT4 translocation to the membrane [96–98]. That is, the specific cycling rate of IR phosphorylation could transmit information through the downstream pathway which is independent of the apparent magnitude of pathway activation. Interestingly, a recent study on the regulation of ERK by EGFR demonstrated this general concept by showing, on a single-cell basis, that the ability of ERK to drive proliferation in epithelial cells was dependent upon the frequency of ERK signaling modulation in those cells in response to EGFR activation [99]. To dissect this connection between frequency and effect on phenotype, the authors of that study implemented a mechanistic model. In sum, to begin to address these basic issues regarding IR regulation by PTPs, the same kind of coupled experimental and modeling approach we recently applied to quantify the kinetics of EGFR dephosphorylation could readily be applied to IR.

Another aspect of IR-mediated signaling amenable to mechanistic modeling is the distinct spatial and temporal regulation of IRS. For example, it has been shown in 3T3 fibroblasts that IRS1 and IRS2 are basally differentially distributed, with IRS1 preferentially concentrated in intracellular membrane compartments and IRS2 preferentially concentrated in the cytosol [100]. In response to exogenous insulin, both substrates are rapidly phosphorylated and translocate to the cytosol within minutes, but IRS1-mediated signaling is more protracted than IRS2-mediated signaling. How these differences arise is not yet fully understood, but hypotheses could be generated and tested based upon the development of associated kinetic models.

Interestingly, the activity of Akt may also be subject to a number of spatiotemporal effects. The extent of this regulation could be influenced by an upstream initiator of Akt activity as determined, for example, by PTP regulation of an upstream initiator (e.g., IR). In 3T3 fibroblasts using a fluorescent reporter of Akt activity, Kunkel et al. [101] identified a more transient activity of Akt in the cytosol versus the nucleus and hypothesized the presence of different phosphatase activity with respect to Akt in these two cellular compartments as an explanation. The same study also found that when the Akt activity reporter was targeted to the plasma membrane, the apparent activity of Akt was increased and the duration of signaling prolonged. This is perhaps not surprising given that the plasma membrane is the intracellular site for Akt phosphorylation by PDK-1. Other mechanisms that may spatially regulate Akt activation within cells include the preferential activation of Akt within lipid raft compartments in the plasma membrane. Indeed Gao et al. [102] found, using a FRET-based approach, that Akt activity is more rapidly induced by growth factor receptors within lipid rafts in membranes. This effect can be so pronounced that chemical disruption of rafts can potently inhibit Akt activation in response to growth factors. It is possible that this apparent preferential activation of Akt within rafts could be related to the differential partitioning between the rafts and non-raft membrane domains of any of the elements which positively or negatively regulate Akt activity. Some evidence in the literature points to preferential exclusion of phosphatases from lipid rafts as one interesting possibility, for example in the regulation of phosphorylation of PAG/Cbp [103].

Beyond the discussion we have had here on spatiotemporal aspects of IR/Akt regulation, there are certainly other examples where the application of some of the modeling approaches previously described would be of use to understand the roles of phosphatases in metabolism. For example, as described in other chapters within this book, the effects of PTP1B expression have been investigated at the whole-body level in mice and separately investigated in the brain, muscle, adipocytes, and liver [104–109]. Through these studies, PTP1B has been identified as an important regulator of body mass and insulin sensitivity, with some important differences in its tissue-specific role. PTP1B may exert its control over metabolism through regulation of insulin receptor (IR) phosphorylation. Like other RTKs, IR is capable of initiating intracellular signaling along multiple parallel pathways, and it is yet unclear what specific pathways may be differentially regulated as a result of differential IR phosphorylation or which of those pathways is most critical for the control of metabolism downstream of IR. Multivariate data-driven methods such as PLSR could be applied to such problems to identify potential ways to rationally interfere with or promote specific pathways for therapeutic purposes.

Conclusion

Computational models can be of great use in addressing the questions of the kind described in the previous section, even when a full set of model parameters may not have been quantitatively determined. Indeed, the very process of constructing a computational model that explains known biological phenomena tends to generate new testable hypotheses even before a model has been completed. For example, even mechanistic computational models lacking a complete set of well-characterized parameters can be readily utilized for local and global model sensitivity analyses to generate a list of testable predictions for which network elements or parameters are most likely to exert the most significant control in pathway regulation. The testing of these basic model predictions can be used in initial steps to refine the model. Ultimately, mechanistic models constructed with a high degree of confidence can then be useful for determining what nodes in the pathway might be most usefully sensitive to therapeutic intervention to either inhibit or promote signaling, for example in metabolic disorders controlled by IR. A similar process of model testing and refinement based on experiment is applicable to the other types of modeling methods described in this chapter. Ultimately, more extensive application of these approaches to the study of phosphatases can be a useful approach for speeding our acquisition of the quantitative insights needed to develop durable therapeutic approaches for diseases characterized by dysregulated metabolism.

Acknowledgements The author gratefully acknowledges Prof. Douglas A. Lauffenburger (MIT) and Prof. Ravi Radhakrishnan (University of Pennsylvania) for helpful feedback on this book chapter.

References

1. Ideker T, Winslow LR, Lauffenburger DA (2006) Bioengineering and systems biology. Ann Biomed Eng 34:1226–1233
2. Lazzara MJ, Lauffenburger DA (2009) Quantitative modeling perspectives on the ErbB system of cell regulatory processes. Exp Cell Res 315:717–725
3. Aldridge BB, Burke JM, Lauffenburger DA, Sorger PK (2006) Physicochemical modelling of cell signalling pathways. Nat Cell Biol 8:1195–1203
4. Kholodenko BN (2006) Cell-signalling dynamics in time and space. Nat Rev Mol Cell Biol 7:165–176
5. Kolitz SE, Lauffenburger DA (2012) Measurement and modeling of signaling at the single-cell level. Biochemistry 51:7433–7443
6. Tonks NK (2006) Protein tyrosine phosphatases: from genes, to function, to disease. Nat Rev Mol Cell Biol 7:833–846
7. Zhang ZY (2002) Protein tyrosine phosphatases: structure and function, substrate specificity, and inhibitor development. Annu Rev Pharmacol Toxicol 42:209–234
8. Fauman EB, Saper MA (1996) Structure and function of the protein tyrosine phosphatases. Trends Biochem Sci 21:413–417
9. Tonks NK, Neel BG (1996) From form to function: signaling by protein tyrosine phosphatases. Cell 87:365–368
10. Ostman A, Hellberg C, Bohmer FD (2006) Protein-tyrosine phosphatases and cancer. Nat Rev Cancer 6:307–320
11. Alonso A, Sasin J, Bottini N, Friedberg I, Friedberg I, Osterman A, Godzik A, Hunter T, Dixon J, Mustelin T (2004) Protein tyrosine phosphatases in the human genome. Cell 117: 699–711
12. Tsou RC, Bence KK (2012) The genetics of PTPN1 and obesity: insights from mouse models of tissue-specific PTP1B deficiency. J Obes 2012:926857
13. Loh K, Fukushima A, Zhang X, Galic S, Briggs D, Enriori PJ, Simonds S, Wiede F, Reichenbach A, Hauser C, Sims NA, Bence KK, Zhang S, Zhang ZY, Kahn BB, Neel BG, Andrews ZB, Cowley MA, Tiganis T (2011) Elevated hypothalamic TCPTP in obesity contributes to cellular leptin resistance. Cell Metab 14:684–699
14. Tsou RC, Zimmer DJ, De Jonghe BC, Bence KK (2012) Deficiency of PTP1B in leptin receptor-expressing neurons leads to decreased body weight and adiposity in mice. Endocrinology 153:4227–4237
15. Banno R, Zimmer D, De Jonghe BC, Atienza M, Rak K, Yang W, Bence KK (2010) PTP1B and SHP2 in POMC neurons reciprocally regulate energy balance in mice. J Clin Invest 120:720–734
16. Tsou RC, Bence KK (2012) Central regulation of metabolism by protein tyrosine phosphatases. Front Neurosci 6:192

17. Karplus M, McCammon JA (2002) Molecular dynamics simulations of biomolecules. Nat Struct Biol 9:646–652
18. Klepeis JL, Lindorff-Larsen K, Dror RO, Shaw DE (2009) Long-timescale molecular dynamics simulations of protein structure and function. Curr Opin Struct Biol 19:120–127
19. Karplus M, Kuriyan J (2005) Molecular dynamics and protein function. Proc Natl Acad Sci U S A 102:6679–6685
20. Purvis J, Liu Y, Ilango V, Radhakrishnan R (2007) Efficacy of tyrosine kinase inhibitors in the mutants of the epidermal growth factor receptor: a multiscale molecular/systems model for phosphorylation and inhibition. In: Proceedings of foundations in systems biology II. IRB, Stuttgart, pp 289–294
21. Liu Y, Purvis J, Shih A, Weinstein J, Agrawal N, Radhakrishnan R (2007) A multiscale computational approach to dissect early events in the Erb family receptor mediated activation, differential signaling, and relevance to oncogenic transformations. Ann Biomed Eng 35:1012–1025
22. Shi F, Telesco SE, Liu Y, Radhakrishnan R, Lemmon MA (2010) ErbB3/HER3 intracellular domain is competent to bind ATP and catalyze autophosphorylation. Proc Natl Acad Sci U S A 107:7692–7697
23. Telesco SE, Shih AJ, Jia F, Radhakrishnan R (2011) A multiscale modeling approach to investigate molecular mechanisms of pseudokinase activation and drug resistance in the HER3/ErbB3 receptor tyrosine kinase signaling network. Mol Biosyst 7:2066–2080
24. Liu B, Bernard B, Wu JH (2006) Impact of EGFR point mutations on the sensitivity to gefitinib: insights from comparative structural analyses and molecular dynamics simulations. Proteins 65:331–346
25. Lee TS, Ma W, Zhang X, Giles F, Cortes J, Kantarjian H, Albitar M (2008) BCR-ABL alternative splicing as a common mechanism for imatinib resistance: evidence from molecular dynamics simulations. Mol Cancer Ther 7:3834–3841
26. Lee TS, Potts SJ, Kantarjian H, Cortes J, Giles F, Albitar M (2008) Molecular basis explanation for imatinib resistance of BCR-ABL due to T315I and P-loop mutations from molecular dynamics simulations. Cancer 112:1744–1753
27. Aloy P, Russell RB (2006) Structural systems biology: modelling protein interactions. Nat Rev Mol Cell Biol 7:188–197
28. Neel BG, Gu H, Pao L (2003) The 'Shp'ing news: SH2 domain-containing tyrosine phosphatases in cell signaling. Trends Biochem Sci 28:284–293
29. Chan G, Kalaitzidis D, Neel BG (2008) The tyrosine phosphatase Shp2 (PTPN11) in cancer. Cancer Metastasis Rev 27:179–192
30. Grossmann KS, Rosario M, Birchmeier C, Birchmeier W (2010) The tyrosine phosphatase Shp2 in development and cancer. Adv Cancer Res 106:53–89
31. Lauriol J, Kontaridis MI (2011) PTPN11-associated mutations in the heart: has LEOPARD changed its RASpots? Trends Cardiovasc Med 21:97–104
32. Edouard T, Montagner A, Dance M, Conte F, Yart A, Parfait B, Tauber M, Salles JP, Raynal P (2007) How do Shp2 mutations that oppositely influence its biochemical activity result in syndromes with overlapping symptoms? Cell Mol Life Sci 64:1585–1590
33. Bentires-Alj M, Paez JG, David FS, Keilhack H, Halmos B, Naoki K, Maris JM, Richardson A, Bardelli A, Sugarbaker DJ, Richards WG, Du J, Girard L, Minna JD, Loh ML, Fisher DE, Velculescu VE, Vogelstein B, Meyerson M, Sellers WR, Neel BG (2004) Activating mutations of the noonan syndrome-associated SHP2/PTPN11 gene in human solid tumors and adult acute myelogenous leukemia. Cancer Res 64:8816–8820
34. Kontaridis MI, Swanson KD, David FS, Barford D, Neel BG (2006) PTPN11 (Shp2) mutations in LEOPARD syndrome have dominant negative, not activating, effects. J Biol Chem 281:6785–6792
35. Hardy S, Julien SG, Tremblay ML (2012) Impact of oncogenic protein tyrosine phosphatases in cancer. Anticancer Agents Med Chem 12:4–18
36. Mattila E, Ivaska J (2011) High-throughput methods in identification of protein tyrosine phosphatase inhibitors and activators. Anticancer Agents Med Chem 11:141–150

37. Wiesmann C, Barr KJ, Kung J, Zhu J, Erlanson DA, Shen W, Fahr BJ, Zhong M, Taylor L, Randal M, McDowell RS, Hansen SK (2004) Allosteric inhibition of protein tyrosine phosphatase 1B. Nat Struct Mol Biol 11:730–737
38. Bilwes AM, den Hertog J, Hunter T, Noel JP (1996) Structural basis for inhibition of receptor protein-tyrosine phosphatase-alpha by dimerization. Nature 382:555–559
39. Jiang G, den Hertog J, Su J, Noel J, Sap J, Hunter T (1999) Dimerization inhibits the activity of receptor-like protein-tyrosine phosphatase-alpha. Nature 401:606–610
40. Park CS, Schneider IC, Haugh JM (2003) Kinetic analysis of platelet-derived growth factor receptor/phosphoinositide 3-kinase/Akt signaling in fibroblasts. J Biol Chem 278:37064–37072
41. Monast CS, Furcht CM, Lazzara MJ (2012) Computational analysis of the regulation of EGFR by protein tyrosine phosphatases. Biophys J 102:2012–2021
42. Hendriks BS, Opresko LK, Wiley HS, Lauffenburger D (2003) Coregulation of epidermal growth factor receptor/human epidermal growth factor receptor 2 (HER2) levels and locations: quantitative analysis of HER2 overexpression effects. Cancer Res 63:1130–1137
43. Hendriks BS, Griffiths GJ, Benson R, Kenyon D, Lazzara M, Swinton J, Beck S, Hickinson M, Beusmans JM, Lauffenburger D, de Graaf D (2006) Decreased internalisation of erbB1 mutants in lung cancer is linked with a mechanism conferring sensitivity to gefitinib. Syst Biol 153:457–466
44. Purvis J, Ilango V, Radhakrishnan R (2008) Role of network branching in eliciting differential short-term signaling responses in the hyper-sensitive epidermal growth factor receptor mutants implicated in lung cancer. Biotechnol Prog 24:540–553
45. Schoeberl B, Eichler-Jonsson C, Gilles ED, Muller G (2002) Computational modeling of the dynamics of the MAP kinase cascade activated by surface and internalized EGF receptors. Nat Biotechnol 20:370–375
46. Birtwistle MR, Hatakeyama M, Yumoto N, Ogunnaike BA, Hoek JB, Kholodenko BN (2007) Ligand-dependent responses of the ErbB signaling network: experimental and modeling analyses. Mol Syst Biol 3:144
47. Hatakeyama M, Kimura S, Naka T, Kawasaki T, Yumoto N, Ichikawa M, Kim JH, Saito K, Saeki M, Shirouzu M, Yokoyama S, Konagaya A (2003) A computational model on the modulation of mitogen-activated protein kinase (MAPK) and Akt pathways in heregulin-induced ErbB signalling. Biochem J 373:451–463
48. Palani S, Sarkar CA (2008) Positive receptor feedback during lineage commitment can generate ultrasensitivity to ligand and confer robustness to a bistable switch. Biophys J 95: 1575–1589
49. Palani S, Sarkar CA (2009) Integrating extrinsic and intrinsic cues into a minimal model of lineage commitment for hematopoietic progenitors. PLoS Comput Biol 5:e1000518
50. Palani S, Sarkar CA (2011) Synthetic conversion of a graded receptor signal into a tunable, reversible switch. Mol Syst Biol 7:480
51. Shah NA, Sarkar CA (2011) Robust network topologies for generating switch-like cellular responses. PLoS Comput Biol 7:e1002085
52. Hendriks BS, Cook J, Burke JM, Beusmans JM, Lauffenburger DA, de Graaf D (2006) Computational modelling of ErbB family phosphorylation dynamics in response to transforming growth factor alpha and heregulin indicates spatial compartmentation of phosphatase activity. Syst Biol 153:22–33
53. Hendriks BS, Orr G, Wells A, Wiley HS, Lauffenburger DA (2005) Parsing ERK activation reveals quantitatively equivalent contributions from epidermal growth factor receptor and HER2 in human mammary epithelial cells. J Biol Chem 280:6157–6169
54. Asthagiri AR, Lauffenburger DA (2001) A computational study of feedback effects on signal dynamics in a mitogen-activated protein kinase (MAPK) pathway model. Biotechnol Prog 17:227–239
55. Heinrich R, Neel BG, Rapoport TA (2002) Mathematical models of protein kinase signal transduction. Mol Cell 9:957–970
56. O'Shaughnessy EC, Palani S, Collins JJ, Sarkar CA (2011) Tunable signal processing in synthetic MAP kinase cascades. Cell 144:119–131

57. Gadkar KG, Varner J, Doyle FJ III (2005) Model identification of signal transduction networks from data using a state regulator problem. Syst Biol 2:17–30
58. Gadkar KG, Gunawan R, Doyle FJ III (2005) Iterative approach to model identification of biological networks. BMC Bioinformatics 6:155
59. Dasika MS, Burgard A, Maranas CD (2006) A computational framework for the topological analysis and targeted disruption of signal transduction networks. Biophys J 91:382–398
60. Klinke DJ II (2009) An empirical Bayesian approach for model-based inference of cellular signaling networks. BMC Bioinformatics 10:371
61. Offterdinger M, Georget V, Girod A, Bastiaens PI (2004) Imaging phosphorylation dynamics of the epidermal growth factor receptor. J Biol Chem 279:36972–36981
62. Reynolds AR, Tischer C, Verveer PJ, Rocks O, Bastiaens PI (2003) EGFR activation coupled to inhibition of tyrosine phosphatases causes lateral signal propagation. Nat Cell Biol 5:447–453
63. Jiang X, Huang F, Marusyk A, Sorkin A (2003) Grb2 regulates internalization of EGF receptors through clathrin-coated pits. Mol Biol Cell 14:858–870
64. Goh LK, Huang F, Kim W, Gygi S, Sorkin A (2010) Multiple mechanisms collectively regulate clathrin-mediated endocytosis of the epidermal growth factor receptor. J Cell Biol 189:871–883
65. Sorkin A, Goh LK (2009) Endocytosis and intracellular trafficking of ErbBs. Exp Cell Res 315:683–696
66. Tarcic G, Boguslavsky SK, Wakim J, Kiuchi T, Liu A, Reinitz F, Nathanson D, Takahashi T, Mischel PS, Ng T, Yarden Y (2009) An unbiased screen identifies DEP-1 tumor suppressor as a phosphatase controlling EGFR endocytosis. Curr Biol 19:1788–1798
67. Araki T, Nawa H, Neel BG (2003) Tyrosyl phosphorylation of Shp2 is required for normal ERK activation in response to some, but not all, growth factors. J Biol Chem 278:41677–41684
68. Bose AK, Janes KA (2013) A high-throughput assay for phosphoprotein-specific phosphatase activity in cellular extracts. Mol Cell Proteomics 12:797–806
69. Murphy LO, Blenis J (2006) MAPK signal specificity: the right place at the right time. Trends Biochem Sci 31:268–275
70. Amit I, Citri A, Shay T, Lu Y, Katz M, Zhang F, Tarcic G, Siwak D, Lahad J, Jacob-Hirsch J, Amariglio N, Vaisman N, Segal E, Rechavi G, Alon U, Mills GB, Domany E, Yarden Y (2007) A module of negative feedback regulators defines growth factor signaling. Nat Genet 39:503–512
71. Haj FG, Verveer PJ, Squire A, Neel BG, Bastiaens PI (2002) Imaging sites of receptor dephosphorylation by PTP1B on the surface of the endoplasmic reticulum. Science 295:1708–1711
72. Haj FG, Markova B, Klaman LD, Bohmer FD, Neel BG (2003) Regulation of receptor tyrosine kinase signaling by protein tyrosine phosphatase-1B. J Biol Chem 278:739–744
73. Eden ER, White IJ, Tsapara A, Futter CE (2010) Membrane contacts between endosomes and ER provide sites for PTP1B-epidermal growth factor receptor interaction. Nat Cell Biol 12:267–272
74. Monteleone MC, Gonzalez Wusener AE, Burdisso JE, Conde C, Caceres A, Arregui CO (2012) ER-bound protein tyrosine phosphatase PTP1B interacts with Src at the plasma membrane/substrate interface. PLoS One 7:e38948
75. Lee HH, Lee HC, Chou CC, Hur SS, Osterday K, Del Alamo JC, Lasheras JC, Chien S (2013) Shp2 plays a crucial role in cell structural orientation and force polarity in response to matrix rigidity. Proc Natl Acad Sci U S A 110:2840–2845
76. Oh ES, Gu H, Saxton TM, Timms JF, Hausdorff S, Frevert EU, Kahn BB, Pawson T, Neel BG, Thomas SM (1999) Regulation of early events in integrin signaling by protein tyrosine phosphatase SHP-2. Mol Cell Biol 19:3205–3215
77. Meng TC, Fukada T, Tonks NK (2002) Reversible oxidation and inactivation of protein tyrosine phosphatases in vivo. Mol Cell 9:387–399

78. Rhee SG, Bae YS, Lee SR, Kwon J (2000) Hydrogen peroxide: a key messenger that modulates protein phosphorylation through cysteine oxidation. Sci STKE 2000:pe1
79. Bae YS, Kang SW, Seo MS, Baines IC, Tekle E, Chock PB, Rhee SG (1997) Epidermal growth factor (EGF)-induced generation of hydrogen peroxide. Role in EGF receptor-mediated tyrosine phosphorylation. J Biol Chem 272:217–221
80. Chiarugi P, Cirri P (2003) Redox regulation of protein tyrosine phosphatases during receptor tyrosine kinase signal transduction. Trends Biochem Sci 28:509–514
81. Deen WM (2012) Analysis of transport phenomena. Oxford University Press, New York
82. Janes KA, Albeck JG, Gaudet S, Sorger PK, Lauffenburger DA, Yaffe MB (2005) A systems model of signaling identifies a molecular basis set for cytokine-induced apoptosis. Science 310:1646–1653
83. Kumar N, Wolf-Yadlin A, White FM, Lauffenburger DA (2007) Modeling HER2 effects on cell behavior from mass spectrometry phosphotyrosine data. PLoS Comput Biol 3:e4
84. Miller-Jensen K, Janes KA, Brugge JS, Lauffenburger DA (2007) Common effector processing mediates cell-specific responses to stimuli. Nature 448:604–608
85. Dworkin M, Mukherjee S, Jayaprakash C, Das J (2012) Dramatic reduction of dimensionality in large biochemical networks owing to strong pair correlations. J R Soc Interface 9:1824–1835
86. Kreeger PK, Wang Y, Haigis KM, Lauffenburger DA (2010) Integration of multiple signaling pathway activities resolves K-RAS/N-RAS mutation paradox in colon epithelial cell response to inflammatory cytokine stimulation. Integr Biol (Camb) 2:202–208
87. Shih AJ, Telesco SE, Radhakrishnan R (2011) Analysis of somatic mutations in cancer: molecular mechanisms of activation in the ErbB family of receptor tyrosine kinases. Cancers (Basel) 3:1195–1231
88. Agrawal NJ, Nukpezah J, Radhakrishnan R (2010) Minimal mesoscale model for protein-mediated vesiculation in clathrin-dependent endocytosis. PLoS Comput Biol 6:e1000926
89. Ramanan V, Agrawal NJ, Liu J, Engles S, Toy R, Radhakrishnan R (2011) Systems biology and physical biology of clathrin-mediated endocytosis. Integr Biol (Camb) 3:803–815
90. Barua D, Faeder JR, Haugh JM (2007) Structure-based kinetic models of modular signaling protein function: focus on Shp2. Biophys J 92:2290–2300
91. Sekar JA, Faeder JR (2012) Rule-based modeling of signal transduction: a primer. Methods Mol Biol 880:139–218
92. Hlavacek WS, Faeder JR, Blinov ML, Posner RG, Hucka M, Fontana W (2006) Rules for modeling signal-transduction systems. Sci STKE 2006:re6
93. Blinov ML, Faeder JR, Goldstein B, Hlavacek WS (2006) A network model of early events in epidermal growth factor receptor signaling that accounts for combinatorial complexity. Biosystems 83:136–151
94. Blinov ML, Faeder JR, Goldstein B, Hlavacek WS (2004) BioNetGen: software for rule-based modeling of signal transduction based on the interactions of molecular domains. Bioinformatics 20:3289–3291
95. Faeder JR, Blinov ML, Hlavacek WS (2009) Rule-based modeling of biochemical systems with BioNetGen. Methods Mol Biol 500:113–167
96. Saltiel AR, Pessin JE (2002) Insulin signaling pathways in time and space. Trends Cell Biol 12:65–71
97. Wiese RJ, Mastick CC, Lazar DF, Saltiel AR (1995) Activation of mitogen-activated protein kinase and phosphatidylinositol 3′-kinase is not sufficient for the hormonal stimulation of glucose uptake, lipogenesis, or glycogen synthesis in 3T3-L1 adipocytes. J Biol Chem 270:3442–3446
98. Guilherme A, Czech MP (1998) Stimulation of IRS-1-associated phosphatidylinositol 3-kinase and Akt/protein kinase B but not glucose transport by beta1-integrin signaling in rat adipocytes. J Biol Chem 273:33119–33122
99. Albeck JG, Mills GB, Brugge JS (2013) Frequency-modulated pulses of ERK activity transmit quantitative proliferation signals. Mol Cell 49:249–261

100. Inoue G, Cheatham B, Emkey R, Kahn CR (1998) Dynamics of insulin signaling in 3T3-L1 adipocytes. Differential compartmentalization and trafficking of insulin receptor substrate (IRS)-1 and IRS-2. J Biol Chem 273:11548–11555
101. Kunkel MT, Ni Q, Tsien RY, Zhang J, Newton AC (2005) Spatio-temporal dynamics of protein kinase B/Akt signaling revealed by a genetically encoded fluorescent reporter. J Biol Chem 280:5581–5587
102. Gao X, Zhang J (2009) Akt signaling dynamics in plasma membrane microdomains visualized by FRET-based reporters. Commun Integr Biol 2:32–34
103. Davidson D, Bakinowski M, Thomas ML, Horejsi V, Veillette A (2003) Phosphorylation-dependent regulation of T-cell activation by PAG/Cbp, a lipid raft-associated transmembrane adaptor. Mol Cell Biol 23:2017–2028
104. Delibegovic M, Zimmer D, Kauffman C, Rak K, Hong EG, Cho YR, Kim JK, Kahn BB, Neel BG, Bence KK (2009) Liver-specific deletion of protein-tyrosine phosphatase 1B (PTP1B) improves metabolic syndrome and attenuates diet-induced endoplasmic reticulum stress. Diabetes 58:590–599
105. Klaman LD, Boss O, Peroni OD, Kim JK, Martino JL, Zabolotny JM, Moghal N, Lubkin M, Kim YB, Sharpe AH, Stricker-Krongrad A, Shulman GI, Neel BG, Kahn BB (2000) Increased energy expenditure, decreased adiposity, and tissue-specific insulin sensitivity in protein-tyrosine phosphatase 1B-deficient mice. Mol Cell Biol 20:5479–5489
106. Elchebly M, Payette P, Michaliszyn E, Cromlish W, Collins S, Loy AL, Normandin D, Cheng A, Himms-Hagen J, Chan CC, Ramachandran C, Gresser MJ, Tremblay ML, Kennedy BP (1999) Increased insulin sensitivity and obesity resistance in mice lacking the protein tyrosine phosphatase-1B gene. Science 283:1544–1548
107. Bence KK, Delibegovic M, Xue B, Gorgun CZ, Hotamisligil GS, Neel BG, Kahn BB (2006) Neuronal PTP1B regulates body weight, adiposity and leptin action. Nat Med 12:917–924
108. Owen C, Czopek A, Agouni A, Grant L, Judson R, Lees EK, McIlroy GD, Goransson O, Welch A, Bence KK, Kahn BB, Neel BG, Mody N, Delibegovic M (2012) Adipocyte-specific protein tyrosine phosphatase 1B deletion increases lipogenesis, adipocyte cell size and is a minor regulator of glucose homeostasis. PLoS One 7:e32700
109. Haj FG, Zabolotny JM, Kim YB, Kahn BB, Neel BG (2005) Liver-specific protein-tyrosine phosphatase 1B (PTP1B) re-expression alters glucose homeostasis of PTP1B-/-mice. J Biol Chem 280:15038–15046

Chapter 3
Protein-Tyrosine Phosphatase 1B Substrates and Control of Metabolism

Yannan Xi and Fawaz G. Haj

Abstract Protein-tyrosine phosphatase 1B (PTP1B) has emerged as an important regulator of several signaling networks that are implicated in human metabolic diseases such as diabetes and obesity. A growing body of evidence demonstrates that PTP1B displays exquisite substrate specificity. In this chapter we review mechanisms that regulate PTP1B–susbtrate interactions and highlight substrates that mediate PTP1B metabolic actions. PTP1B–substrate interactions are modulated by PTP1B subcellular location and numerous posttranslational modifications that regulate its activity such as oxidation, nitrosylation, sulfhydration, sumoylation, phosphorylation, and proteolysis. The metabolic actions of PTP1B are mediated by key physiological substrates that regulate insulin and leptin signaling, cell–cell communication, and endoplasmic reticulum (ER) stress response.

Abbreviations

ER	Endoplasmic reticulum
AMPK	AMP-activated protein kinase
ASO	Antisense oligonucleotide
ATF6	Activating transcription factor 6
BRET	Bioluminescence resonance energy transfer
EGFR	Epidermal growth factor receptor
eIF2α	Eukaryotic translation initiation factor 2α

Y. Xi, B.Sc.
Department of Nutrition, University of California Davis, Davis, CA 95616, USA

F.G. Haj, M.Sc., D.Phil. (✉)
Department of Nutrition, University of California Davis, Davis, CA 95616, USA

Department of Internal Medicine, University of California Davis, Sacramento, CA 95817, USA
e-mail: fghaj@ucdavis.edu

K.K. Bence (ed.), *Protein Tyrosine Phosphatase Control of Metabolism*,
DOI 10.1007/978-1-4614-7855-3_3, © Springer Science+Business Media New York 2013

ERK	Extracellular signal-regulated kinase
FRET	Fluorescence resonance energy transfer
Gab	Grb2-associated binder
GH	Growth hormone
H_2S	Hydrogen sulfide
HFD	High fat diet
INM	Inner nuclear membrane
IR	Insulin receptor
IRE1α	Inositol requiring enzyme 1α
IRS	Insulin receptor substrate
JAK	Janus kinase
KO	Knockout
LepR	Leptin receptor
MVB	Multivesicular bodies
NO	Nitric oxide
PDGFR	Platelet-derived growth factor receptor
PERK	PKR-like ER-resident kinase
PI3K	Phosphatidylinositol 3-kinase
PIAS	Protein inhibitor of activated STAT1
PM	Plasma membrane
POMC	Pro-opiomelanocortin
PPARγ	Peroxisome proliferator-activated receptor γ
PTP1B	Protein-tyrosine phosphatase 1B
RNS	Reactive nitrogen species
ROS	Reactive oxygen species
RTK	Receptor tyrosine kinase
SH	Src homology
STAT	Signal transducer and activator of transcription
SUMO	Small ubiquitin-like modifier
UPR	Unfolded protein response
XBP1	X-box binding protein 1
ZO1	Zonula occludens 1

Regulation of PTP1B–Substrate Interaction

PTP1B function is regulated through a variety of mechanisms that include its subcellular location and posttranslational modifications that modulate its catalytic activity. PTP1B has a hydrophobic domain that tethers it to the cytoplasmic face of the ER and restricts its access to substrates (Fig. 3.1). In addition, several posttranslational modifications, sometimes working in tandem, regulate PTP1B activity including oxidation, nitrosylation, sulfhydration, sumoylation, phosphorylation, and proteolysis (Fig. 3.1). The growing understanding of PTP1B–substrate interaction and regulation of PTP1B catalytic activity can be utilized to develop novel therapeutic strategies for metabolic diseases.

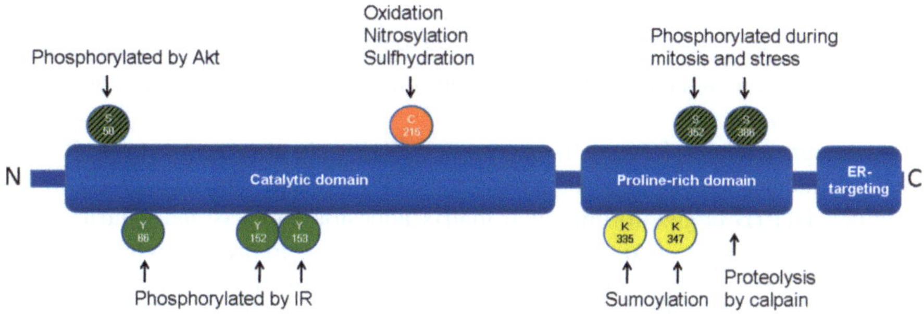

Fig. 3.1 PTP1B structural domains and modes of regulation. Schematic representation of PTP1B domains: N-terminal catalytic domain, proline-rich domain, and a C-terminal ER-targeting domain. PTP1B is regulated by serine phosphorylation at S50, S352, and S386 (*green with black stripes*), tyrosine phosphorylation at Y66, Y152, and Y153 (*green*), oxidation, S-nitrosylation, and sulfhydration of the catalytic site C215 (*orange*); sumoylation at K335 and K347 (*yellow*); and proteolysis by calpain

Subcellular Location

An important constraint on the function of PTP1B and its ability to gain access to substrates is its subcellular location. One of the earliest discoveries about PTP1B was its localization on the cytoplasmic face of the ER by means of a hydrophobic (35-residue) C-terminal anchor sequence [1, 2]. This raised a conundrum since biochemical and genetic studies established that PTP1B dephosphorylates several receptor tyrosine kinases (RTKs), including the activated epidermal growth factor receptor (EGFR), platelet-derived growth factor receptor (PDGFR), and insulin receptor (IR) [3–10]. This topological dilemma was addressed using fluorescence resonance energy transfer (FRET) which demonstrated that PTP1B-catalyzed dephosphorylation required endocytosis of the EGFR and PDGFR and occurred at specific sites on the surface of the ER [4]. Most of the activated RTKs interact with PTP1B after internalization, establishing that RTK activation and inactivation are spatially and temporally segregated within cells [4]. Subsequent studies further confirmed these findings and extended them to the IR [6, 10, 11]. The ER localization of PTP1B is ideally suited to prevent tyrosine phosphorylation of immature RTKs during biosynthesis. Indeed, studies investigating PTP1B–IR interaction using FRET or bioluminescence resonance energy transfer (BRET) demonstrate basal interaction between these proteins, and this can be enhanced by treatment with tunicamycin (an inhibitor of N-glycosylation and RTK cell surface maturation) [6]. Thus, PTP1B can access RTKs when both proteins are in the ER, during RTK biosynthesis. In addition to its ability to dephosphorylate RTKs, PTP1B has been identified as a potential regulator of RTK endocytosis. PTP1B–EGFR interaction occurs through direct membrane contact between multivesicular bodies (MVB) and the ER, with PTP1B activity promoting the sequestration of EGFR to MVB internal vesicles [12]. Consistent with these findings, additional studies identified the endosomal protein STAM2, which regulates sorting of activated RTKs for degradation, as a PTP1B substrate [13]. Together, these studies demonstrate PTP1B involvement in the dephosphorylation and endocytosis of RTKs, two key mechanisms of RTK signal attenuation and termination, and highlight the multifaceted impact of PTP1B on RTK signaling.

Although activated RTKs gain access to PTP1B only after endocytosis, PTP1B also can interact with some plasma membrane (PM)-bound substrates [14, 15]. For example, PTP1B targets forming cell-matrix adhesion contacts and contributes to the stabilization of focal adhesions. This process appears to involve dynamic extension of the ER via a microtubule-dependent process [16]. PTP1B also can access substrates at regions of cell–cell contact [17–19], although how these interactions are regulated remains largely unexplored. These findings raise an apparent paradox: why can PTP1B dephosphorylate some substrates after endocytosis, whereas others can be targeted on the PM? Quantitative cellular imaging and mathematical modeling of protein mobility demonstrate that the ER network comes in close proximity to the PM at apparently specialized regions of cell–cell contact, enabling PTP1B to engage substrates at these sites [20]. At these regions, the ER network appears to be organized thereby creating specialized zones of ER–PM interaction at cell–cell contacts in which PTP1B may access the PM, demonstrating that the ER plays a dynamic role in regulating signaling at regions of cell–cell contacts via PTP1B. Collectively, these studies demonstrate that PTP1B subcellular location plays an important role in its ability to access substrates.

Posttranslational Modifications

In addition to subcellular location which restricts PTP1B access to substrates, multiple lines of evidence provide insights into modes of PTP1B regulation by posttranslational modifications. These include oxidation, nitrosylation, sulfyhydration, sumoylation, phosphorylation, and proteolysis (Fig. 3.1).

Oxidation

PTP1B, similar to other members of the classical PTP subfamily, contains a conserved "signature motif" within its active site, wherein the invariant cysteine (Cys215) is essential for catalysis [21]. Due to the unique chemical environment of the active site, this cysteine displays a low PKa (4.5–5.5) which enhances its nucleophilic attack on substrate phosphotyrosines [22], but renders it susceptible to oxidation by reactive oxygen species (ROS) [23]. Biochemical and crystallographic studies demonstrate that oxidation of the active site cysteine abrogates its nucleophilic properties, thus inhibiting PTP1B activity. ROS, particularly H_2O_2 functions as an intracellular second messenger, and RTK activation leads to a transient burst of H_2O_2 production which is needed for full receptor phosphorylation and downstream signaling [24]. Therefore, one function of stimulus-induced ROS production is to transiently inactivate PTP1B, which usually exerts inhibitory constraint on the system, to initiate a signaling response to the stimulus. Indeed, biochemical studies demonstrate that PTP1B is reversibly oxidized at its active site and its enzymatic activity is attenuated in response to epidermal growth factor [25] and insulin [26] stimulation. Notably, quantitative imaging studies reveal that PTP1B activity is spatially regulated in the cell, thereby creating distinct microenvironments that enable RTK signal propagation and signal termination [27]. In addition, crystallographic analysis of PTP1B revealed that oxidation of the active site cysteine rapidly converts the sulphenic acid (S-OH) form of PTP1B to a cyclic sulphenamide [28, 29]. This is accompanied by profound changes in the architecture of the active site, in which residues that are usually buried adopt solvent-exposed positions [28, 29]. These structural changes that are readily reversible under physiological conditions help protect the enzyme from higher order, irreversible oxidation and facilitate reduction back to the active form. Taken together, biochemical and crystallographic studies provide mechanistic insights into PTP1B oxidation; however, several outstanding questions remain including how oxidation specificity is achieved. One element of specificity is likely localization of the phosphatase in close proximity to the site(s) of ROS production. For example, Nox4, which regulates ROS production in response to insulin [30], colocalizes with PTP1B at intracellular membranes [31]. It is worth noting that studies on PTP1B oxidation utilized cells that were exposed to an oxygen-rich environment, and additional studies are needed to investigate PTP1B oxidation under lower oxygen concentrations that are comparable to those encountered *in vivo*.

Nitrosylation

The free radical nitric oxide (NO) and its derivatives, which constitute the reactive nitrogen species (RNS), are the products of NO synthases [32] and function as second messengers that regulate cellular signaling [33]. *S*-nitrosylation of cysteine residues is a posttranslational modification that regulates the activity of key enzymes and influences various cellular functions [34, 35]. Mass spectroscopy-based studies reveal that NO inactivates PTP1B by *S*-nitrosylation of its active site cysteine [36]. Importantly, *S*-nitrosylation prevents the active site cysteine from subsequent oxidation when subjected to oxidative stress, and thus provide a protective mechanism to prevent ROS-induced irreversible oxidation of PTP1B [37]. Thus, PTP1B nitrosylation is another posttranslational modification that regulates reversible inactivation and ensures appropriate signaling response to the stimulus.

Sulfhydration

Hydrogen sulfide (H_2S) is a signaling molecule that regulates a broad spectrum of physiological processes at subtoxic concentrations [38]. The effects of H_2S are mediated through the sulfhydration of specific cysteine residues in target proteins [39]. PTP1B is reversibly inactivated by endogenously generated H_2S during the ER stress response through sulfhydration of its active site cysteine [40]. Sulfhydration of PTP1B *in vitro* occurs at H_2S concentrations that are encountered *in vivo* [41]. In addition, the rate of PTP1B inactivation by H_2S is 22.4 ± 1.8 M^{-1} s^{-1} which is comparable to its rate of inactivation by H_2O_2 (10 ± 1.4 M^{-1} s^{-1}) and NO (2.1 ± 0.5 M^{-1} s^{-1}) [40]. Decreased H_2S production leads to decreased tyrosine phosphorylation of PKR-like ER-resident kinase (PERK), a PTP1B substrate, and reduces PERK activation in response to ER stress. Importantly, treatment with a small molecule PTP1B inhibitor (MSI-1436) abrogates the effects of suppressing H_2S production on the ER stress response [40]. These studies establish that sulfyhydration of PTP1B, which occurs at high stoichiometry in cells under ER stress conditions, reversibly inactivates it and provides a mechanism for modulating ER stress and the metabolic response.

Sumoylation

Small ubiquitin-like modifier (SUMO) proteins are important regulators of various facets of cellular function. SUMO modification is implicated in protein stability and interaction, nuclear-cytosolic transport, and progression through the cell cycle [42, 43]. PTP1B interacts with a protein inhibitor of activated STAT1 (PIAS), a SUMO E3 ligase that promotes sumoylation of PTP1B in fibroblast cell lines [44]. PTP1B is sumoylated on at least two C-terminal lysine residues and the ER-targeting domain of PTP1B is required for maximum sumoylation [44]. Insulin-induced sumoylation of PTP1B results in transient reduction in enzymatic activity and inhibits the negative effect of PTP1B on insulin signaling [44]. In addition, PTP1B appears to localize on the inner nuclear membrane (INM) where it is heavily sumoylated in a cell cycle-dependent manner and interacts with emerin (an INM protein that regulates nuclear architecture) [45]. It is plausible that PTP1B regulates emerin tyrosine phosphorylation and may have nuclear functions that are modulated by sumoylation. Additional studies implicate PTP1B sumoylation in adipocyte differentiation [46]. Indeed, PTP1B knockout (KO) and wild type-reconstituted preadipocytes differentiate into adipocytes while sumoylation-resistant PTP1B mutant-reconstituted preadipocytes exhibit marked attenuation in differentiation and lipid accumulation [46]. The differentiation deficit in sumoylation-resistant reconstituted preadipocytes can be reversed by treatment with the PPARγ activator troglitazone [46]. Therefore, sumoylation is a yet another posttranslational modification that regulates PTP1B activation.

Phosphorylation

Phosphorylation is a key posttranslational modification that regulates enzyme function and assembly of protein complexes. PTP1B is regulated by both tyrosine and serine phosphorylation at multiple sites (Fig. 3.1); however, the effects of phosphorylation on enzyme activation appear controversial in some cases. Insulin stimulates tyrosine phosphorylation of PTP1B (at Tyr 66, 152, and 153) [47] and results in increased phosphatase activity [48]. On the other hand, *in vivo* studies demonstrate that insulin injection in mice induces PTP1B tyrosine phosphorylation and concomitant decrease in its activity in skeletal muscle and adipose tissue [49]. The reason(s) for these discrepant results is currently not known. In addition to tyrosine phosphorylation, PTP1B is phosphorylated on serine residues. During mitosis, PTP1B is phosphorylated on multiple serine residues (including Ser 352 and 386), although this does not appear to alter enzymatic activity [50]. Similarly, stress-induced PTP1B phosphorylation involves both Ser 352 and 386 [51]. Moreover, phosphorylation of PTP1B at Ser 50 by Akt attenuates PTP1B enzymatic activity and its ability to dephosphorylate the IR [52], possibly as part of a positive feedback mechanism to potentiate insulin signaling.

Proteolysis

Proteolysis is a common posttranslational modification that regulates PTP activity. Calpain-mediated cleavage of the C-terminal, ER-anchoring domain of PTP1B in activated platelets generates an activated, soluble enzyme [53]. Disruption of calpain1 in mice results in significant defects in platelet aggregation and clot retraction [54]. In addition, calpain1-PTP1B double KO mice reveal that PTP1B is a physiological target of calpain1 in platelets [55]. Notably, the reversibly oxidized PTP1B is more vulnerable to calpain proteolysis [56], demonstrating how seemingly different posttranslational modifications work in tandem to regulate PTP1B activation. Whether calpain-dependent cleavage promotes PTP1B catalytic activity or targets the phosphatase for complete proteolysis still needs clarification and may depend on the stimulus and/or cellular context.

PTP1B-Substrate Identification

Several properties factor into PTP1B interactions with substrates including binding motifs and the structure of its catalytic domain. PTP1B contains binding motifs that modulate its interactions with some proteins. For example, the proline-rich domain at the C-terminus (amino acids: 278–401) mediate association with several Src homology 3 (SH3) domain-containing substrates such as $p130^{Cas}$ [57]. In addition to binding motifs, the catalytic domain of PTP1B, while closely related to other PTPs, has intrinsic specificity and preference for certain substrates. Crystallographic and peptide binding studies define the molecular basis of PTP1B interaction with one of its key substrates, the IR, and demonstrate that the sequence E/D-pY-pY-R/K is important for optimal recognition of IR by PTP1B [58]. Importantly, PTP1B displays ~ 70-fold higher affinity for tandem pTyr-containing peptides from the IR activation loop compared with mono-pTyr derivatives [58]. Remarkably, this motif can be utilized to predict physiological PTP1B substrates. Indeed, JAK (janus kinase) subfamily of kinases contain this motif in their activation loop and JAK2 and TYK2 are physiological substrates of PTP1B [59]. Although the motif is important for optimal substrate recognition, it is not a requirement for PTP1B substrates as several well-established substrates do not contain the tandem phosphotyrosine sequence.

Identification of PTP1B substrates is an essential step towards a better understanding of its physiological functions. Wild-type PTP1B has a catalytic constant (k_{cat}) of ~2×10^3 molecules/min [60] rendering its interaction with substrates transient and difficult to detect using standard biochemical and imaging approaches. Insights gained from PTP1B structural studies helped define several residues that are important for substrate recognition and catalysis. This led to the generation of a mutant enzyme that retains substrate binding but is catalytically impaired, a so-called substrate trapping mutant [61]. The residue that is mutated in PTP1B to generate the substrate trapping mutant is the invariant catalytic acid (Asp181) that is conserved in all members of the PTP family. Expression of PTP1B (D181A) mutant in cells forms stable complexes with tyrosine-phosphorylated substrates and enables their identification. Indeed, the substrate trapping approach has been successfully utilized to identify several physiologically relevant PTP1B substrates [61–64]. Its application also revealed that members of the PTP family display exquisite substrate specificity in a cellular context. The combined use of the substrate trapping approach with other methodologies such as cell imaging and quantitative proteomics holds great promise in identifying and characterizing novel PTP1B substrates. Moreover, the contribution of the candidate substrate to PTP1B physiological functions can be further interrogated using gene deletion studies *in vivo*.

PTP1B Substrates and Metabolic Regulation

Metabolic syndrome and type 2 diabetes are complex disorders that are associated with obesity, sedentary life style, aging, and genetic predisposition [65, 66]. The increased prevalence of these diseases presents serious health challenges and highlights the urgent need to elucidate the underlying molecular mechanisms in order to develop effective therapeutic interventions. Ample evidence from biochemical, structural, and genetic studies establish PTP1B as a modulator of metabolic signaling in mammals. These remarkable findings highlight the potential use of PTP1B inhibitors for the treatment of obesity, diabetes, and the metabolic syndrome. Next we will review the key findings that implicate PTP1B and its substrates in insulin and leptin signaling, cell–cell communication, and ER stress, thereby highlighting the important role of PTP1B in regulating the metabolic state (Fig. 3.2).

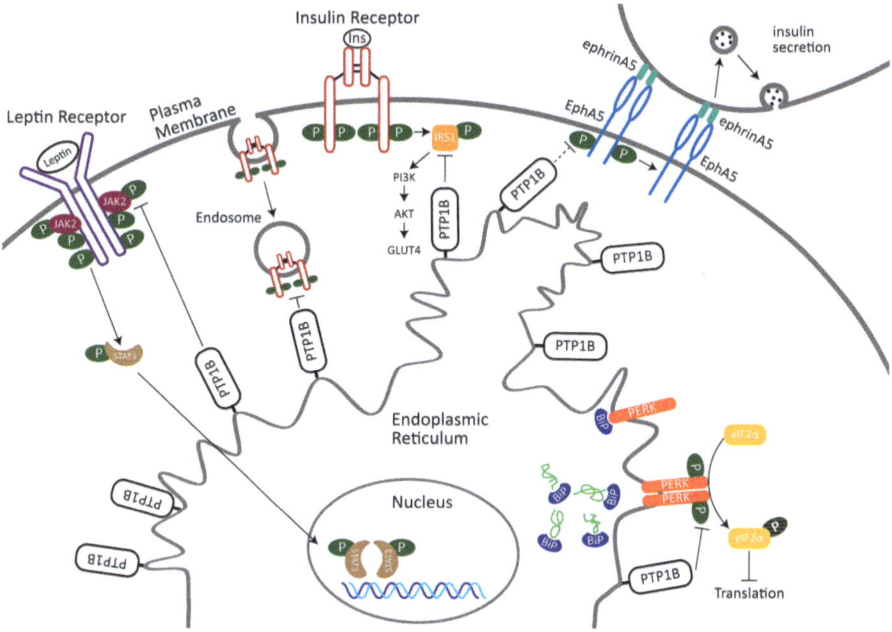

Fig. 3.2 PTP1B–substrates interaction and metabolic signaling. PTP1B is a regulator of insulin and leptin signaling, cell–cell communication, and ER stress. Insulin induces trans-phosphorylation and activation of the IR, leading to phosphorylation of IRS1 and subsequent activation of PI3K and AKT. PTP1B attenuates insulin signaling by dephosphorylating IRS1 and the endocytosed IR. PTP1B attenuates leptin signaling by dephosphorylating JAK2. Leptin receptor activation results in the recruitment and phosphorylation of JAK2 and phosphorylation of STAT3. Phosphorylated STAT3 dimerizes and translocates to the nucleus where it regulates the transcription of target genes. PTP1B regulates signaling at regions of cell–cell contact. PTP1B is anchored to the cytoplasmic face of the ER via a hydrophobic domain. The ER network comes in close proximity to the plasma membrane at apparently specialized regions of cell–cell contact (*top, right corner*), enabling PTP1B to engage substrates at these sites. EphA5-ephrinA5 bidirectional signaling regulates basal and glucose-stimulated insulin secretion. PTP1B is a direct regulator of ER stress response. Under basal conditions, PERK is maintained inactive by its association with the molecular chaperone BiP through its luminal domain. When improperly folded proteins (*green*) accumulate in the ER, BiP dissociates from PERK which undergoes homo-oligomerization and trans-phosphorylation resulting in activation. PERK phosphorylates eIF2α, a modification that blocks initiation of translation. PTP1B dephosphorylates PERK at Tyr 615 and attenuates its activation. Tyrosine-phosphorylated residues are indicated by *green ovals* and serine-phosphorylated residue indicated by *green oval* with *black stripes*. *Solid lines* indicate established interactions and *dotted line* indicates putative interaction

Insulin Signaling

Insulin is secreted from pancreatic β cells into the portal circulation and acts as a major regulator of glucose homeostasis by means of a complex and highly integrated network of signaling events [67]. Upon binding to the IR, insulin induces trans-phosphorylation of several tyrosine residues, leading to the recruitment and phosphorylation of insulin receptor substrates (IRSs) and Grb2-associated binder

(Gab) family proteins (Fig. 3.2). These serve as docking sites for Src homology 2 (SH2) domain-containing signal relay molecules, such as the p85 regulatory subunit of phosphatidylinositol 3-kinase (PI3K). PI3K activity is stimulated by its binding to IRS proteins [68], leading to activation of downstream targets such as Akt [69]. Insulin also activates ERK1/2, which phosphorylates various substrates to promote gene expression [70]. Research on insulin signaling using tissue-specific gene deletion in mice as well as other approaches provides critical insights into the tissue-specific effects of insulin and demonstrates that impaired insulin signaling is central to the development of the metabolic syndrome [71].

PTP1B is an established regulator of insulin signaling, supported by significant evidence from biochemical, genetic, and inhibitor studies. Collectively, these studies demonstrate that PTP1B is a *bona fide* IR phosphatase and that PTP1B inhibits insulin signaling by dephosphorylating the IR and possibly IRS1. Early microinjection studies of PTP1B protein into *Xenopus oocytes* illustrate a role for this phosphatase in antagonizing insulin signaling and regulating IR phosphorylation [72, 73]. Subsequent studies using neutralizing antibodies and overexpression approaches further demonstrate a role for PTP1B in IR phosphorylation and signaling [74–76]. In addition, substrate trapping methodology reveals direct association between PTP1B and IR [47, 77, 78] and IRS1 [46, 79]. Although these studies strongly implicated PTP1B in regulating IR signaling, the most compelling evidence was provided by the development and characterization of PTP1B KO mice.

Major insights into the physiological role of PTP1B in insulin signaling were obtained by targeted disruption of PTP1B gene (*Ptpn1*) in mice [8]. Whole-body PTP1B KO mice develop normally and have comparable life span to control littermates. However, PTP1B KO mice exhibit enhanced and/or prolonged IR phosphorylation in liver and muscle. In line with these findings, PTP1B KO mice exhibit increased systemic insulin sensitivity and enhanced glucose tolerance compared with controls [8]. Interestingly, when fed a high fat diet, PTP1B KO mice are resistant to weight gain and remain lean, whereas control mice rapidly gain weight and develop insulin resistance [8]. These findings were confirmed and extended in an independently generated mouse line (where PTP1B exon 1 was targeted for disruption rather than exon 5/6) [9]. Similarly, PTP1B KO mice also exhibit improved insulin sensitivity with enhanced insulin signaling in skeletal muscle but not in adipose tissue [9]. Furthermore, these mice exhibit increased basal metabolic rate and total energy expenditure [9]. Of note, despite the potential of PTP1B to regulate growth factor RTKs signaling, PTP1B KO mice do not display phenotypes associated with aberrant receptor signaling and do not show predisposition to cancer [8, 9]. Although ablation of PTP1B expression leads to hyperphosphorylation of EGFR and PDGFR, presumably compensatory mechanism(s) prevents hyperactivation of downstream signaling triggered by these receptors [5].

Tissue-specific expression and deletion studies in mice provide additional insights into the tissue-specific metabolic functions of PTP1B. Transgenic overexpression of PTP1B in muscle leads to impaired activation of the IR following insulin stimulation and decreased muscle glucose uptake [80]. In addition, adenovirus-mediated PTP1B reexpression in the liver of PTP1B KO mice leads to reduced insulin sensitivity and decreased insulin-stimulated IR tyrosine phosphorylation [81]. IR phospho-specific antibodies reveal preferential dephosphorylation of hepatic IR Tyr 1162/1163 by PTP1B *in vivo* [81]. Moreover, mice with neuronal-specific PTP1B deletion exhibit reduced body mass and adiposity due to decreased food intake and increased energy expenditure [82]. Similar effects are observed when PTP1B is deleted in pro-opiomelanocortin (POMC) neurons in the hypothalamus [83] and in leptin receptor (LepR)-expressing neurons [84]. While the underlying molecular mechanism(s) remains to be elucidated, neuronal PTP1B deletion reduces hypothalamic AMPK activity and this is likely to be a contributor to weight regulation by PTP1B [85]. Furthermore, mice with muscle-specific PTP1B-deficiency exhibit body weight and adiposity that are comparable to control mice on either regular chow or HFD [86]. Nevertheless, muscle-specific PTP1B KO mice exhibit improved systemic insulin sensitivity that is most likely caused by increased phosphorylation of the IR. Similarly, liver-specific PTP1B KO mice exhibit increased insulin sensitivity independent of changes in adiposity [87].

The metabolic phenotype of PTP1B-deficient mice generated a surge of interest in developing PTP1B-specific inhibitors. While this led to the generation of very potent inhibitors *in vitro*, they have thus far not progressed beyond this preclinical stage largely due to intracellular delivery and specificity problems [88–92]. However, antisense oligonucleotides (ASO) provide a significant advantage compared with other drugs: exquisite specificity. PTP1B ASO (ISIS 113715; targeting nucleotides 861–880) reduced PTP1B mRNA and protein in liver and fat (with no effect on skeletal muscle) in murine models of obesity (*ob*/*ob* and *db*/*db* mice) [93]. Importantly, ASO treatment enhanced insulin signaling in the liver and was effective in improving systemic insulin sensitivity [93]. Further, ASO administration to monkeys reduces PTP1B expression in liver and adipose and improves insulin sensitivity [94]. Together, these studies validate PTP1B as a physiological regulator of insulin signaling and a therapeutic target.

Leptin Signaling

Leptin is a hormone that is produced by adipose tissue and acts in the brain to regulate feeding and energy expenditure [95]. Elevated concentrations of leptin indicate energy abundance to the brain leading to suppression of food intake and increased energy expenditure [96]. LepR is a type I cytokine receptor that associates with the tyrosine kinase JAK2 [96, 97]. Leptin stimulation activates JAK2 leading to autophosphorylation and phosphorylation of LepR which then recruits signal relay molecules, such as signal transducer and activator of transcription 3 (STAT3) (Fig. 3.2) [98]. Phosphorylated STAT3 dimerizes and translocates to the nucleus where it regulates the transcription of target genes [96, 98]. Leptin-deficient (*ob*/*ob*) mice are hyperphagic and develop severe obesity and insulin resistance [99].

In vitro studies identified JAK2 as a PTP1B substrate in fibroblasts upon interferon stimulation [59]. However, the physiological relevance of PTP1B–JAK2 interaction is demonstrated in PTP1B KO mice. A lack of PTP1B is associated with elevated JAK2 phosphorylation and activation of JAK2 and its target STAT3 within the hypothalamus. In addition, substrate trapping experiments demonstrate that leptin-activated JAK2 is a direct PTP1B substrate and that PTP1B suppresses phosphorylation of JAK2 activation site tyrosines (Tyr 1007/1008) [101]. Further, mice with compound *ob/ob*-PTP1B deletion exhibit increased resting metabolic rate, decreased weight gain, and have enhanced response towards leptin-mediated weight loss, demonstrating the importance of intact leptin signaling to the phenotypes associated with PTP1B deficiency [100, 101]. Interestingly, however, *ob/ob*-PTP1B KO mice weigh slightly less than *ob/ob* mice, suggesting that PTP1B may regulate obesity via a leptin-independent signaling pathway. In this regard, PTP1B also regulates growth hormone (GH)-induced JAK2 phopshorylation [102, 103]. GH is an important regulator of energy metabolism that acts through JAK-STAT pathway. Therefore, increasing signaling through this pathway might contribute to the reduced obesity in PTP1B KO and further underscores the complex role of PTP1B in energy metabolism.

Cell–Cell Communication

Cell–cell communication within any tissue is an important aspect for proper organ function and for maintaining homeostasis. ER-anchored PTP1B can access PM substrates at regions of cell–cell contact and regulate cell–cell communication [17–19]. Recent studies reveal that the ER network comes in close proximity to the PM at regions of cell–cell contact, thereby enabling PTP1B to engage PM substrates at these sites (Fig. 3.2) [20]. Proteins that reside at cell–cell contacts, such as Zonula Occludens (ZO-1) and p120 catenin, are hyper-phosphorylated in PTP1B-deficient fibroblasts and represent candidate PTP1B substrates [63]. In addition, members of the Eph RTK family have been identified as novel PTP1B substrates [20, 104]. Eph RTK and their PM-bound ephrin ligands play multiple roles including regulation of cell–cell communication [105]. Eph and ephrins are divided into A and B subclasses, with most EphAs binding to ephrin As (A1-A5) and most EphBs binding to ephrin Bs (B1-B3) [105]. A unique feature of the Eph-ephrin system is bidirectional signaling that originates from Eph receptor (forward signaling) and ephrin ligands (reverse signaling). Thus, Eph-ephrin signaling is well suited for regulating cell–cell communication (Fig. 3.2). Cell imaging and biochemical studies reveal direct interaction between PTP1B and EphA3 and demonstrate that the interaction can occur at the PM at areas of EphA3/ephrin-mediated cell–cell contact [104]. In addition, EphA2 tyrosine phosphorylation is increased at regions of cell–cell contact upon PTP1B inhibition in fibroblasts, suggesting that EphA2 is a putative PTP1B substrate at these sites [20]. Together, these studies represent a growing body of evidence that implicates PTP1B in the regulation of EphA phosphorylation and signaling. While the functional implications of PTP1B–Eph interaction remain to be determined, it is likely to encompass cell–cell communication.

Cell–cell communication is vital for metabolic regulation and its failure contributes to disease pathogenesis. For example, the importance of cell–cell interactions in pancreatic islet function and insulin secretion has long been known. Pancreatic β cell failure is a common pathological component in the progression of both type 1 and type 2 diabetes [106, 107]. β cells dynamically respond to fluctuations in blood glucose concentrations with the regulated secretion of insulin [108]. Cell–cell communication between β cells ensures suppression of basal insulin secretion and enhancement of glucose-stimulated insulin secretion [109]. This enables the secretion of low amounts of insulin during starvation and higher levels during feeding. EphA-ephrin A signaling is important for β cell communication and insulin secretion [110]. EphA5 signaling inhibits, while ephrin A5 signaling promotes glucose-stimulated insulin secretion. Importantly, EphA5 tyrosine phosphorylation decreases after glucose stimulation and alleviates the inhibitory effects on insulin secretion [110]. Thus, under basal conditions EphA5 signaling inhibits insulin secretion, but upon glucose stimulation, dephosphorylation of EphA5 attenuates its signaling and favors ephrin A5 signaling leading to insulin secretion. The PTP(s) that regulates EphA5 phosphorylation and signaling in pancreatic islets remain to be determined. Of note, PTP1B is expressed in pancreatic islets and a compound KO mouse model reveals that global PTP1B deficiency mitigates the severe diabetes caused by insulin receptor substrate 2 (IRS2) deletion [111]. In this model, PTP1B deficiency increases islet area, enhances glucose tolerance, and delays diabetes suggesting a potential role for PTP1B in endocrine pancreatic function *in vivo*. Additional studies are required to determine if PTP1B regulates pancreatic EphA5 phosphorylation and β cell–cell communication. It is noteworthy that some drugs used to treat type 2 diabetes (such as incretins) affect cell–cell interactions. Hence, it is plausible that molecules which modulate pancreatic cell–cell communication could be viable drug targets that need to be explored.

Endoplasmic Reticulum Stress

The ER is highly responsive to nutrient and energy status of the cell and plays an essential role in folding and maturation of newly synthesized proteins. Several studies demonstrate that ER dysfunction is a significant contributor to metabolic disease [112, 113]. When the folding capacity of the ER is exceeded, unfolded proteins accumulate within the ER lumen and perturb ER homeostasis, a process termed ER stress [114]. Cells use adaptive mechanisms to counter the deleterious effects of ER stress known as the unfolded protein response (UPR) [115]. UPR consists of three branches that are controlled by the ER transmembrane proteins PERK, inositol requiring enzyme 1α (IRE1α), and activating transcription factor 6 (ATF6) [113, 116, 117]. PERK phosphorylates the α-subunit of translation initiation factor 2 (eIF2α), a modification that blocks initiation of mRNA translation in response to ER stress (Fig. 3.2) [118–120]. PERK is a serine/threonine kinase whose autophosphorylation within its activation loop is essential for its activation and eIF2α phosphorylation [119]. In addition, PERK is a tyrosine-phosphorylated protein [121] and its activity is regulated by tyrosine phosphorylation; autophosphorylation of PERK at Tyr 615 is required for maximal kinase activity [122]. IRE1α activation leads to the unconventional splicing of X-box binding protein 1 (XBP1) mRNA and the transcription of genes encoding ER chaperones [123–125]. The third canonical branch of ER stress signaling includes the ATF6 transcription factors [126, 127]. These UPR arms synergize to attenuate stress [114], but if the compensatory mechanisms fail, induction of UPR can lead to elimination of stressed cells by apoptosis [128, 129].

PTP1B is implicated in the regulation of ER stress signaling. PTP1B-deficient mouse embryonic fibroblasts exhibit impaired IRE1α signaling and attenuated ER stress-induced apoptosis [130]. In addition, liver-specific PTP1B deficiency protects mice against HFD-induced ER stress [87, 131]. In contrast, PTP1B overexpression in insulinoma MIN6 β-cells mitigates chemical-induced PERK/eIF2α signaling, and PTP1B deficiency increases ER stress-induced cell death [132]. Similarly, PTP1B deficiency in adipocytes and adipose tissue leads to upregulation of PERK-eIF2α phosphorylation and sensitizes adipocytes to chemical-induced ER stress [64]. The reason(s) for these discrepancies is currently not clear. Conceivably, PTP1B has distinct substrates and/or regulates distinct arms of ER stress signaling in different tissues and in response to various challenges. Alternatively, PTP1B may affect the same pathways in different tissues, but the effects of those pathways may differ in a tissue-specific manner.

The effects of PTP1B on the ER stress response can be general (caused by indirect regulation of one or more sub-arm of ER stress signaling pathway) and/or specific (caused by direct regulation of a key target(s)). Notably, two recent studies reveal direct regulation of ER stress response by PTP1B and demonstrate that PERK is a PTP1B substrate [40, 64]. In response to chemical-induced ER stress in fibroblasts, PTP1B is reversibly inhibited by sulfhydration [40]. Importantly, PTP1B inhibition directly promotes PERK Tyr 615 phosphorylation and activity during the response to ER stress [40]. In line with these findings, PTP1B-deficient adipocytes and adipose tissue exhibit increased PERK activation and phosphorylation [64]. Moreover, substrate trapping reveals direct interaction between PTP1B and PERK and identifies PERK Tyr 615 as a mediator of this association [64]. Additional studies are required to decipher the tissue/stimulus-dependent regulation of ER stress signaling by PTP1B and its contribution to metabolic regulation.

Conclusions and Perspectives

Metabolic homeostasis requires integration of complex signaling networks which, when dysregulated, contribute to the metabolic syndrome and related disorders. The increased prevalence of metabolic syndrome highlights the urgent need for therapeutic interventions. PTP1B presents a favorable drug target as its inhibition might have beneficial effects for the treatment of obesity and type 2 diabetes. Several

PTP1B substrates are key components of fundamentally important metabolic signal transduction pathways, and it is likely that additional substrates will be identified. Insights from PTP1B–substrate interactions and tissue-specific PTP1B functions will have practical consequences for generating PTP1B inhibitors with the appropriate combination of bioavailability, specificity, and affinity. While PTP1B inhibition is a viable option that should be pursued for therapeutic intervention, it is prudent to consider multipronged approaches to combat insulin resistance and metabolic dysregulation.

References

1. Frangioni JV, Beahm PH, Shifrin V, Jost CA, Neel BG (1992) The nontransmembrane tyrosine phosphatase PTP-1B localizes to the endoplasmic reticulum via its 35 amino acid C-terminal sequence. Cell 68(3):545–560
2. Woodford-Thomas TA, Rhodes JD, Dixon JE (1992) Expression of a protein tyrosine phosphatase in normal and v-src-transformed mouse 3T3 fibroblasts. J Cell Biol 117(2):401–414
3. Liu F, Chernoff J (1997) Protein tyrosine phosphatase 1B interacts with and is tyrosine phosphorylated by the epidermal growth factor receptor. Biochem J 327(Pt 1):139–145
4. Haj FG, Verveer PJ, Squire A, Neel BG, Bastiaens PI (2002) Imaging sites of receptor dephosphorylation by PTP1B on the surface of the endoplasmic reticulum. Science 295(5560):1708–1711
5. Haj FG, Markova B, Klaman LD, Bohmer FD, Neel BG (2003) Regulation of receptor tyrosine kinase signaling by protein tyrosine phosphatase-1B. J Biol Chem 278(2):739–744
6. Boute N, Boubekeur S, Lacasa D, Issad T (2003) Dynamics of the interaction between the insulin receptor and protein tyrosine-phosphatase 1B in living cells. EMBO Rep 4(3):313–319
7. Dadke S, Kusari J, Chernoff J (2000) Down-regulation of insulin signaling by protein-tyrosine phosphatase 1B is mediated by an N-terminal binding region. J Biol Chem 275(31): 23642–23647
8. Elchebly M, Payette P, Michaliszyn E, Cromlish W, Collins S, Loy AL et al (1999) Increased insulin sensitivity and obesity resistance in mice lacking the protein tyrosine phosphatase-1B gene. Science 283(5407):1544–1548
9. Klaman LD, Boss O, Peroni OD, Kim JK, Martino JL, Zabolotny JM et al (2000) Increased energy expenditure, decreased adiposity, and tissue-specific insulin sensitivity in protein-tyrosine phosphatase 1B-deficient mice. Mol Cell Biol 20(15):5479–5489
10. Romsicki Y, Reece M, Gauthier JY, Asante-Appiah E, Kennedy BP (2004) Protein tyrosine phosphatase-1B dephosphorylation of the insulin receptor occurs in a perinuclear endosome compartment in human embryonic kidney 293 cells. J Biol Chem 279(13):12868–12875
11. Cromlish WA, Tang M, Kyskan R, Tran L, Kennedy BP (2006) PTP1B-dependent insulin receptor phosphorylation/residency in the endocytic recycling compartment of CHO-IR cells. Biochem Pharmacol 72(10):1279–1292
12. Eden ER, White IJ, Tsapara A, Futter CE (2010) Membrane contacts between endosomes and ER provide sites for PTP1B-epidermal growth factor receptor interaction. Nat Cell Biol 12(3):267–272
13. Stuible M, Abella JV, Feldhammer M, Nossov M, Sangwan V, Blagoev B et al (2010) PTP1B targets the endosomal sorting machinery: dephosphorylation of regulatory sites on the ESCRT component STAM2. J Biol Chem 285(31):23899–23907
14. Xu G, Craig AW, Greer P, Miller M, Anastasiadis PZ, Lilien J et al (2004) Continuous association of cadherin with beta-catenin requires the non-receptor tyrosine-kinase Fer. J Cell Sci 117(Pt 15):3207–3219

15. Anderie I, Schulz I, Schmid A (2007) Direct interaction between ER membrane-bound PTP1B and its plasma membrane-anchored targets. Cell Signal 19(3):582–592
16. Hernandez MV, Sala MG, Balsamo J, Lilien J, Arregui CO (2006) ER-bound PTP1B is targeted to newly forming cell-matrix adhesions. J Cell Sci 119(Pt 7):1233–1243
17. Balsamo J, Leung T, Ernst H, Zanin MK, Hoffman S, Lilien J (1996) Regulated binding of PTP1B-like phosphatase to N-cadherin: control of cadherin-mediated adhesion by dephosphorylation of beta-catenin. J Cell Biol 134(3):801–813
18. Pathre P, Arregui C, Wampler T, Kue I, Leung TC, Lilien J et al (2001) PTP1B regulates neurite extension mediated by cell-cell and cell-matrix adhesion molecules. J Neurosci Res 63(2):143–150
19. Nakamura Y, Patrushev N, Inomata H, Mehta D, Urao N, Kim HW et al (2008) Role of protein tyrosine phosphatase 1B in vascular endothelial growth factor signaling and cell-cell adhesions in endothelial cells. Circ Res 102(10):1182–1191
20. Haj FG, Sabet O, Kinkhabwala A, Wimmer-Kleikamp S, Roukos V, Han HM et al (2012) Regulation of signaling at regions of cell-cell contact by endoplasmic reticulum-bound protein-tyrosine phosphatase 1B. PLoS One 7(5):e36633
21. Andersen JN, Mortensen OH, Peters GH, Drake PG, Iversen LF, Olsen OH et al (2001) Structural and evolutionary relationships among protein tyrosine phosphatase domains. Mol Cell Biol 21(21):7117–7136
22. Denu JM, Dixon JE (1998) Protein tyrosine phosphatases: mechanisms of catalysis and regulation. Curr Opin Chem Biol 2(5):633–641
23. Salmeen A, Barford D (2005) Functions and mechanisms of redox regulation of cysteine-based phosphatases. Antioxid Redox Signal 7(5–6):560–577
24. Rhee SG, Bae YS, Lee SR, Kwon J (2000) Hydrogen peroxide: a key messenger that modulates protein phosphorylation through cysteine oxidation. Sci STKE 2000(53):pe1
25. Lee SR, Kwon KS, Kim SR, Rhee SG (1998) Reversible inactivation of protein-tyrosine phosphatase 1B in A431 cells stimulated with epidermal growth factor. J Biol Chem 273(25):15366–15372
26. Meng TC, Buckley DA, Galic S, Tiganis T, Tonks NK (2004) Regulation of insulin signaling through reversible oxidation of the protein-tyrosine phosphatases TC45 and PTP1B. J Biol Chem 279(36):37716–37725
27. Yudushkin IA, Schleifenbaum A, Kinkhabwala A, Neel BG, Schultz C, Bastiaens PI (2007) Live-cell imaging of enzyme-substrate interaction reveals spatial regulation of PTP1B. Science 315(5808):115–119
28. Salmeen A, Andersen JN, Myers MP, Meng TC, Hinks JA, Tonks NK et al (2003) Redox regulation of protein tyrosine phosphatase 1B involves a sulphenyl-amide intermediate. Nature 423(6941):769–773
29. Van Montfort RL, Congreve M, Tisi D, Carr R, Jhoti H (2003) Oxidation state of the active-site cysteine in protein tyrosine phosphatase 1B. Nature 423(6941):773–777
30. Mahadev K, Motoshima H, Wu X, Ruddy JM, Arnold RS, Cheng G et al (2004) The NAD(P)H oxidase homolog Nox4 modulates insulin-stimulated generation of H2O2 and plays an integral role in insulin signal transduction. Mol Cell Biol 24(5):1844–1854
31. Martyn KD, Frederick LM, von Loehneysen K, Dinauer MC, Knaus UG (2006) Functional analysis of Nox4 reveals unique characteristics compared to other NADPH oxidases. Cell Signal 18(1):69–82
32. Hess DT, Matsumoto A, Kim SO, Marshall HE, Stamler JS (2005) Protein S-nitrosylation: purview and parameters. Nat Rev Mol Cell Biol 6(2):150–166
33. Lane P, Hao G, Gross SS (2001) S-nitrosylation is emerging as a specific and fundamental posttranslational protein modification: head-to-head comparison with O-phosphorylation. Sci STKE 2001(86):re1
34. Gu Z, Kaul M, Yan B, Kridel SJ, Cui J, Strongin A et al (2002) S-nitrosylation of matrix metalloproteinases: signaling pathway to neuronal cell death. Science 297(5584):1186–1190
35. Mannick JB, Hausladen A, Liu L, Hess DT, Zeng M, Miao QX et al (1999) Fas-induced caspase denitrosylation. Science 284(5414):651–654

36. Chen YY, Huang YF, Khoo KH, Meng TC (2007) Mass spectrometry-based analyses for identifying and characterizing S-nitrosylation of protein tyrosine phosphatases. Methods 42(3):243–249
37. Chen YY, Chu HM, Pan KT, Teng CH, Wang DL, Wang AH et al (2008) Cysteine S-nitrosylation protects protein-tyrosine phosphatase 1B against oxidation-induced permanent inactivation. J Biol Chem 283(50):35265–35272
38. Gadalla MM, Snyder SH (2010) Hydrogen sulfide as a gasotransmitter. J Neurochem 113(1):14–26
39. Mustafa AK, Gadalla MM, Sen N, Kim S, Mu W, Gazi SK et al (2009) H2S signals through protein S-sulfhydration. Sci Signal 2(96):ra72
40. Krishnan N, Fu C, Pappin DJ, Tonks NK (2012) H2S-induced sulfhydration of the phosphatase PTP1B and its role in the endoplasmic reticulum stress response. Sci Signal 303(9):C916–C923
41. Baskar R, Bian J (2011) Hydrogen sulfide gas has cell growth regulatory role. Eur J Pharmacol 656(1–3):5–9
42. Wilkinson KA, Henley JM (2010) Mechanisms, regulation and consequences of protein SUMOylation. Biochem J 428(2):133–145
43. Sarge KD, Park-Sarge OK (2011) SUMO and its role in human diseases. Int Rev Cell Mol Biol 288:167–183
44. Dadke S, Cotteret S, Yip SC, Jaffer ZM, Haj F, Ivanov A et al (2007) Regulation of protein tyrosine phosphatase 1B by sumoylation. Nat Cell Biol 9(1):80–85
45. Yip SC, Cotteret S, Chernoff J (2012) Sumoylated protein tyrosine phosphatase 1B localizes to the inner nuclear membrane and regulates the tyrosine phosphorylation of emerin. J Cell Sci 125(Pt 2):310–316
46. Matsuo K, Bettaieb A, Nagata N, Matsuo I, Keilhack H, Haj FG (2011) Regulation of brown fat adipogenesis by protein tyrosine phosphatase 1B. PloS one 6(1):e16446
47. Bandyopadhyay D, Kusari A, Kenner KA, Liu F, Chernoff J, Gustafson TA et al (1997) Protein-tyrosine phosphatase 1B complexes with the insulin receptor in vivo and is tyrosine-phosphorylated in the presence of insulin. J Biol Chem 272(3):1639–1645
48. Dadke S, Kusari A, Kusari J (2001) Phosphorylation and activation of protein tyrosine phosphatase (PTP) 1B by insulin receptor. Mol Cell Biochem 221(1–2):147–154
49. Tao J, Malbon CC, Wang HY (2001) Insulin stimulates tyrosine phosphorylation and inactivation of protein-tyrosine phosphatase 1B in vivo. J Biol Chem 276(31):29520–29525
50. Flint AJ, Gebbink MF, Franza BR Jr, Hill DE, Tonks NK (1993) Multi-site phosphorylation of the protein tyrosine phosphatase, PTP1B: identification of cell cycle regulated and phorbol ester stimulated sites of phosphorylation. EMBO J 12(5):1937–1946
51. Shifrin VI, Davis RJ, Neel BG (1997) Phosphorylation of protein-tyrosine phosphatase PTP-1B on identical sites suggests activation of a common signaling pathway during mitosis and stress response in mammalian cells. J Biol Chem 272(5):2957–2962
52. Ravichandran LV, Chen H, Li Y, Quon MJ (2001) Phosphorylation of PTP1B at Ser(50) by Akt impairs its ability to dephosphorylate the insulin receptor. Mol Endocrinol 15(10): 1768–1780
53. Frangioni JV, Oda A, Smith M, Salzman EW, Neel BG (1993) Calpain-catalyzed cleavage and subcellular relocation of protein phosphotyrosine phosphatase 1B (PTP-1B) in human platelets. EMBO J 12(12):4843–4856
54. Azam M, Andrabi SS, Sahr KE, Kamath L, Kuliopulos A, Chishti AH (2001) Disruption of the mouse mu-calpain gene reveals an essential role in platelet function. Mol Cell Biol 21(6):2213–2220
55. Kuchay SM, Kim N, Grunz EA, Fay WP, Chishti AH (2007) Double knockouts reveal that protein tyrosine phosphatase 1B is a physiological target of calpain-1 in platelets. Mol Cell Biol 27(17):6038–6052
56. Trumpler A, Schlott B, Herrlich P, Greer PA, Bohmer FD (2009) Calpain-mediated degradation of reversibly oxidized protein-tyrosine phosphatase 1B. FEBS J 276(19):5622–5633

57. Liu F, Hill DE, Chernoff J (1996) Direct binding of the proline-rich region of protein tyrosine phosphatase 1B to the Src homology 3 domain of p130(Cas). J Biol Chem 271(49): 31290–31295
58. Salmeen A, Andersen JN, Myers MP, Tonks NK, Barford D (2000) Molecular basis for the dephosphorylation of the activation segment of the insulin receptor by protein tyrosine phosphatase 1B. Mol Cell 6(6):1401–1412
59. Myers MP, Andersen JN, Cheng A, Tremblay ML, Horvath CM, Parisien JP et al (2001) TYK2 and JAK2 are substrates of protein-tyrosine phosphatase 1B. J Biol Chem 276(51):47771–47774
60. Zhang M, Van Etten RL, Stauffacher CV (1994) Crystal structure of bovine heart phosphotyrosyl phosphatase at 2.2-A resolution. Biochemistry 33(37):11097–11105
61. Flint AJ, Tiganis T, Barford D, Tonks NK (1997) Development of "substrate-trapping" mutants to identify physiological substrates of protein tyrosine phosphatases. Proc Natl Acad Sci U S A 94(5):1680–1685
62. Stuible M, Dube N, Tremblay ML (2008) PTP1B regulates cortactin tyrosine phosphorylation by targeting Tyr446. J Biol Chem 283(23):15740–15746
63. Mertins P, Eberl HC, Renkawitz J, Olsen JV, Tremblay ML, Mann M et al (2008) Investigation of protein tyrosine phosphatase 1B function by quantitative proteomics. Mol Cell Proteomics 7(9):1763–1777
64. Bettaieb A, Matsuo K, Matsuo I, Wang S, Melhem R, Koromilas AE et al (2012) Protein tyrosine phosphatase 1B deficiency potentiates PERK/eIF2alpha signaling in brown adipocytes. PLoS One 7(4):e34412
65. Spiegelman BM, Flier JS (2001) Obesity and the regulation of energy balance. Cell 104(4):531–543
66. Biddinger SB, Kahn CR (2006) From mice to men: insights into the insulin resistance syndromes. Annu Rev Physiol 68:123–158
67. Taniguchi CM, Emanuelli B, Kahn CR (2006) Critical nodes in signalling pathways: insights into insulin action. Nat Rev Mol Cell Biol 7(2):85–96
68. Luo J, Cantley LC (2005) The negative regulation of phosphoinositide 3-kinase signaling by p85 and it's implication in cancer. Cell Cycle 4(10):1309–1312
69. Avruch J (1998) Insulin signal transduction through protein kinase cascades. Mol Cell Biochem 182(1–2):31–48
70. Pouyssegur J, Volmat V, Lenormand P (2002) Fidelity and spatio-temporal control in MAP kinase (ERKs) signalling. Biochem Pharmacol 64(5–6):755–763
71. Rask-Madsen C, Kahn CR (2012) Tissue-specific insulin signaling, metabolic syndrome, and cardiovascular disease. Arterioscler Thromb Vasc Biol 32(9):2052–2059
72. Cicirelli MF, Tonks NK, Diltz CD, Weiel JE, Fischer EH, Krebs EG (1990) Microinjection of a protein-tyrosine-phosphatase inhibits insulin action in Xenopus oocytes. Proc Natl Acad Sci U S A 87(14):5514–5518
73. Tonks NK, Cicirelli MF, Diltz CD, Krebs EG, Fischer EH (1990) Effect of microinjection of a low-Mr human placenta protein tyrosine phosphatase on induction of meiotic cell division in Xenopus oocytes. Mol Cell Biol 10(2):458–463
74. Ahmad F, Li PM, Meyerovitch J, Goldstein BJ (1995) Osmotic loading of neutralizing antibodies demonstrates a role for protein-tyrosine phosphatase 1B in negative regulation of the insulin action pathway. J Biol Chem 270(35):20503–20508
75. Kenner KA, Anyanwu E, Olefsky JM, Kusari J (1996) Protein-tyrosine phosphatase 1B is a negative regulator of insulin- and insulin-like growth factor-I-stimulated signaling. J Biol Chem 271(33):19810–19816
76. Chen H, Wertheimer SJ, Lin CH, Katz SL, Amrein KE, Burn P et al (1997) Protein-tyrosine phosphatases PTP1B and syp are modulators of insulin-stimulated translocation of GLUT4 in transfected rat adipose cells. J Biol Chem 272(12):8026–8031
77. Walchli S, Curchod ML, Gobert RP, Arkinstall S, Hooft van Huijsduijnen R (2000) Identification of tyrosine phosphatases that dephosphorylate the insulin receptor. A brute force approach based on "substrate-trapping" mutants. J Biol Chem 275(13):9792–9796

78. Seely BL, Staubs PA, Reichart DR, Berhanu P, Milarski KL, Saltiel AR et al (1996) Protein tyrosine phosphatase 1B interacts with the activated insulin receptor. Diabetes 45(10): 1379–1385
79. Goldstein BJ, Bittner-Kowalczyk A, White MF, Harbeck M (2000) Tyrosine dephosphorylation and deactivation of insulin receptor substrate-1 by protein-tyrosine phosphatase 1B. Possible facilitation by the formation of a ternary complex with the Grb2 adaptor protein. J Biol Chem 275(6):4283–4289
80. Zabolotny JM, Haj FG, Kim YB, Kim HJ, Shulman GI, Kim JK et al (2004) Transgenic overexpression of protein-tyrosine phosphatase 1B in muscle causes insulin resistance, but overexpression with leukocyte antigen-related phosphatase does not additively impair insulin action. J Biol Chem 279(23):24844–24851
81. Haj FG, Zabolotny JM, Kim YB, Kahn BB, Neel BG (2005) Liver-specific protein-tyrosine phosphatase 1B (PTP1B) re-expression alters glucose homeostasis of PTP1B-/-mice. J Biol Chem 280(15):15038–15046
82. Bence KK, Delibegovic M, Xue B, Gorgun CZ, Hotamisligil GS, Neel BG et al (2006) Neuronal PTP1B regulates body weight, adiposity and leptin action. Nat Med 12(8):917–924
83. Banno R, Zimmer D, De Jonghe BC, Atienza M, Rak K, Yang W et al (2010) PTP1B and SHP2 in POMC neurons reciprocally regulate energy balance in mice. J Clin Investig 120(3):720–734
84. Tsou RC, Zimmer DJ, De Jonghe BC, Bence KK (2012) Deficiency of PTP1B in leptin receptor-expressing neurons leads to decreased body weight and adiposity in mice. Endocrinology 153(9):4227–4237
85. Xue B, Pulinilkunnil T, Murano I, Bence KK, He H, Minokoshi Y et al (2009) Neuronal protein tyrosine phosphatase 1B deficiency results in inhibition of hypothalamic AMPK and isoform-specific activation of AMPK in peripheral tissues. Mol Cell Biol 29(16):4563–4573
86. Delibegovic M, Bence KK, Mody N, Hong EG, Ko HJ, Kim JK et al (2007) Improved glucose homeostasis in mice with muscle-specific deletion of protein-tyrosine phosphatase 1B. Mol Cell Biol 27(21):7727–7734
87. Delibegovic M, Zimmer D, Kauffman C, Rak K, Hong EG, Cho YR et al (2009) Liver-specific deletion of protein-tyrosine phosphatase 1B (PTP1B) improves metabolic syndrome and attenuates diet-induced endoplasmic reticulum stress. Diabetes 58(3):590–599
88. Hooft van Huijsduijnen R, Bombrun A, Swinnen D (2002) Selecting protein tyrosine phosphatases as drug targets. Drug Discov Today 7(19):1013–1019
89. Lee S, Wang Q (2007) Recent development of small molecular specific inhibitor of protein tyrosine phosphatase 1B. Med Res Rev 27(4):553–573
90. Zhang S, Zhang ZY (2007) PTP1B as a drug target: recent developments in PTP1B inhibitor discovery. Drug Discov Today 12(9–10):373–381
91. Vintonyak VV, Antonchick AP, Rauh D, Waldmann H (2009) The therapeutic potential of phosphatase inhibitors. Curr Opin Chem Biol 13(3):272–283
92. Blaskovich MA (2009) Drug discovery and protein tyrosine phosphatases. Curr Med Chem 16(17):2095–2176
93. Zinker BA, Rondinone CM, Trevillyan JM, Gum RJ, Clampit JE, Waring JF et al (2002) PTP1B antisense oligonucleotide lowers PTP1B protein, normalizes blood glucose, and improves insulin sensitivity in diabetic mice. Proc Natl Acad Sci U S A 99(17): 11357–11362
94. Swarbrick MM, Havel PJ, Levin AA, Bremer AA, Stanhope KL, Butler M et al (2009) Inhibition of protein tyrosine phosphatase-1B with antisense oligonucleotides improves insulin sensitivity and increases adiponectin concentrations in monkeys. Endocrinology 150(4):1670–1679
95. Zhang Y, Proenca R, Maffei M, Barone M, Leopold L, Friedman JM (1994) Positional cloning of the mouse obese gene and its human homologue. Nature 372(6505):425–432
96. Ahima RS, Flier JS (2000) Leptin. Annu Rev Physiol 62:413–437
97. Tartaglia LA (1997) The leptin receptor. J Biol Chem 272(10):6093–6096

98. Ahima RS, Osei SY (2004) Leptin signaling. Physiol Behav 81(2):223–241
99. Pelleymounter MA, Cullen MJ, Baker MB, Hecht R, Winters D, Boone T et al (1995) Effects of the obese gene product on body weight regulation in ob/ob mice. Science 269(5223):540–543
100. Cheng A, Uetani N, Simoncic PD, Chaubey VP, Lee-Loy A, McGlade CJ et al (2002) Attenuation of leptin action and regulation of obesity by protein tyrosine phosphatase 1B. Dev cell 2(4):497–503
101. Zabolotny JM, Bence-Hanulec KK, Stricker-Krongrad A, Haj F, Wang Y, Minokoshi Y et al (2002) PTP1B regulates leptin signal transduction in vivo. Dev Cell 2(4):489–495
102. Pasquali C, Curchod ML, Walchli S, Espanel X, Guerrier M, Arigoni F et al (2003) Identification of protein tyrosine phosphatases with specificity for the ligand-activated growth hormone receptor. Mol Endocrinol 17(11):2228–2239
103. Gu F, Dube N, Kim JW, Cheng A, Ibarra-Sanchez Mde J, Tremblay ML et al (2003) Protein tyrosine phosphatase 1B attenuates growth hormone-mediated JAK2-STAT signaling. Mol Cell Biol 23(11):3753–3762
104. Nievergall E, Janes PW, Stegmayer C, Vail ME, Haj FG, Teng SW et al (2010) PTP1B regulates Eph receptor function and trafficking. J Cell Biol 191(6):1189–1203
105. Pasquale EB (2008) Eph-ephrin bidirectional signaling in physiology and disease. Cell 133(1):38–52
106. Mathis D, Vence L, Benoist C (2001) beta-cell death during progression to diabetes. Nature 414(6865):792–798
107. Butler AE, Janson J, Bonner-Weir S, Ritzel R, Rizza RA, Butler PC (2003) Beta-cell deficit and increased beta-cell apoptosis in humans with type 2 diabetes. Diabetes 52(1):102–110
108. Henquin JC (2000) Triggering and amplifying pathways of regulation of insulin secretion by glucose. Diabetes 49(11):1751–1760
109. Ravier MA, Guldenagel M, Charollais A, Gjinovci A, Caille D, Sohl G et al (2005) Loss of connexin36 channels alters beta-cell coupling, islet synchronization of glucose-induced Ca2+ and insulin oscillations, and basal insulin release. Diabetes 54(6):1798–1807
110. Konstantinova I, Nikolova G, Ohara-Imaizumi M, Meda P, Kucera T, Zarbalis K et al (2007) EphA-Ephrin-A-mediated beta cell communication regulates insulin secretion from pancreatic islets. Cell 129(2):359–370
111. Kushner JA, Haj FG, Klaman LD, Dow MA, Kahn BB, Neel BG et al (2004) Islet-sparing effects of protein tyrosine phosphatase-1b deficiency delays onset of diabetes in IRS2 knockout mice. Diabetes 53(1):61–66
112. Ozcan U, Cao Q, Yilmaz E, Lee AH, Iwakoshi NN, Ozdelen E et al (2004) Endoplasmic reticulum stress links obesity, insulin action, and type 2 diabetes. Science 306(5695): 457–461
113. Hummasti S, Hotamisligil GS (2010) Endoplasmic reticulum stress and inflammation in obesity and diabetes. Circ Res 107(5):579–591
114. Schroder M, Kaufman RJ (2005) The mammalian unfolded protein response. Annu Rev Biochem 74:739–789
115. Kaufman RJ, Scheuner D, Schroder M, Shen X, Lee K, Liu CY et al (2002) The unfolded protein response in nutrient sensing and differentiation. Nat Rev Mol Cell Biol 3(6):411–421
116. Ron D, Walter P (2007) Signal integration in the endoplasmic reticulum unfolded protein response. Nat Rev Mol Cell Biol 8(7):519–529
117. Hotamisligil GS (2010) Endoplasmic reticulum stress and the inflammatory basis of metabolic disease. Cell 140(6):900–917
118. Shi Y, Vattem KM, Sood R, An J, Liang J, Stramm L et al (1998) Identification and characterization of pancreatic eukaryotic initiation factor 2 alpha-subunit kinase, PEK, involved in translational control. Mol Cell Biol 18(12):7499–7509
119. Harding HP, Zhang Y, Ron D (1999) Protein translation and folding are coupled by an endoplasmic-reticulum-resident kinase. Nature 397(6716):271–274
120. Novoa I, Zhang Y, Zeng H, Jungreis R, Harding HP, Ron D (2003) Stress-induced gene expression requires programmed recovery from translational repression. EMBO J 22(5): 1180–1187

121. Ma Y, Lu Y, Zeng H, Ron D, Mo W, Neubert TA (2001) Characterization of phosphopeptides from protein digests using matrix-assisted laser desorption/ionization time-of-flight mass spectrometry and nanoelectrospray quadrupole time-of-flight mass spectrometry. Rapid Commun Mass Spectrom 15(18):1693–1700
122. Su Q, Wang S, Gao HQ, Kazemi S, Harding HP, Ron D et al (2008) Modulation of the eukaryotic initiation factor 2 alpha-subunit kinase PERK by tyrosine phosphorylation. J Biol Chem 283(1):469–475
123. Sidrauski C, Walter P (1997) The transmembrane kinase Ire1p is a site-specific endonuclease that initiates mRNA splicing in the unfolded protein response. Cell 90(6):1031–1039
124. Yoshida H, Matsui T, Yamamoto A, Okada T, Mori K (2001) XBP1 mRNA is induced by ATF6 and spliced by IRE1 in response to ER stress to produce a highly active transcription factor. Cell 107(7):881–891
125. Calfon M, Zeng H, Urano F, Till JH, Hubbard SR, Harding HP et al (2002) IRE1 couples endoplasmic reticulum load to secretory capacity by processing the XBP-1 mRNA. Nature 415(6867):92–96
126. Ye J, Rawson RB, Komuro R, Chen X, Dave UP, Prywes R et al (2000) ER stress induces cleavage of membrane-bound ATF6 by the same proteases that process SREBPs. Mol Cell 6(6):1355–1364
127. Okada T, Yoshida H, Akazawa R, Negishi M, Mori K (2002) Distinct roles of activating transcription factor 6 (ATF6) and double-stranded RNA-activated protein kinase-like endoplasmic reticulum kinase (PERK) in transcription during the mammalian unfolded protein response. Biochem J 366(Pt 2):585–594
128. Zinszner H, Kuroda M, Wang X, Batchvarova N, Lightfoot RT, Remotti H et al (1998) CHOP is implicated in programmed cell death in response to impaired function of the endoplasmic reticulum. Genes Dev 12(7):982–995
129. Nishitoh H, Matsuzawa A, Tobiume K, Saegusa K, Takeda K, Inoue K et al (2002) ASK1 is essential for endoplasmic reticulum stress-induced neuronal cell death triggered by expanded polyglutamine repeats. Genes Dev 16(11):1345–1355
130. Gu F, Nguyen DT, Stuible M, Dube N, Tremblay ML, Chevet E (2004) Protein-tyrosine phosphatase 1B potentiates IRE1 signaling during endoplasmic reticulum stress. J Biol Chem 279(48):49689–49693
131. Agouni A, Mody N, Owen C, Czopek AJ, Zimmer D, Bentires-Alj M et al (2011) Liver-specific deletion of protein tyrosine phosphatase (PTP) 1B improves obesity- and pharmacologically-induced endoplasmic reticulum stress. Biochem J 438(2):369–378
132. Bettaieb A, Liu S, Xi Y, Nagata N, Matsuo K, Matsuo I et al (2011) Differential regulation of endoplasmic reticulum stress by protein tyrosine phosphatase 1B and T cell protein tyrosine phosphatase. J Biol Chem 286(11):9225–9235

Chapter 4
PTP1B and TCPTP in CNS Signaling and Energy Balance

Kendra K. Bence and Tony Tiganis

Abstract The regulation of energy balance is under tight homeostatic control. Biological mechanisms have evolved over time to ensure adequate nutritional status and appropriate body composition in response to metabolic and environmental stimuli. The central nervous system (CNS) plays an important role in the regulation of body weight and in the control of normal glucose homeostasis. Several key areas of the CNS are involved in energy balance, including the nuclei of the hypothalamus, hindbrain, and limbic (reward) centers of the brain. Within these brain regions critical cellular signaling pathways have been identified that mediate a multitude of metabolic processes, including feeding, body weight gain/loss, energy expenditure, core temperature regulation, peripheral insulin sensitivity, and liver metabolism. Two such pathways are the leptin and insulin signaling pathways. Rapid reversible phosphorylation events within these key CNS signaling pathways are critical to the tight regulation of energy balance control, and disruption of these events can contribute to the pathogenesis of the metabolic syndrome. Protein tyrosine phosphatases, or PTPs, catalyze the dephosphorylation of phosphorylated tyrosyl residues and thus are important regulators of intracellular signaling pathways. In this chapter, the contributions of protein tyrosine phosphatase 1B (PTP1B) and its closest homologue, T cell PTP (TCPTP), to CNS control of energy balance will be highlighted.

K.K. Bence, Ph.D. (✉)
Department of Animal Biology, School of Veterinary Medicine, University of Pennsylvania, Philadelphia, PA 19104-6046, USA
e-mail: kbence@vet.upenn.edu

T. Tiganis, Ph.D., B.Sc.
Department of Biochemistry and Molecular Biology, Monash University, Clayton, VIC 3800, Australia

K.K. Bence (ed.), *Protein Tyrosine Phosphatase Control of Metabolism*,
DOI 10.1007/978-1-4614-7855-3_4,

Abbreviations

AgRP	Agouti-related peptide
ARC	Arcuate nucleus
BAT	Brown adipose tissue
Cga	Glycoprotein hormone alpha-subunit
CNS	Central nervous system
DMH	Dorsomedial hypothalamus
ER	Endoplasmic reticulum
GH	Growth hormone
icv	Intracerebroventricular
IR	Insulin receptor
IRS-1	Insulin receptor substrate-1
JAK-2	Janus-activated kinase 2
LepRb	Leptin receptor
LH	Lateral hypothalamus
MEFs	Mouse embryonic fibroblasts
MTII	Melanotan II
NPY	Neuropeptide Y
NTS	Nucleus of the solitary tract
PI3K	Phosphatidylinositol 3-kinase
POMC	Proopiomelanocortin
PTP1B	Protein tyrosine phosphatase 1B
PVN	Paraventricular nucleus
SHP-2	Src homology phosphatase-2
SOCS3	Suppressor of cytokine signaling 3
TCPTP	T cell PTP
VMH	Ventromedial hypothalamus
VTA	Ventral tegmental area
α-MSH	α-melanocyte-stimulating hormone

PTP1B-Deficient Mice Are Lean

PTP1B is a non-receptor PTP encoded by the *PTPN1* gene, which has been implicated in the regulation of growth factor signaling, cell cycle control, oncogenic signaling, and the response to cellular stress [1]. The first hint that PTP1B may be an important regulator of body weight and insulin sensitivity came when two independent groups generated and characterized *Ptpn1–/–* mice. Elchebly and colleagues found that *Ptpn1–/–* mice had lower blood glucose and serum insulin levels compared to wild type controls, as well as enhanced insulin sensitivity [2]. When placed on a high-fat diet, *Ptpn1–/–* mice gained less weight than controls and remained more insulin-sensitive. Klaman and colleagues also found that *Ptpn1–/–* mice were lean and resistant to weight gain on a high-fat diet. The decreased adiposity was shown to be due to an elevated metabolic rate and total energy expenditure [3]. Since PTP1B was already known to be an insulin receptor (IR) phosphatase [4, 5], the phenotype of enhanced insulin sensitivity was not unexpected. However, the decreased adiposity and resistance to diet-induced obesity were intriguing and sparked considerable attention from both academia and the pharmaceutical industry. Although the basis of the obesity resistance was unknown at the time, it was ultimately ascribed to enhanced central leptin sensitivity.

Leptin, a Master Regulator of Body Weight

Leptin is an adipocyte-secreted hormone that acts on leptin-receptor (LepRb) expressing neurons in the hypothalamus and elsewhere in the brain to suppress food intake and increase energy expenditure. Leptin in the circulation typically correlates with adiposity, thus communicating information to the brain regarding nutrient status and allowing for precise control of energy balance [reviewed in [6, 7]]. Leptin also modulates many other processes via the brain, including fertility/reproduction, immune function, and glucose homeostasis [reviewed in [6–9]]. In the arcuate nucleus (ARC) of the hypothalamus leptin elicits its effects by (1) altering the frequency of action potentials [depolarizing anorexigenic proopiomelanocortin (POMC) expressing neurons] to affect synaptic transmission, and (2) altering gene transcription; leptin promotes the expression of POMC [the precursor of α-melanocyte-stimulating hormone (α-MSH)] in POMC neurons and inhibits the expression of NPY (neuropeptide Y) and AgRP (agouti-related peptide) in orexigenic NPY/AgRP orexigenic neurons [[10–14] and reviewed in [6, 7]]. α-MSH signals via melanocortin 4 receptors (MC4R) on downstream target neurons in hypothalamic regions such as the paraventricular nucleus (PVN) that affect food intake and energy expenditure, whereas AgRP antagonizes α-MSH binding to MC4R [reviewed in [6, 7]]. The LepRb is also expressed in other neuronal populations in the CNS to affect body weight, including other hypothalamic nuclei such as the PVN, the ventromedial hypothalamus (VMH), the dorsomedial hypothalamus (DMH), and the lateral hypothalamus (LH) [reviewed in [6, 7]], as well as other regions of the brain including the nucleus of the solitary tract (NTS) of the hindbrain [15, 16] and regions involved in reward-related behavior including dopaminergic neurons in the ventral tegmental area (VTA) [17–20].

The LepRb is a member of the type I cytokine receptor family; a single *Lepr* gene produces several LepR isoforms via alternative splicing [21]. The "long isoform" of the receptor, LepRb, is the only isoform that can couple to the associated protein tyrosine kinase JAK-2 (Janus-activated kinase 2) via membrane-proximal Box1 and Box2 motifs present in LepRb intracellular domain [22]. Since the LepRb does not have intrinsic kinase activity, the association with JAK-2 is critical to activate downstream signaling components. Leptin activation of JAK-2 promotes phosphorylation of LepRb on three critical intracellular tyrosine residues, Y985, Y1077, and Y1138 [reviewed in [6, 7, 23]]. Phosphorylation of Y985 recruits Src homology phosphatase-2 (SHP-2) leading to ERK1/2 activation, and suppressor of cytokine signaling 3 (SOCS3), which mediates negative feedback inhibition of leptin signaling. Phosphorylation of Y1077 recruits the transcription factor STAT5, while phospho-Y1138 recruits STAT3 leading to transcriptional changes that alter energy balance. Activation of JAK-2 also promotes signaling via additional effector cascades, including the phosphatidylinositol 3-kinase (PI3K)/Akt pathway, which acutely suppresses food intake and depolarizes POMC neurons and may underlie the acute effects of leptin on food intake [24–27].

A Lean Mouse in Search of a Mechanism

Shortly after the obesity resistance phenotype of *Ptpn1–/–* mice was described, several groups established that JAK-2 could serve a direct substrate of PTP1B in vitro and in vivo. The sequence (E/D)-pY-pY-(R/K), which had previously been shown to be integral to the interaction of PTP1B with the active site of the IR [28], was found to be a more global consensus PTP1B substrate recognition motif [29–31]. In particular, PTP1B was shown to dephosphorylate the corresponding Y1007/Y1008 activation loop autophosphorylation site of JAK-2 [29–31]. In HEK293 cells stimulated with IFN-γ, Y1007/Y1008 phosphorylated JAK-2 formed a complex with the substrate-trapping D181A mutant of PTP1B [29]. Furthermore, the Y1007/Y1008 site within the JAK-2 activation loop was hyperphosphorylated in *Ptpn1–/–* mouse embryonic fibroblasts (MEFs) compared to PTP1B+/+ MEFs [29, 30]. In *Ptpn1–/–* MEFs reexpressing PTP1B-D181A, JAK-2 co-immunoprecipitated with PTP1B-D181A in leptin-stimulated cells; this interaction was disrupted in the presence of the competitive PTP inhibitor sodium orthovanadate [30]. Similarly, in 293LA cells (a human kidney cell line stably overexpressing JAK-2), pull-down assays showed that GST-PTP1B-D181A co-immunoprecipitated JAK-2 from leptin-stimulated cells [31]. Overexpression of PTP1B in COS-7 or CHO cells engineered to express the LepRb resulted in a dose-dependent reduction in leptin-induced JAK-2 and STAT3 phosphorylation [30, 32]. Finally, overexpression of PTP1B in a mouse hypothalamic cell line (GT1-7) resulted in a dose-dependent decrease of leptin-induced JAK-2 and STAT3 phosphorylation, as well as suppression of leptin-induced changes in gene expression [33].

Evidence that PTP1B is an important regulator of leptin signaling under physiological conditions comes from in vivo studies using mouse models of PTP1B-deficiency. Importantly, PTP1B mRNA expression is enriched in areas of the brain that are known sites of LepRb expression, including the nuclei of the hypothalamus and the hippocampus [30, 34]. As such, leptin-induced phosphorylation of STAT3 is increased in the hypothalamus of *Ptpn1–/–* mice compared to wild type controls, and *Ptpn1–/–* mice show enhanced leptin-induced suppression of food intake and body weight compared to controls [30, 31]. The primary importance of the hypothalamus in mediating the beneficial effects of PTP1B-deficiency is highlighted by the finding that chemical ablation of neurons within the mediobasal hypothalamus of *Ptpn1–/–* mice using gold thioglucose results in hyperphagia and elevated body weight [30]. Combined these in vitro and in vivo studies placed PTP1B in the limelight as a critical regulator of body weight and an important negative regulator of leptin signaling under physiological conditions.

Mind Over Matter: Central Role for PTP1B in the Control of Energy Balance

Since PTP1B is a relatively ubiquitously expressed PTP, the relevant sites of PTP1B's metabolic effects were not initially clear. In order to definitively assess the tissue and cell-type specific functions of PTP1B, mice with selective PTP1B-deficiency were generated and characterized. Notably, these studies established the brain as the primary site mediating PTP1B's effects on body weight and adiposity [35]. Deficiency of PTP1B in the muscle (MCK-*Ptpn1–/–*) [36], liver (Albumin-*Ptpn1–/–*) [37, 38], or adipocytes (Adiponectin-*Ptpn1–/–*) [39] of mice does not result in a body weight phenotype; the phenotypes of these mice are reviewed in elsewhere in this book. In contrast, mice with neuronal PTP1B-deficiency (Nestin-*Ptpn1–/–*) display reduced body weight and adiposity when fed a high-fat diet compared to wild type controls [35]. This phenotype is likely due to a combination of reduced food intake, elevated energy expenditure, and increased locomotor activity seen in these mice. Similar to whole-body *Ptpn1–/–* mice, Nestin-*Ptpn1–/–* mice display enhanced leptin sensitivity and elevated hypothalamic pSTAT3 levels [35]. Further support for PTP1B's central role in regulating energy balance comes from inhibitor and antisense studies. Rats pretreated with a selective cell-permeable PTP1B inhibitor via intracerebroventricular (*icv*) delivery into the third ventricle display enhanced leptin-induced suppression of food intake in response to *icv* leptin [40]. Selective reduction of PTP1B expression in rat hypothalamus by *icv* infusion of antisense oligonucleotides specific for PTP1B results in decreased body weight and adiposity, enhanced hypothalamic leptin and insulin signaling, and improved glucose homeostasis in obese diabetic rats [41]. Combined these genetic and pharmacologic studies provide strong evidence that the majority of the energy balance effects resulting from PTP1B-deficiency are mediated within the central nervous system.

A subset of hypothalamic neurons has been shown to play an important role in the control of energy balance. Mice lacking LepRbs specifically within POMC neurons are mildly obese and hyperleptinemic, indicating that these neurons contribute to leptin's overall effects on body weight [42]. In addition, restoration of LepRs within POMC neurons in LepRb-deficient (*db/db*) mice partially rescues hyperphagia and obesity of *db/db* mice [43]. Consistent with the notion that leptin signaling within this limited subset of neurons is important to body mass control, POMC-*Ptpn1–/–* mice have reduced body weight and adiposity, and elevated energy expenditure, when placed on a high-fat (55 % fat) diet from weaning [44]. These mice display enhanced leptin sensitivity and also display enhanced peripheral insulin sensitivity and glucose homeostasis when matched for body weight and adiposity. Notably, these data were the first indication that central PTP1B may exert control over peripheral insulin sensitivity. POMC-*Ptpn1–/–* mice also have more brown adipose tissue (BAT) and elevated plasma triiodothyronine levels in response to a 4-day cold challenge, as well as reduced spontaneous physical activity following 1-day of cold exposure compared to controls [45]. Combined these data show that PTP1B in POMC neurons plays a role in the control of body weight, leptin sensitivity, glucose homeostasis, and homeostatic responses to cold exposure.

POMC is expressed within cells of the anterior pituitary, the arcuate nucleus of the hypothalamus, and in the hindbrain. A metabolic role for PTP1B in the pituitary has been ruled out since mice with anterior pituitary-specific deletion (glycoprotein hormone alpha-subunit (Cga)-*Ptpn1–/–*) do not show alterations in body weight or glucose homeostasis [44]. The most well-characterized populations of POMC neurons are localized to the arcuate nucleus; however, leptin-responsive neurons are also found in the NTS of the hindbrain [46–48]. Studies utilizing hindbrain delivery (via the fourth ventricle) of leptin or the melanocortin 3/4 receptor agonist melanotan II (MTII) demonstrate that POMC-*Ptpn1–/–* mice are indeed hypersensitive to hindbrain leptin- and MTII-induced suppression of body weight and food intake, and leptin- and MTII-stimulation of spontaneous physical activity, suggesting that this subpopulation of POMC neurons may contribute to PTP1B's metabolic effects [49].

Leptin-Dependent and -Independent Effects of PTP1B-Deficiency

The improved metabolic phenotype of PTP1B-deficient mice (both whole body and brain-specific) is primarily attributed to improved leptin sensitivity as outlined above. If all metabolic improvements are mediated via leptin signaling, then mice deficient in both PTP1B and leptin (*ob/ob* mice) might be expected to have a similar metabolic phenotype compared to *ob/ob* single mutants, due to a lack of circulating leptin. In fact, *ob/ob*:*Ptpn1–/–* mice exhibit suppressed body weight gain in comparison with *ob/ob* single mutants [31], and *db/db*:*Ptpn1–/–* double mutants exhibit improved plasma lipids compared to *db/db* mice [50], suggesting that PTP1B's metabolic effects may not be solely due to its regulation of leptin signaling. Recently, mice lacking PTP1B in LepRb-expressing cells (LepRb-*Ptpn1–/–*) were generated and compared to global *Ptpn1–/–* mice in order to assess whether any metabolic improvements of PTP1B-deficiency are mediated outside of this cell type. These studies show that in mice fed a high-fat diet from weaning, the body weights of LepRb-*Ptpn1–/–* and global *Ptpn1–/–* mice are virtually identical, suggesting that the lack of PTP1B within LepRb-expressing cells is primarily responsible for the decreased body weight. As expected, LepRb-*Ptpn1–/–* mice show the same extent of leptin hypersensitivity as global *Ptpn1–/–* mice. However, there are subtle yet important differences in some phenotypes of LepRb-*Ptpn1–/–* mice. On a low fat chow diet, LepRb-*Ptpn1–/–* mice weigh less than global *Ptpn1–/–* mice, but have relatively more adiposity than *Ptpn1–/–* mice; in addition, LepRb-*Ptpn1–/–* mice have reduced food intake whereas global *Ptpn1–/–* mice actually have slightly elevated food intake. These differences raise the intriguing possibility that PTP1B may have distinct functions outside of leptin-receptor expressing cells. Importantly, these studies do not distinguish effects of PTP1B on the LepRb signaling pathway from other receptors that may be co-expressed with the LepRb, including the insulin receptor.

In the CNS, insulin acts as a catabolic hormone promoting a state of negative energy balance [51, 52]. As discussed elsewhere in this book, PTP1B can dephosphorylate the IR [2–5, 53] and IR substrate-1 (IRS-1) [54] and PTP1B negatively regulates insulin signaling in liver and skeletal muscle to affect glucose homeostasis [2, 3, 36, 37]. Moreover, *icv* administration of PTP1B antisense oligonucleotides into the third ventricle of rats enhances insulin signaling and insulin-induced satiety [41]; this at least raises the possibility that some of the improved metabolic effects of PTP1B deficiency may be due to enhanced central insulin signaling. In addition, since PTP1B acts on JAK-2, a downstream kinase in the leptin signaling pathway, it is also possible that PTP1B might negatively regulate other type I cytokine signaling pathways within the brain to affect energy balance. For example, PTP1B has been shown to regulate IL-6 signaling in the liver and in smooth muscle cells [55, 56], and IL-6 is a known central regulator of body weight [57, 58]. PTP1B may also have novel targets in the energy balance centers of the brain. These interesting possibilities remain to be examined.

T Cell Protein Tyrosine Phosphatase

T Cell Protein Tyrosine Phosphatase (TCPTP) (encoded by *PTPN2*) is a classical tyrosine-specific PTP, so called because it was originally cloned from a T cell cDNA library [59]. Although TCPTP is particularly abundant in T cells and hematopoietic cells [60, 61], it is nonetheless a ubiquitous enzyme that is expressed in all cell types and tissues [reviewed in [62, 63]]. Two variants of TCPTP are expressed that arise from alternative splicing of *PTPN2* message: a 48 kDa form, which like PTP1B is targeted to the endoplasmic reticulum (ER) by a hydrophobic C-terminus, and a 45 kDa variant that lacks the hydrophobic C-terminus and is targeted to the nucleus by a bipartite nuclear localization sequence [59, 64–67]. The two variants are expressed differentially in murine tissues and cell types, with only the 45 kDa form being detected in T cells [61], the 48 kDa TCPTP predominating in total brain homogenates and both forms being expressed equally in hypothalamus [68]. Although the 48 kDa form is likely to act on tyrosine phosphorylated receptor PTK substrates after they are activated and undergo endocytosis, as proposed for PTP1B [29, 69–73], the 45 kDa variant can dephosphorylate substrates in both the nucleus and cytoplasm. Despite an apparent exclusive nuclear localization in resting cells, the 45 kDa TCPTP can exit the nucleus in response to a variety of stimuli to access substrates in the cytoplasm and at the plasma membrane [74–78]. In this regard, the nuclear locale of the 45 kDa TCPTP may allow for the temporal and spatial control of TCPTP function. Cytoplasmic TCPTP substrates include receptor PTKs such as the IR [53, 77, 79], non-receptor PTKs such as c-Src, Fyn, Lck [61, 68, 78, 80] and

JAK-1 and -3 [81], and PTK substrates such as the adaptor protein p52^{Shc} [75]. The only known nuclear substrates for TCPTP include members of the STAT family of transcription factors, including STAT-1, -3, -5, and -6 [68, 79, 80, 82–85], that translocate to the nucleus when tyrosine phosphorylated. In particular, TCPTP has been shown to regulate STAT3 signaling in a variety of tissues in vivo, including the liver where it regulates the STAT3-mediated repression of gluconeogenesis [79], and the hypothalamus, where TCPTP regulates leptin signaling and energy homeostasis [68].

TCPTP-Null Mice: More Than Just an Immune Phenotype?

Mice that are globally deficient for TCPTP (*Ptpn2–/–*) on a BALB/c-129SJ background [60] die soon after birth from severe anemia, hematopoietic defects, and the development of progressive systemic inflammatory disease, characterized by increases in circulating pro-inflammatory cytokines and lymphocytic infiltrates in non-lymphoid tissues [60, 86]. This is in stark contrast to the phenotype of *Ptpn1–/–* mice [2, 3] that have no overt alteration in life span and exhibit decreased adiposity and increased insulin sensitivity [2, 3]. *Ptpn2–/–* mice (BALB/c-129SJ) exhibit growth retardation, a hunched posture, piloerection, decreased mobility, and diarrhea, succumbing by 3–5 weeks of age [60]. The overt morbidity and mortality and the hematopoietic defects and anemia in *Ptpn2–/–* mice may be attributable to bone marrow stromal cell abnormalities [60, 87]. However, more recent studies point towards the morbidity and mortality also being strain-dependent. Wiede et al. [88] have reported that mice with a global deficiency in TCPTP on a C57BL/6 background exhibit growth retardation, but live longer and do not display the outward signs of morbidity that characterize *Ptpn2–/–* (BALB/c) mice [88]. Mice on the C57BL/6 background strain are proportionately smaller, consistent with the growth retardation being developmental in nature [88]. Interestingly, mice that lack TCPTP exclusively in neuronal and glial cells (*Nes*-Cre;*Ptpn2*$^{lox/lox}$: Nestin-*Ptpn2–/–*) are also runted [68]. Nestin-*Ptpn2–/–* mice have increased hypothalamic STAT5 phosphorylation and decreased circulating growth hormone (GH), which amongst other things affects postnatal growth [68]. Hypothalamic STAT5 signaling in hypothalamic neurons can repress GH releasing hormone to inhibit GH release from the anterior pituitary [89–92] and this may account for the growth retardation in Nestin-*Ptpn2–/–* mice [68]; whether increased hypothalamic STAT5 signaling and decreased GH release also account for the runted appearance of *Ptpn2–/–* mice remains to be seen. In addition to being runted, *Ptpn2–/–* (C57BL/6) mice exhibit decreased relative adiposity [88], consistent with TCPTP affecting energy balance. This interpretation is in keeping with recent studies using Nestin-*Ptpn2–/–* mice that have highlighted TCPTP's role in the central control of leptin sensitivity and body weight [68]. Thus, the phenotype associated with global TCPTP deficiency likely reflects not only the impact of TCPTP deficiency on the immune compartment, but also alterations in overarching homeostatic systems, including those involved in the central control of growth and energy homeostasis.

PTP1B and TCPTP: Antagonizing JAK2 and STAT3

JAK-2 is a *bona fide* substrate for PTP1B [29–31]. Indeed, PTP1B selectively recognizes JAK-2 and not the LepRb or STAT3 as substrates in response to leptin [29–31]. In keeping with PTP1B acting at the level of JAK-2, the leptin-induced and JAK-2-mediated suppression of AMPK activity is enhanced in mice lacking PTP1B in neuronal and glial cells [93]. In contrast to PTP1B, TCPTP acts on the leptin pathway at the level of STAT3 [68]. TCPTP overexpression in LEPR-B-expressing CHO cells attenuates leptin-induced STAT3 Y705 phosphorylation without affecting JAK-2 Y1007/Y1008 phosphorylation or leptin-induced PI3K/Akt signaling (as assessed by Akt Ser-473 phosphorylation), whereas stable shRNA-mediated TCPTP knockdown in SK-N-SH-SY5Y neuroblastoma cells enhances leptin-induced STAT3 phosphorylation, but not JAK-2 Y1007/Y1008 phosphorylation, or downstream Ras/MAPK signaling [68]. In contrast to the ER-restricted PTP1B, which is capable of dephosphorylating JAK-2 at the plasma membrane and/or in the cytoplasm after LepRb endocytosis [29, 69–73], the nuclear form of TCPTP acts on STAT3 in the nucleus [68]. Thus, PTP1B and TCPTP coordinately contribute to the regulation of leptin signaling, acting at the level of JAK-2 and STAT3 in the cytoplasm and nucleus, respectively.

TCPTP Regulates Leptin Sensitivity

As discussed there is a large body of evidence implicating PTP1B in the regulation of leptin sensitivity and in the central control of body weight. In keeping with the regulation of leptin signaling in cells in vitro, studies characterizing mice lacking TCPTP in neuronal cells have also borne out an important role for TCPTP in the neuronal control of body weight [68]. Nestin-*Ptpn2–/–* mice exhibit enhanced leptin sensitivity characterized by increased STAT3 signaling in the ARC and VMH [68]. As in vitro, TCPTP's role in leptin signaling in vivo appears to be restricted to the regulation of the STAT3 pathway, with TCPTP deficiency enhancing the transcriptional effects on POMC and AgRP, but not NPY [68]. In mice, TCPTP neuronal cell deficiency enhanced the leptin-induced suppression of food intake and body weight and resulted in increased energy expenditure, decreased bone mineral density, and lower plasma leptin levels [68], all consistent with enhanced central leptin signaling. Paradoxically, relative fat mass was not attenuated, but rather increased and ambulatory activity decreased, rather than being increased in Nestin-*Ptpn2–/–* mice [68], pointing towards TCPTP regulating additional neuronal pathways in the hypothalamus and possibly elsewhere in the CNS. GH-induced hypothalamic STAT5 signaling was enhanced in TCPTP-nestin-Cre mice, circulating GH and IGF-1 levels decreased and mice were 10–20 % smaller than their floxed counterparts, or nestin-Cre controls [68]. Thus the increase in adiposity might be explained by TCPTP deficiency exacerbating the hypothalamic GH-induced and STAT-5 mediated repression of GH production/secretion from the anterior pituitary to increase peripheral adiposity and decrease growth [89–92]. Although it is possible that the effects on growth might have in turn impacted overall leptin sensitivity, it is important to note that the key features of the effects of TCPTP deficiency on leptin sensitivity (including the promotion of STAT3 signaling and the effects on body weight and energy expenditure) could be recapitulated by the *icv* administration of a specific TCPTP inhibitor [68, 94]. Although further studies are necessary to delineate TCPTP's role in different leptin-responsive neurons in the hypothalamus and elsewhere in the brain, taken together the studies described above provide compelling evidence for TCPTP having an important role in the central control of energy balance, growth, and body weight.

PTP1B, TCPTP, and Cellular Leptin Resistance

The anorectic and metabolic responses to leptin are diminished in the obese state [6, 7]. Several underlying mechanisms are thought to contribute to the decreased leptin responsiveness including alterations in the blood brain barrier and gliosis that impose physical barriers to leptin and cause alterations in synaptic plasticity and neuronal cell death, as well as neuronal cell-intrinsic mechanisms that directly attenuate leptin signaling [6, 7, 95–99]. Elevated hypothalamic SOCS3 and PTP1B protein levels in obesity are thought to contribute to the attenuation of the leptin signal and the development of cellular leptin resistance [6, 7]. Circulating factors that are elevated in obesity and type 2 diabetes including leptin, inflammatory cytokines and free fatty acids have been shown to induce the expression of PTP1B in cells [100–104, 106]. In vivo, chronic *icv* leptin infusion (2 weeks) induces PTP1B expression in the hypothalami of rats [105], whereas in mice, TNF has been shown to acutely induce PTP1B expression in both hypothalami and peripheral tissues through NFκB recruitment to the *Ptpn1* promoter [106]. Therefore, the hyperleptinemia and inflammation associated with the development of obesity are key contributors to the increased hypothalamic PTP1B expression and the promotion of cellular leptin resistance.

As for PTP1B, hypothalamic TCPTP levels are also increased in obese mice [68]. The degree of increase is similar to that noted for PTP1B and SOCS3, consistent with TCPTP also being an important contributor to disease pathology [68]. However, unlike PTP1B whose expression can be induced by factors such as free fatty acids and TNF [103, 104, 106], the increase in TCPTP appears to be mediated by leptin alone. TCPTP protein expression is induced by leptin in LEPR-B-expressing cells in vitro and intraperitoneal leptin administration rapidly increases hypothalamic *Ptpn2* mRNA and TCPTP protein in vivo [68]. Furthermore, hypothalamic TCPTP is not elevated in leptin-deficient obese *ob/ob* mice, consistent with leptin being essential for the elevated hypothalamic TCPTP in obesity [68]. In keeping with this, the increase in hypothalamic TCPTP in high fat-fed mice coincides with an increase in circulating leptin levels. Moreover, the increase in hypothalamic TCPTP expression in high fat-fed mice follows that of SOCS3,

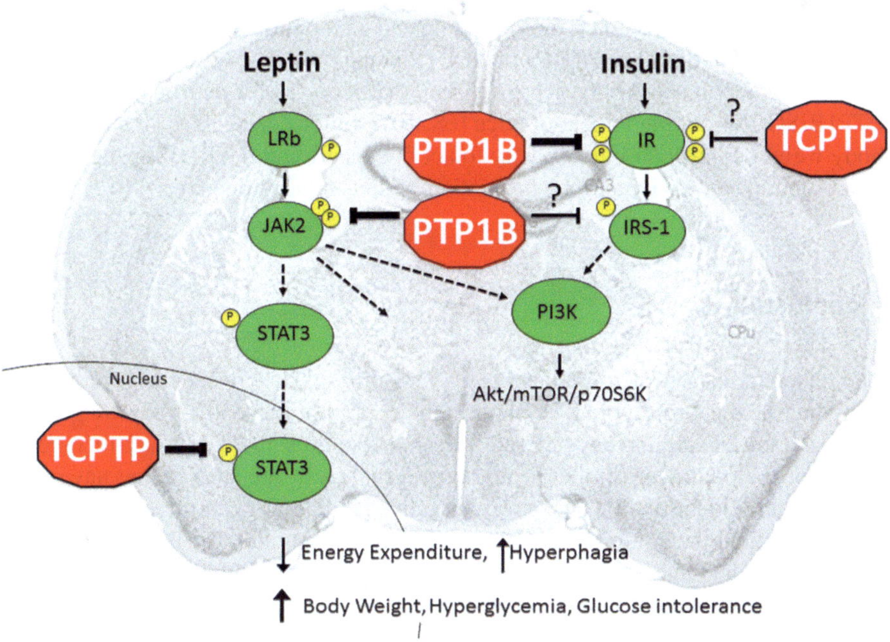

Fig. 4.1 The roles of neuronal PTP1B and TCPTP on the development of central leptin and insulin resistance

PTP1B, and the onset of inflammation [68]. Therefore, leptin-induced increases in hypothalamic TCPTP potentially exacerbate the development of cellular leptin resistance initiated by inflammation and the induction of hypothalamic PTP1B and SOCS3 expression.

The increases in hypothalamic PTP1B and TCPTP are consistent with these two PTPs being causally involved with the development of cellular leptin resistance and obesity. Although the contributions of PTP1B to leptin resistance are well established, the role of TCPTP has only recently come to light. As for PTP1B knockouts, mice that lack TCPTP in neuronal cells gain less weight when fed a high-fat diet [68]. The decreased weight gain is similar to that reported previously for PTP1B neuronal cell-specific knockout mice. Importantly, high fat-fed Nestin-*Ptpn2–/–* mice exhibit enhanced leptin-induced STAT3 signaling and leptin sensitivity (as measured by the effects of leptin on body weight) accompanied by elevated energy expenditure and improved whole-body glucose tolerance and insulin sensitivity [68]. Thus, the decreased weight gain in Nestin-*Ptpn2–/–* mice is likely to be attributable to TCPTP deficiency enhancing leptin sensitivity. It remains unclear if PTP1B and TCPTP exert their effects on leptin sensitivity by acting in the same neurons. However, given their respective roles in JAK2 versus STAT3 inactivation, one could envisage that this would be the case (see Fig. 4.1). Indeed, deletion of both PTP1B and TCPTP in neuronal cells (generated using the *Nes*-Cre transgene) results in an additive enhancement of leptin sensitivity and an overall improvement in glucose homeostasis and a more pronounced attenuation of diet-induced obesity than the deletion of either PTP1B or TCPTP alone [68]. The effect on diet-induced obesity and leptin sensitivity is greater than that reported for neuronal cell-specific PTP1B/SOCS3 double knockout mice [68, 107] underscoring TCPTP's key role in body weight control and its capacity to act in concert with PTP1B for the attenuation of leptin sensitivity and the promotion of diet-induced obesity.

Conclusion

The characterization of mice lacking PTP1B more than a decade ago brought to light the role this prototypic PTP plays in the control of adiposity, body weight, and glucose homeostasis and engendered widespread interest in PTP1B as a therapeutic target for the treatment of obesity and type 2 diabetes. The studies that have since followed have identified PTP1B as key negative regulator of central leptin signaling and highlighted the contributions of elevated hypothalamic PTP1B to the development of diet- and age-associated cellular leptin resistance and obesity. Delineating the precise neuronal populations in which PTP1B exerts its effects and determining its potential to regulate additional neuronal signaling pathways remains an exciting area for future research. Beyond the regulation of leptin signaling in POMC neurons in the ARC [44] and the NTS of the hindbrain [49], it will be important to assess the role of PTP1B in other leptin-responsive neurons of the melanocortin system, including AgRP neurons and second-order neurons emanating from the PVN. In addition, it will be necessary to assess PTP1B's role in non-AgRP GABAergic leptin-responsive neurons in the hypothalamus (ARC, DMH, or LH) that control the tone of POMC neurons [108] and in leptin-responsive dopaminergic neurons of the VTA that affect food and drug reward [17–20]. Furthermore, given the established role of PTP1B in IR regulation in the periphery [2, 3, 37, 109], it will be interesting to assess the role of PTP1B in insulin signaling in different regions of the brain including AgRP and POMC neurons of the ARC, where insulin action regulates peripheral glucose and fat metabolism [8, 110–116], and steroidogenic factor 1-expressing neurons in the VMH, where insulin signaling attenuates leptin sensitivity and promotes obesity [117]. The recent studies by Loh et al. [68] have highlighted the important role of TCPTP in the attenuation of leptin sensitivity and the promotion of diet-induced cellular leptin resistance. Moreover, these studies have shown that the deletion of TCPTP, in addition to PTP1B in neuronal cells, results in greater leptin sensitivity and resistance to diet-induced obesity. The enhanced leptin sensitivity and resistance to diet-induced obesity in the double knockout mice has been attributed to the concerted alleviation of JAK2 and STAT3 dephosphorylation by the otherwise elevated hypothalamic PTP1B and TCPTP, respectively. However, it remains to be seen whether PTP1B and TCPTP do indeed exert their effects on leptin sensitivity by acting in the same neuronal populations. Furthermore, TCPTP has been implicated in the dephosphorylation of the IR and shown to act in a coordinated manner in the same cell with PTP1B to regulate the duration and intensity of IR signaling, respectively [53, 77, 79]. Thus, TCPTP may also function with PTP1B to attenuate hypothalamic insulin signaling and increases in hypothalamic TCPTP might contribute to the development of central insulin resistance, but this also remains to be established. Similarly it remains to be determined if PTP1B and TCPTP contribute to central leptin signaling and the development of cellular leptin resistance in primates. Assuming that they do, the striking attenuation of diet-induced obesity reported by Loh et al. [68] in mice lacking both PTP1B and TCPTP in neuronal cells raises the exciting possibility that the combined inhibition of these two PTPs in the hypothalamus may provide an effective strategy to alleviate diet-induced cellular leptin resistance to combat obesity. Although most PTP1B inhibitors also at least partially inhibit TCPTP, the important challenge will be to avoid TCPTP inhibition in the periphery and to synthesize inhibitors that can cross the blood brain barrier. Preclinical and clinical studies have established the effectiveness of intranasal insulin or leptin delivery in the promotion satiety and enhancement of glucose homeostasis [118–121]; the combined intranasal delivery of PTP1B/TCPTP inhibitors may provide additive benefit.

Acknowledgements T.T. is a National Health and Medical Research Council (NHMRC) of Australia Principal Research Fellow and supported by grants from the NHMRC. K.B. is an Associate Professor at the University of Pennsylvania, School of Veterinary Medicine, and is supported by the National Institutes of Health (NIDDK).

References

1. Dube N, Tremblay ML (2004) Beyond the metabolic function of PTP1B. Cell Cycle 3:550–553
2. Elchebly M, Payette P, Michaliszyn E, Cromlish W, Collins S, Loy AL, Normandin D, Cheng A, Himms-Hagen J, Chan CC, Ramachandran C, Gresser MJ, Tremblay ML, Kennedy BP (1999) Increased insulin sensitivity and obesity resistance in mice lacking the protein tyrosine phosphatase-1B gene. Science 283:1544–1548
3. Klaman LD, Boss O, Peroni OD, Kim JK, Martino JL, Zabolotny JM, Moghal N, Lubkin M, Kim YB, Sharpe AH, Stricker-Krongrad A, Shulman GI, Neel BG, Kahn BB (2000) Increased energy expenditure, decreased adiposity, and tissue-specific insulin sensitivity in protein-tyrosine phosphatase 1B-deficient mice. Mol Cell Biol 20:5479–5489
4. Kenner KA, Anyanwu E, Olefsky JM, Kusari J (1996) Protein-tyrosine phosphatase 1B is a negative regulator of insulin- and insulin-like growth factor-I-stimulated signaling. J Biol Chem 271:19810–19816
5. Ahmad F, Li PM, Meyerovitch J, Goldstein BJ (1995) Osmotic loading of neutralizing antibodies demonstrates a role for protein-tyrosine phosphatase 1B in negative regulation of the insulin action pathway. J Biol Chem 270:20503–20508
6. Myers MG, Cowley MA, Munzberg H (2008) Mechanisms of leptin action and leptin resistance. Annu Rev Physiol 70:537–556
7. Myers MG Jr, Heymsfield SB, Haft C, Kahn BB, Laughlin M, Leibel RL, Tschop MH, Yanovski JA (2012) Challenges and opportunities of defining clinical leptin resistance. Cell Metab 15:150–156

8. Marino JS, Xu Y, Hill JW (2011) Central insulin and leptin-mediated autonomic control of glucose homeostasis. Trends Endocrinol Metab 22:275–285
9. Xu Y, Elmquist JK, Fukuda M (2011) Central nervous control of energy and glucose balance: focus on the central melanocortin system. Ann N Y Acad Sci 1243:1–14
10. Schwartz MW, Seeley RJ, Campfield LA, Burn P, Baskin DG (1996) Identification of targets of leptin action in rat hypothalamus. J Clin Invest 98:1101–1106
11. Schwartz MW, Seeley RJ, Woods SC, Weigle DS, Campfield LA, Burn P, Baskin DG (1997) Leptin increases hypothalamic pro-opiomelanocortin mRNA expression in the rostral arcuate nucleus. Diabetes 46:2119–2123
12. Mizuno TM, Kleopoulos SP, Bergen HT, Roberts JL, Priest CA, Mobbs CV (1998) Hypothalamic pro-opiomelanocortin mRNA is reduced by fasting and [corrected] in ob/ob and db/db mice, but is stimulated by leptin. Diabetes 47:294–297
13. Elias CF, Aschkenasi C, Lee C, Kelly J, Ahima RS, Bjorbaek C, Flier JS, Saper CB, Elmquist JK (1999) Leptin differentially regulates NPY and POMC neurons projecting to the lateral hypothalamic area. Neuron 23:775–786
14. Cowley MA, Smart JL, Rubinstein M, Cerdan MG, Diano S, Horvath TL, Cone RD, Low MJ (2001) Leptin activates anorexigenic POMC neurons through a neural network in the arcuate nucleus. Nature 411:480–484
15. Hayes MR, Skibicka KP, Leichner TM, Guarnieri DJ, DiLeone RJ, Bence KK, Grill HJ (2010) Endogenous leptin signaling in the caudal nucleus tractus solitarius and area postrema is required for energy balance regulation. Cell Metab 11:77–83
16. Scott MM, Williams KW, Rossi J, Lee CE, Elmquist JK (2011) Leptin receptor expression in hindbrain Glp-1 neurons regulates food intake and energy balance in mice. J Clin Invest 121:2413–2421
17. Figlewicz DP, MacDonald Naleid A, Sipols AJ (2007) Modulation of food reward by adiposity signals. Physiol Behav 91:473–478
18. Fulton S, Woodside B, Shizgal P (2000) Modulation of brain reward circuitry by leptin. Science 287:125–128
19. Fulton S, Pissios P, Manchon RP, Stiles L, Frank L, Pothos EN, Maratos-Flier E, Flier JS (2006) Leptin regulation of the mesoaccumbens dopamine pathway. Neuron 51:811–822
20. Hommel JD, Trinko R, Sears RM, Georgescu D, Liu ZW, Gao XB, Thurmon JJ, Marinelli M, DiLeone RJ (2006) Leptin receptor signaling in midbrain dopamine neurons regulates feeding. Neuron 51:801–810
21. Lee GH, Proenca R, Montez JM, Carroll KM, Darvishzadeh JG, Lee JI, Friedman JM (1996) Abnormal splicing of the leptin receptor in diabetic mice. Nature 379:632–635
22. Kloek C, Haq AK, Dunn SL, Lavery HJ, Banks AS, Myers MG Jr (2002) Regulation of Jak kinases by intracellular leptin receptor sequences. J Biol Chem 277:41547–41555
23. Villanueva EC, Myers MG Jr (2008) Leptin receptor signaling and the regulation of mammalian physiology. Int J Obes (Lond) 32(Suppl 7):S8–S12
24. Hill JW, Williams KW, Ye C, Luo J, Balthasar N, Coppari R, Cowley MA, Cantley LC, Lowell BB, Elmquist JK (2008) Acute effects of leptin require PI3K signaling in hypothalamic proopiomelanocortin neurons in mice. J Clin Invest 118:1796–1805
25. Xu Y, Hill JW, Fukuda M, Gautron L, Sohn JW, Kim KW, Lee CE, Choi MJ, Lauzon DA, Dhillon H, Lowell BB, Zigman JM, Zhao JJ, Elmquist JK (2010) PI3K signaling in the ventromedial hypothalamic nucleus is required for normal energy homeostasis. Cell Metab 12:88–95
26. Niswender KD, Morrison CD, Clegg DJ, Olson R, Baskin DG, Myers MG Jr, Seeley RJ, Schwartz MW (2003) Insulin activation of phosphatidylinositol 3-kinase in the hypothalamic arcuate nucleus: a key mediator of insulin-induced anorexia. Diabetes 52:227–231
27. Dagon Y, Hur E, Zheng B, Wellenstein K, Cantley LC, Kahn BB (2012) p70S6 kinase phosphorylates AMPK on serine 491 to mediate leptin's effect on food intake. Cell Metab 16:104–112
28. Salmeen A, Andersen JN, Myers MP, Tonks NK, Barford D (2000) Molecular basis for recognition and dephosphorylation of the activation segment of the insulin receptor by protein tyrosine phosphatase 1B. Mol Cell 6:1401–1412

29. Myers MP, Andersen JN, Cheng A, Tremblay ML, Horvath CM, Parisien JP, Salmeen A, Barford D, Tonks NK (2001) TYK2 and JAK2 are substrates of protein-tyrosine phosphatase 1B. J Biol Chem 276:47771–47774
30. Zabolotny JM, Bence-Hanulec KK, Stricker-Krongrad A, Haj F, Wang Y, Minokoshi Y, Kim YB, Elmquist JK, Tartaglia LA, Kahn BB, Neel BG (2002) PTP1B regulates leptin signal transduction in vivo. Dev Cell 2:489–495
31. Cheng A, Uetani N, Simoncic PD, Chaubey VP, Lee-Loy A, McGlade CJ, Kennedy BP, Tremblay ML (2002) Attenuation of leptin action and regulation of obesity by protein tyrosine phosphatase 1B. Dev Cell 2:497–503
32. Lund IK, Hansen JA, Andersen HS, Moller NP, Billestrup N (2005) Mechanism of protein tyrosine phosphatase 1B-mediated inhibition of leptin signalling. J Mol Endocrinol 34:339–351
33. Kaszubska W, Falls HD, Schaefer VG, Haasch D, Frost L, Hessler P, Kroeger PE, White DW, Jirousek MR, Trevillyan JM (2002) Protein tyrosine phosphatase 1B negatively regulates leptin signaling in a hypothalamic cell line. Mol Cell Endocrinol 195:109–118
34. Guan KL, Haun RS, Watson SJ, Geahlen RL, Dixon JE (1990) Cloning and expression of a protein-tyrosine-phosphatase. Proc Natl Acad Sci U S A 87:1501–1505
35. Bence KK, Delibegovic M, Xue B, Gorgun CZ, Hotamisligil GS, Neel BG, Kahn BB (2006) Neuronal PTP1B regulates body weight, adiposity and leptin action. Nat Med 12:917–924
36. Delibegovic M, Bence KK, Mody N, Hong EG, Ko HJ, Kim JK, Kahn BB, Neel BG (2007) Improved glucose homeostasis in mice with muscle-specific deletion of protein-tyrosine phosphatase 1B. Mol Cell Biol 27:7727–7734
37. Delibegovic M, Zimmer D, Kauffman C, Rak K, Hong EG, Cho YR, Kim JK, Kahn BB, Neel BG, Bence KK (2009) Liver-specific deletion of protein-tyrosine phosphatase 1B (PTP1B) improves metabolic syndrome and attenuates diet-induced ER stress. Diabetes 58:590–599
38. Agouni A, Mody N, Owen C, Czopek A, Zimmer D, Bentires-Alj M, Bence KK, Delibegovic M (2011) Liver-specific deletion of protein tyrosine phosphatase (PTP) 1B improves obesity- and pharmacologically induced endoplasmic reticulum stress. Biochem J 438:369–378
39. Owen C, Czopek A, Agouni A, Grant L, Judson R, Lees EK, McIlroy GD, Goransson O, Welch A, Bence KK, Kahn BB, Neel BG, Mody N, Delibegovic M (2012) Adipocyte-specific protein tyrosine phosphatase 1B deletion increases lipogenesis, adipocyte cell size and is a minor regulator of glucose homeostasis. PLoS One 7:e32700
40. Morrison CD, White CL, Wang Z, Lee SY, Lawrence DS, Cefalu WT, Zhang ZY, Gettys TW (2007) Increased hypothalamic protein tyrosine phosphatase 1B contributes to leptin resistance with age. Endocrinology 148:433–440
41. Picardi PK, Calegari VC, Prada Pde O, Moraes JC, Araujo E, Marcondes MC, Ueno M, Carvalheira JB, Velloso LA, Saad MJ (2008) Reduction of hypothalamic protein tyrosine phosphatase improves insulin and leptin resistance in diet-induced obese rats. Endocrinology 149:3870–3880
42. Balthasar N, Coppari R, McMinn J, Liu SM, Lee CE, Tang V, Kenny CD, McGovern RA, Chua SC Jr, Elmquist JK, Lowell BB (2004) Leptin receptor signaling in POMC neurons is required for normal body weight homeostasis. Neuron 42:983–991
43. Huo L, Gamber K, Greeley S, Silva J, Huntoon N, Leng XH, Bjorbaek C (2009) Leptin-dependent control of glucose balance and locomotor activity by POMC neurons. Cell Metab 9:537–547
44. Banno R, Zimmer D, De Jonghe BC, Atienza M, Rak K, Yang W, Bence KK (2010) PTP1B and SHP2 in POMC neurons reciprocally regulate energy balance in mice. J Clin Invest 120:720–734
45. De Jonghe BC, Hayes MR, Banno R, Skibicka KPPD, Zimmer DJ, Bowen KAPD, Leichner TM, Alhadeff AL, Kanoski SEPD, Cyr NE, Nillni EA, Grill HJ, Bence KK (2011) Deficiency of PTP1B in POMC neurons leads to alterations in energy balance and homeostatic response to cold exposure. Am J Physiol Endocrinol Metab 300(6):E1002–E1011
46. Ellacott KL, Halatchev IG, Cone RD (2006) Characterization of leptin-responsive neurons in the caudal brainstem. Endocrinology 147:3190–3195

47. Bronstein DM, Schafer MK, Watson SJ, Akil H (1992) Evidence that beta-endorphin is synthesized in cells in the nucleus tractus solitarius: detection of POMC mRNA. Brain Res 587:269–275
48. Palkovits M, Eskay RL (1987) Distribution and possible origin of beta-endorphin and ACTH in discrete brainstem nuclei of rats. Neuropeptides 9:123–137
49. De Jonghe BC, Hayes MR, Kanoski SE, Zimmer DJ, Grill HJ, Bence KK (2012) Food intake reductions and increases in energetic responses by hindbrain leptin and melanotan II are enhanced in mice with POMC-specific PTP1B deficiency. Am J Physiol Endocrinol Metab 303(5):E644–E651
50. Ali MI, Ketsawatsomkron P, Belin de Chantemele EJ, Mintz JD, Muta K, Salet C, Black SM, Tremblay ML, Fulton DJ, Marrero MB, Stepp DW (2009) Deletion of protein tyrosine phosphatase 1b improves peripheral insulin resistance and vascular function in obese, leptin-resistant mice via reduced oxidant tone. Circ Res 105:1013–1022
51. Schwartz MW, Woods SC, Porte D Jr, Seeley RJ, Baskin DG (2000) Central nervous system control of food intake. Nature 404:661–671
52. Porte D Jr, Baskin DG, Schwartz MW (2005) Insulin signaling in the central nervous system: a critical role in metabolic homeostasis and disease from C. elegans to humans. Diabetes 54:1264–1276
53. Galic S, Hauser C, Kahn BB, Haj FG, Neel BG, Tonks NK, Tiganis T (2005) Coordinated regulation of insulin signaling by the protein tyrosine phosphatases PTP1B and TCPTP. Mol Cell Biol 25:819–829
54. Goldstein BJ, Bittner-Kowalczyk A, White MF, Harbeck M (2000) Tyrosine dephosphorylation and deactivation of insulin receptor substrate-1 by protein-tyrosine phosphatase 1B. Possible facilitation by the formation of a ternary complex with the Grb2 adaptor protein. J Biol Chem 275:4283–4289
55. Revuelta-Cervantes J, Mayoral R, Miranda S, Gonzalez-Rodriguez A, Fernandez M, Martin-Sanz P, Valverde AM (2011) Protein tyrosine phosphatase 1B (PTP1B) deficiency accelerates hepatic regeneration in mice. Am J Pathol 178:1591–1604
56. Chang Y (2011) A central role of PTP1B in hyperinsulinemia-enhanced IL-6 signaling in dedifferentiated vascular smooth muscle cells. J Diabet Metabol 2:118
57. Wallenius K, Wallenius V, Sunter D, Dickson SL, Jansson JO (2002) Intracerebroventricular interleukin-6 treatment decreases body fat in rats. Biochem Biophys Res Commun 293:560–565
58. Wallenius V, Wallenius K, Ahren B, Rudling M, Carlsten H, Dickson SL, Ohlsson C, Jansson JO (2002) Interleukin-6-deficient mice develop mature-onset obesity. Nat Med 8:75–79
59. Cool DE, Tonks NK, Charbonneau H, Walsh KA, Fischer EH, Krebs EG (1989) cDNA isolated from a human T-cell library encodes a member of the protein-tyrosine-phosphatse family. Proc Natl Acad Sci U S A 86:5257–5261
60. You-Ten KE, Muise ES, Itie A, Michaliszyn E, Wagner J, Jothy S, Lapp WS, Tremblay ML (1997) Impaired bone marrow microenvironment and immune function in T cell protein tyrosine phosphatase-deficient mice. J Exp Med 186:683–693
61. Wiede F, Shields BJ, Chew SH, Kyparissoudis K, van Vliet C, Galic S, Tremblay ML, Russell SM, Godfrey DI, Tiganis T (2011) T cell protein tyrosine phosphatase attenuates T cell signaling to maintain tolerance in mice. J Clin Invest 121:4758–4774
62. Tiganis T, Bennett AM (2007) Protein tyrosine phosphatase function: the substrate perspective. Biochem J 402:1–15
63. Tiganis T (2013) PTP1B and TCPTP—nonredundant phosphatases in insulin signaling and glucose homeostasis. FEBS J 280(2):445–458
64. Mosinger B Jr, Tillmann U, Westphal H, Tremblay ML (1992) Cloning and characterization of a mouse cDNA encoding a cytoplasmic protein-tyrosine-phosphatase. Proc Natl Acad Sci U S A 89:499–503
65. Tillmann U, Wagner J, Boerboom D, Westphal H, Tremblay ML (1994) Nuclear localization and cell cycle regulation of a murine protein tyrosine phosphatase. Mol Cell Biol 14:3030–3040
66. Lorenzen JA, Dadabay CY, Fischer EH (1995) COOH-Terminal sequence motifs target the T-cell protein tyrosine phosphatase to the ER and nucleus. J Cell Biol 131:631–643

67. Tiganis T, Flint AJ, Adam SA, Tonks NK (1997) Association of the T-cell protein tyrosine phosphatase with nuclear import factor p97. J Biol Chem 272:21548–21557
68. Loh K, Fukushima A, Zhang X, Galic S, Briggs D, Enriori PJ, Simonds S, Wiede F, Reichenbach A, Hauser C, Sims NA, Bence KK, Zhang S, Zhang ZY, Kahn BB, Neel BG, Andrews ZB, Cowley MA, Tiganis T (2011) Elevated hypothalamic TCPTP in obesity contributes to cellular leptin resistance. Cell Metab 14:684–699
69. Haj FG, Verveer PJ, Squire A, Neel BG, Bastiaens PI (2002) Imaging sites of receptor dephosphorylation by PTP1B on the surface of the endoplasmic reticulum. Science 295:1708–1711
70. Nievergall E, Janes PW, Stegmayer C, Vail ME, Haj FG, Teng SW, Neel BG, Bastiaens PI, Lackmann M (2010) PTP1B regulates Eph receptor function and trafficking. J Cell Biol 191:1189–1203
71. Eden ER, White IJ, Tsapara A, Futter CE (2010) Membrane contacts between endosomes and ER provide sites for PTP1B-epidermal growth factor receptor interaction. Nat Cell Biol 12:267–272
72. Haj FG, Sabet O, Kinkhabwala A, Wimmer-Kleikamp S, Roukos V, Han HM, Grabenbauer M, Bierbaum M, Antony C, Neel BG, Bastiaens PI (2012) Regulation of signaling at regions of cell-cell contact by endoplasmic reticulum-bound protein-tyrosine phosphatase 1B. PLoS One 7:e36633
73. Monteleone MC, Gonzalez Wusener AE, Burdisso JE, Conde C, Caceres A, Arregui CO (2012) ER-bound protein tyrosine phosphatase PTP1B interacts with Src at the plasma membrane/substrate interface. PLoS One 7:e38948
74. Tiganis T, Kemp BE, Tonks NK (1999) The protein tyrosine phosphatase TCPTP regulates epidermal growth factor receptor-mediated and phosphatidylinositol 3-kinase-dependent signalling. J Biol Chem 274:27768–27775
75. Tiganis T, Bennett AM, Ravichandran KS, Tonks NK (1998) Epidermal growth factor receptor and the adaptor protein $p52^{Shc}$ are specific substrates of T-cell protein tyrosine phosphatase. Mol Cell Biol 18:1622–1634
76. Lam MH, Michell BJ, Fodero-Tavoletti MT, Kemp BE, Tonks NK, Tiganis T (2001) Cellular stress regulates the nucleocytoplasmic distribution of the protein tyrosine phosphatase TCPTP. J Biol Chem 276:37700–37707
77. Galic S, Klingler-Hoffmann M, Fodero-Tavoletti MT, Puryer MA, Meng TC, Tonks NK, Tiganis T (2003) Regulation of insulin receptor signaling by the protein tyrosine phosphatase TCPTP. Mol Cell Biol 23:2096–2108
78. van Vliet C, Bukczynska PE, Puryer MA, Sadek CM, Shields BJ, Tremblay ML, Tiganis T (2005) Selective regulation of tumor necrosis factor-induced Erk signaling by Src family kinases and the T cell protein tyrosine phosphatase. Nat Immunol 6:253–260
79. Fukushima A, Loh K, Galic S, Fam B, Shields B, Wiede F, Tremblay ML, Watt MJ, Andrikopoulos S, Tiganis T (2010) T-cell protein tyrosine phosphatase attenuates STAT3 and insulin signaling in the liver to regulate gluconeogenesis. Diabetes 59:1906–1914
80. Shields BJ, Hauser C, Bukczynska PE, Court NW, Tiganis T (2008) DNA replication stalling attenuates tyrosine kinase signaling to suppress S phase progression. Cancer Cell 14:166–179
81. Simoncic PD, Lee-Loy A, Barber DL, Tremblay ML, McGlade CJ (2002) The T cell protein tyrosine phosphatase is a negative regulator of janus family kinases 1 and 3. Curr Biol 12: 446–453
82. Aoki N, Matsuda T (2002) A nuclear protein tyrosine phosphatase TC-PTP is a potential negative regulator of the PRL-mediated signaling pathway: dephosphorylation and deactivation of signal transducer and activator of transcription 5a and 5b by TC-PTP in nucleus. Mol Endocrinol 16:58–69
83. Yamamoto T, Sekine Y, Kashima K, Kubota A, Sato N, Aoki N, Matsuda T (2002) The nuclear isoform of protein-tyrosine phosphatase TC-PTP regulates interleukin-6-mediated signaling pathway through STAT3 dephosphorylation. Biochem Biophys Res Commun 297:811–817
84. ten Hoeve J, Ibarra-Sanchez MJ, Fu Y, Zhu W, Tremblay M, David M, Shuai K (2002) Identification of a nuclear Stat1 protein tyrosine phosphatase. Mol Cell Biol 22:5662–5668

85. Lu X, Chen J, Sasmono RT, Hsi ED, Sarosiek KA, Tiganis T, Lossos IS (2007) T-cell protein tyrosine phosphatase, distinctively expressed in activated-B-cell-like diffuse large B-cell lymphomas, is the nuclear phosphatase of STAT6. Mol Cell Biol 27:2166–2179
86. Heinonen KM, Nestel FP, Newell EW, Charette G, Seemayer TA, Tremblay ML, Lapp WS (2004) T Cell Protein Tyrosine Phosphatase deletion results in progressive systemic inflammatory disease. Blood 103(9):3457–3464
87. Bourdeau A, Dube N, Heinonen KM, Theberge JF, Doody KM, Tremblay ML (2007) TC-PTP-deficient bone marrow stromal cells fail to support normal B lymphopoiesis due to abnormal secretion of interferon-{gamma}. Blood 109:4220–4228
88. Wiede F, Hui Chew S, van Vliet C, Poulton IJ, Kyparissoudis K, Sasmono T, Loh K, Tremblay ML, Godfrey DI, Sims NA, Tiganis T (2012) Strain-dependent differences in bone development, myeloid hyperplasia, morbidity and mortality in Ptpn2-deficient mice. Plos One 7:e36703
89. Minami S, Kamegai J, Hasegawa O, Sugihara H, Okada K, Wakabayashi I (1993) Expression of growth hormone receptor gene in rat hypothalamus. J Neuroendocrinol 5:691–696
90. Burton KA, Kabigting EB, Clifton DK, Steiner RA (1992) Growth hormone receptor messenger ribonucleic acid distribution in the adult male rat brain and its colocalization in hypothalamic somatostatin neurons. Endocrinology 131:958–963
91. Lichanska AM, Waters MJ (2008) How growth hormone controls growth, obesity and sexual dimorphism. Trends Genet 24:41–47
92. Bennett E, McGuinness L, Gevers EF, Thomas GB, Robinson IC, Davey HW, Luckman SM (2005) Hypothalamic STAT proteins: regulation of somatostatin neurones by growth hormone via STAT5b. J Neuroendocrinol 17:186–194
93. Xue B, Pulinilkunnil T, Murano I, Bence KK, He H, Minokoshi Y, Asakura K, Lee A, Haj F, Furukawa N, Catalano KJ, Delibegovic M, Balschi JA, Cinti S, Neel BG, Kahn BB (2009) Neuronal protein tyrosine phosphatase 1B deficiency results in inhibition of hypothalamic AMPK and isoform-specific activation of AMPK in peripheral tissues. Mol Cell Biol 29:4563–4573
94. Zhang S, Chen L, Luo Y, Gunawan A, Lawrence DS, Zhang ZY (2009) Acquisition of a potent and selective TC-PTP inhibitor via a stepwise fluorophore-tagged combinatorial synthesis and screening strategy. J Am Chem Soc 131:13072–13079
95. El-Haschimi K, Pierroz DD, Hileman SM, Bjorbaek C, Flier JS (2000) Two defects contribute to hypothalamic leptin resistance in mice with diet-induced obesity. J Clin Invest 105:1827–1832
96. Caro JF, Kolaczynski JW, Nyce MR, Ohannesian JP, Opentanova I, Goldman WH, Lynn RB, Zhang PL, Sinha MK, Considine RV (1996) Decreased cerebrospinal-fluid/serum leptin ratio in obesity: a possible mechanism for leptin resistance. Lancet 348:159–161
97. Schwartz MW, Peskind E, Raskind M, Boyko EJ, Porte D Jr (1996) Cerebrospinal fluid leptin levels: relationship to plasma levels and to adiposity in humans. Nat Med 2:589–593
98. Horvath TL, Sarman B, Garcia-Caceres C, Enriori PJ, Sotonyi P, Shanabrough M, Borok E, Argente J, Chowen JA, Perez-Tilve D, Pfluger PT, Bronneke HS, Levin BE, Diano S, Cowley MA, Tschop MH (2010) Synaptic input organization of the melanocortin system predicts diet-induced hypothalamic reactive gliosis and obesity. Proc Natl Acad Sci U S A 107:14875–14880
99. Thaler JP, Yi CX, Schur EA, Guyenet SJ, Hwang BH, Dietrich MO, Zhao X, Sarruf DA, Izgur V, Maravilla KR, Nguyen HT, Fischer JD, Matsen ME, Wisse BE, Morton GJ, Horvath TL, Baskin DG, Tschop MH, Schwartz MW (2012) Obesity is associated with hypothalamic injury in rodents and humans. J Clin Invest 122:153–162
100. Taghibiglou C, Rashid-Kolvear F, Van Iderstine SC, Le-Tien H, Fantus IG, Lewis GF, Adeli K (2002) Hepatic very low density lipoprotein-ApoB overproduction is associated with attenuated hepatic insulin signaling and overexpression of protein-tyrosine phosphatase 1B in a fructose-fed hamster model of insulin resistance. J Biol Chem 277:793–803
101. Lam NT, Lewis JT, Cheung AT, Luk CT, Tse J, Wang J, Bryer-Ash M, Kolls JK, Kieffer TJ (2004) Leptin increases hepatic insulin sensitivity and protein tyrosine phosphatase 1B expression. Mol Endocrinol 18:1333–1345
102. Obata T, Maegawa H, Kashiwagi A, Pillay TS, Kikkawa R (1998) High glucose-induced abnormal epidermal growth factor signaling. J Biochem 123:813–820

103. Shao J, Gao Y, Yuan Z (1998) [Free fatty acids promoting PTP1B expression in rat skeletal muscle and hepatic cells]. Zhonghua Yi Xue Za Zhi 78:753–755
104. Parvaneh L, Meshkani R, Bakhtiyari S, Mohammadtaghvaie N, Gorganifiruzjaee S, Taheripak G, Golestani A, Foruzandeh M, Larijani B, Taghikhani M (2010) Palmitate and inflammatory state additively induce the expression of PTP1B in muscle cells. Biochem Biophys Res Commun 396:467–471
105. Berthou F, Rouch C, Gertler A, Gerozissis K, Taouis M (2011) Chronic central leptin infusion differently modulates brain and liver insulin signaling. Mol Cell Endocrinol 337:89–95
106. Zabolotny JM, Kim YB, Welsh LA, Kershaw EE, Neel BG, Kahn BB (2008) Protein-tyrosine phosphatase 1B expression is induced by inflammation in vivo. J Biol Chem 283:14230–14241
107. Briancon N, McNay DE, Maratos-Flier E, Flier JS (2010) Combined neural inactivation of SOCS-3 and PTP-1B reveal additive, synergistic, and factor-specific roles in the regulation of body energy balance. Diabetes 59:3074–3084
108. Vong L, Ye C, Yang Z, Choi B, Chua S Jr, Lowell BB (2011) Leptin action on GABAergic neurons prevents obesity and reduces inhibitory tone to POMC neurons. Neuron 71:142–154
109. Haj FG, Zabolotny JM, Kim YB, Kahn BB, Neel BG (2005) Liver specific protein-tyrosine phosphatase 1B (PTP1B) Re-expression alters glucose homeostasis of PTP1B-/-mice. J Biol Chem 280:15038–15046
110. Gelling RW, Morton GJ, Morrison CD, Niswender KD, Myers MG Jr, Rhodes CJ, Schwartz MW (2006) Insulin action in the brain contributes to glucose lowering during insulin treatment of diabetes. Cell Metab 3:67–73
111. Koch L, Wunderlich FT, Seibler J, Konner AC, Hampel B, Irlenbusch S, Brabant G, Kahn CR, Schwenk F, Bruning JC (2008) Central insulin action regulates peripheral glucose and fat metabolism in mice. J Clin Invest 118:2132–2147
112. Plum L, Belgardt BF, Bruning JC (2006) Central insulin action in energy and glucose homeostasis. J Clin Invest 116:1761–1766
113. Bruning JC, Gautam D, Burks DJ, Gillette J, Schubert M, Orban PC, Klein R, Krone W, Muller-Wieland D, Kahn CR (2000) Role of brain insulin receptor in control of body weight and reproduction. Science 289:2122–2125
114. Obici S, Feng Z, Karkanias G, Baskin DG, Rossetti L (2002) Decreasing hypothalamic insulin receptors causes hyperphagia and insulin resistance in rats. Nat Neurosci 5:566–572
115. Inoue H, Ogawa W, Asakawa A, Okamoto Y, Nishizawa A, Matsumoto M, Teshigawara K, Matsuki Y, Watanabe E, Hiramatsu R, Notohara K, Katayose K, Okamura H, Kahn CR, Noda T, Takeda K, Akira S, Inui A, Kasuga M (2006) Role of hepatic STAT3 in brain-insulin action on hepatic glucose production. Cell Metab 3:267–275
116. Scherer T, O'Hare J, Diggs-Andrews K, Schweiger M, Cheng B, Lindtner C, Zielinski E, Vempati P, Su K, Dighe S, Milsom T, Puchowicz M, Scheja L, Zechner R, Fisher SJ, Previs SF, Buettner C (2011) Brain insulin controls adipose tissue lipolysis and lipogenesis. Cell Metab 13:183–194
117. Klockener T, Hess S, Belgardt BF, Paeger L, Verhagen LA, Husch A, Sohn JW, Hampel B, Dhillon H, Zigman JM, Lowell BB, Williams KW, Elmquist JK, Horvath TL, Kloppenburg P, Bruning JC (2011) High-fat feeding promotes obesity via insulin receptor/PI3K-dependent inhibition of SF-1 VMH neurons. Nat Neurosci 14:911–918
118. Schulz C, Paulus K, Johren O, Lehnert H (2012) Intranasal leptin reduces appetite and induces weight loss in rats with diet-induced obesity (DIO). Endocrinology 153:143–153
119. Hallschmid M, Higgs S, Thienel M, Ott V, Lehnert H (2012) Postprandial administration of intranasal insulin intensifies satiety and reduces intake of palatable snacks in women. Diabetes 61:782–789
120. Benedict C, Brede S, Schioth HB, Lehnert H, Schultes B, Born J, Hallschmid M (2011) Intranasal insulin enhances postprandial thermogenesis and lowers postprandial serum insulin levels in healthy men. Diabetes 60:114–118
121. Jauch-Chara K, Friedrich A, Rezmer M, Melchert UH, G Scholand-Engler H, Hallschmid M, Oltmanns KM (2012) Intranasal insulin suppresses food intake via enhancement of brain energy levels in humans. Diabetes 61:2261–2268

Chapter 5
PTP1B in the Periphery: Regulating Insulin Sensitivity and ER Stress

Mirela Delibegovic and Nimesh Mody

Abstract Protein tyrosine phosphatase 1B (PTP1B) is a major regulator of body mass and insulin sensitivity and plays an important role in signal transduction of many important metabolic pathways such as insulin, leptin, integrin, growth hormone, and endoplasmic reticulum (ER) stress response signaling. In this article we review current literature on the divergent role of PTP1B in different tissues in ER stress response signal transduction and its control of insulin sensitivity in vivo. Recent evidence suggests that PTP1B is a direct player in the ER stress response pathway potentially due to direct dephosphorylation of PKR-like ER kinase (PERK) on its tyrosine 619 site. Considering that the ER stress response pathway is involved in the pathophysiology of insulin and leptin resistance, aging, neurodegenerative disorders, cancer, and other diseases, it implicates PTP1B inhibitors as a viable therapeutic target in the treatment and/or prevention of a number of these diseases.

Abbreviations

ATF4	Activating transcription factor 4
ATF6	Activating transcription factor 6
BIP	Binding immunoglobulinprotein
CHOP	CAAT/enhancer-binding protein homologous protein
EDEM	ER degradation-enhancing α-mannosidase-like protein

M. Delibegovic, Ph.D. (✉) • N. Mody, Ph.D.
Institute of Medical Sciences, University of Aberdeen,
Foresterhill Health Campus, Aberdeen AB25 2ZD, Aberdeen, UK
e-mail: m.delibegovic@abdn.ac.uk

K.K. Bence (ed.), *Protein Tyrosine Phosphatase Control of Metabolism*,
DOI 10.1007/978-1-4614-7855-3_5,

eIF2α	Translation initiation factor 2α
ER	Endoplasmic reticulum stress
ERAD	ER-associated degradation
ERDJ4	ER-localized DnaJ homologue
GADD34	Growth arrest and DNA damage 34
GRP94	Glucose-regulated protein 94
GSK3	Glycogen synthase kinase 3
IR	Insulin receptor
IRE1	Inositol-requiring enzyme 1
IRS	Insulin receptor substrate
mTOR	Mammalian target of rapamycin
PERK	PKR-like ER kinase
PI3K	Phosphatidylinositol 3-kinase
PKB	Protein kinase B
PTP1B	Protein tyrosine phosphatase 1B
UPR	Unfolded protein response
XBP1	X box-binding protein 1

Introduction

Insulin is a pancreatic hormone that controls energy metabolism in tissues such as liver, muscle, and white adipose tissue (WAT) [1]. Insulin exerts its effects by binding to the insulin receptors (IR) located on the plasma membrane, thus inducing autophosphorylation of the receptor on tyrosine residues, which stimulates its tyrosine kinase activity (Fig. 5.1). The activated IR phosphorylates substrates including the insulin receptor substrate (IRS) proteins 1–4, Shc, Cbl, and Gab-1 [2]. Upon phosphorylation, the IRS proteins act as docking sites for several Src homology region 2 (SH2) domain containing proteins, including the p85 regulatory subunit of phosphatidylinositol 3-kinase (PI3K), which results in PI3K activation [3]. PI3K activation subsequently leads to activation of protein kinase B (Akt/PKB), which has diverse intracellular targets, including glycogen synthase kinase 3 (GSK3) and the mammalian target of rapamycin (mTOR). Importantly, Akt/PKB is required to promote translocation of GLUT4 to the plasma membrane, and consequently, increase glucose uptake. Termination of the signal involves inactivation of the IR kinase by dephosphorylation on three tyrosine residues which are located in the activation loop of the receptor [4].

Protein tyrosine phosphatase 1B (PTP1B) is a widely expressed, prototypical non-transmembrane tyrosine phosphatase that can dephosphorylate several receptor tyrosine kinases (RTKs), such as IR, epidermal growth factor receptor (EGFR), platelet-derived growth factor receptor (PDGFR), and insulin-like growth factor 1 receptor (IGF1-R) [5–7]. A 35-residue hydrophobic region at the C-terminus targets PTP1B to the endoplasmic reticulum (ER) [8], where its catalytic domain is exposed to the cytoplasm, and even though PTP1B can be released from the ER under certain

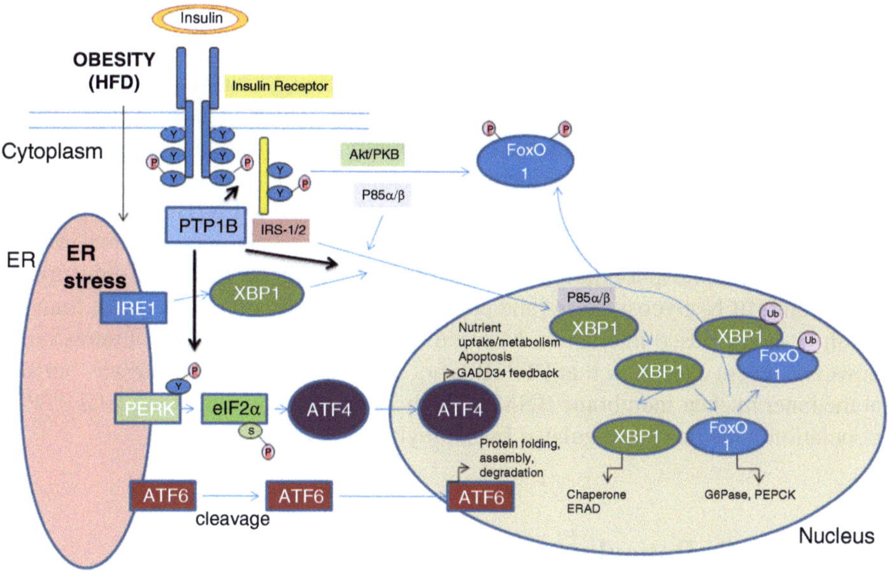

Fig. 5.1 ER stress response signaling, insulin receptor signaling, and PTP1B. Insulin binds to the insulin receptors (IR) located on the plasma membrane, inducing autophosphorylation of the receptor on tyrosine residues. This leads to induction of a signaling cascade downstream of IR, such as phosphorylation of IRS proteins, activation of PI3K, and Akt/PKB. IRE1, PERK, and ATF6 monitor the accumulation of misfolded proteins inside the ER and upon induction of ER stress, such as that caused by obesity (genetic or dietary), IRE1 endonuclease activity facilitates the splicing of XBP-1 mRNA (XBP1s). XBP-1s binding of one of the PI3K regulatory subunits, p85α or p85β, enhances nuclear entry of this transcription factor and induction of UPR target genes. This results in an increase in the expression of chaperones and ERAD proteins, therefore improving ER-folding capacity and maintaining insulin sensitivity. Moreover, XBP1, in an insulin signaling-independent manner, can translocate into the nucleus and bind the transcription factor FoxO1 to promote its proteosomal degradation. PTP1B directly dephosphorylates the IR and is also suggested to dephosphorylate the Tyr615 site on PERK, thereby directly regulating components of ER stress response machinery. In addition, hepatic nuclear translocation of p85a-XBP1s complex is increased in liver-specific PTP1B knockout mice suggesting that PTP1B directly or indirectly, through general improvements in whole-body insulin sensitivity, plays a role in this process

conditions, the full length, ER-localized protein is the predominant form found in most cell types [7]. In some cells, PTP1B gets cleaved by the protease calpain which cuts off its hydrophobic tail thus releasing a more active, soluble form of the enzyme [9–11]. One interesting question that has been puzzling scientists has been the question of how an ER-resident phosphatase can dephosphorylate RTKs which are localized on the plasma membrane or the endocytic vesicles [7, 12, 13]. Using imaging fluorescence resonance energy transfer (FRET) between transiently expressed RTK-green fluorescent protein (GFP) fusions and sulfoindocyanine (Cy3)-conjugated monoclonal antibodies to PTP1B, and stimulation with growth factors, the sites of interaction between PTP1B and RTK (EGFR and PDGFR) dephosphorylation were mapped [12]. ER-anchored PTP1B was found to specifically dephosphorylate endocytosed RTKs, contributing to RTK signal termination.

Furthermore, use of bioluminescence resonance energy transfer (BRET) allowed investigation of the interaction between PTP1B and the IR in real time, observing the interaction between the two that occurs rapidly upon insulin stimulation (within 30 s) [1]; it was also shown that there is an insulin-independent interaction between PTP1B and the IR that requires PTP1B to be localized to the ER. However, previous studies found that PTP1B localization at the ER was not required for insulin-stimulated IR dephosphorylation by PTP1B in 3T3-L1 adipocytes [14]. Tunicamycin, an inhibitor of N-glycosylation (and potent inducer of ER stress), has been found to result in an increased interaction between the IR and PTP1B [1, 15]. Interestingly, however, recent data show that PTP1B is not only located at the ER surface, but also at the inner nuclear membrane (INM), which allows this fraction of the total PTP1B population to be heavily regulated by sumoylation [11, 16].

Endoplasmic Reticulum and ER Stress

The endoplasmic reticulum (ER) is an organelle required for synthesis and metabolism of metabolites and biologically active proteins [17]. This is a highly regulated process consisting of chaperones, signaling molecules, and a network of degradation machinery that maintain cellular homeostasis. In response to stress and alterations in cellular homeostasis, such as the accumulation of unfolded protein aggregates in response to a viral load, high glucose levels, oxidative stress, and similar, the ER signaling machinery triggers the inhibition of protein synthesis and up-regulation of genes whose products are involved in protein folding, cell cycle exit, and/or apoptosis [18]. This leads to stimulation of a pathway termed the unfolded protein response (UPR), which promotes the return of the ER to its normal physiological state.

UPR is mediated initially by three molecules, PKR-like ER kinase (PERK), activating transcription factor 6 (ATF6), and Inositol-requiring enzyme 1 (IRE1) (Fig. 5.1) [18, 19]. The ER luminal domain of PERK, IRE1, and ATF6 interacts with the ER chaperone GRP78 (or binding immunoglobulinprotein [BIP]) (glucose-regulated protein). Activated PERK phosphorylates translation initiation factor 2α (eIF2α), which decreases the rate of general protein translation and protein load on the ER [20, 21]. Phosphorylation of eIF2α paradoxically increases translation of activating transcription factor 4 (ATF4) mRNA to produce a transcription factor that activates expression of several UPR target genes [20, 22]. Expression profiling studies have found that PERK, eIF2α, and ATF4 are required for expression of genes involved in amino acid biosynthesis and transport, anti-oxidative stress, and apoptosis, e.g., growth arrest and DNA damage 34 (GADD34), and CAAT/enhancer-binding protein homologous protein (CHOP) [23, 24]. Activation of the ER protein kinase IRE1 triggers its endoribonuclease activity to induce cleavage of X box-binding protein 1 (XBP-1) mRNA [25], which is then ligated by an uncharacterized RNA ligase and translated to produce spliced XBP-1 protein (Fig. 5.1) [26].

This spliced XBP-1 protein is a highly active transcription factor and one of the key regulators of ER-folding capacity [27]. Some of the genes identified that require the IRE1/XBP-1 pathway are components of the ER-associated degradation (ERAD) machinery, such as ER degradation-enhancing α-mannosidase-like protein (EDEM) [28, 29] and major ER chaperones, e.g., GRP78/BIP, GRP94 and the endoplasmic reticulum-localized DnaJ homologue (ERDJ4) [27, 28]. At the same time, ATF6 is released from GRP78 and transits to the Golgi body where it is cleaved to release a transcriptionally active ATF6 fragment [30], which acts in concert with spliced XBP-1 protein to induce expression of genes encoding protein chaperones and components of the ERAD machinery [31, 32].

As stated above, the UPR deals with adverse effects of ER stress and enhances cell survival. However, prolonged ER stress leads to apoptosis, inflammation, and lipid accumulation and is commonly referred to as the "ER stress response" [33]. Obesity, whether caused by lifestyle factors or genetic deficiency, results in conditions that increase demand on the ER resulting in chronic activation of the pathway, subsequently leading to development of pathophysiological conditions and development of the ER stress response (Fig. 5.1). This is particularly clear in the liver, adipose tissue, and pancreas, where changes in tissue architecture, increases in protein synthesis, and perturbations in cellular energy fluxes occur [34]. Recent studies demonstrated that ER stress is increased in adipose and liver tissues in both dietary and genetic obesity in mice [35–38], although it is not clear if this also occurs in human obesity. Examination of the ER stress markers in adipose tissue of obese and obese/diabetic volunteers revealed no differences in ER stress induction in comparison to the lean human volunteers in this tissue [39]. ER stress and the UPR are linked to major inflammatory and stress-signaling networks via several distinct mechanisms, including the activation of JNK-AP-1 and IKB kinase nuclear factor κB (IKK-NFκB) pathways and the production of reactive oxygen species (ROS). Interestingly, these are also the pathways and mechanisms that play a central role in obesity-induced inflammation and metabolic abnormalities [19]. For example, JNK activation by IRE-1α during ER stress is one key pathway to increased inflammation. In the nucleus, JNK up-regulates the expression of inflammatory genes through activation of the AP-1 transcription factor complexes; thus the beneficial metabolic effects observed in the JNK1-deficient mouse are suggested to be mediated, at least in part, through the suppression of inflammatory cytokines, such as TNF-α, IL-6, and MCP-1, compared with wild-type mice on a high-fat diet (HFD). IRE-1α can also activate the IKK-NFκB pathway, which is critical in the induction of multiple inflammatory genes such as TNF-α and IL-6 and is also implicated in insulin resistance. The NFκB pathway may also be activated through PERK signaling, as PERK-mediated phosphorylation of eIF2α results in the inhibition of translation of the inhibitor of NFκB (IkB) protein, the major negative regulator of NFκB, thus allowing the activation of NFκB and the induction of its pro-inflammatory targets [19]. Modulation of ER-folding capacity through gain- and loss-of-function studies with X box-binding protein (XBP-1) showed a close link between ER function and insulin action in vitro and in vivo [35]. Interestingly, PTP1B$^{-/-}$ primary and immortalized cells show impaired IRE1α signaling, decreased JNK activation, and a defect in XBP-1 splicing [40], revealing an unanticipated role of PTP1B in modulation of ER stress signaling.

Tissue-Specific Role of PTP1B in Regulation of Insulin Sensitivity, Body Mass/Adiposity, and Energy Homeostasis

As discussed above, studies integrating crystallographic, kinetic, and PTP1B peptide-binding data have revealed highly specific interactions between PTP1B and IR activation loop tyrosine residues [41], and PTP1B has been shown in vitro to dephosphorylate phosphotyrosine residues on IRS1 with high specificity [42]. However, it was not until the generation of PTP1B global knockout mice (PTP1B$^{-/-}$) that key evidence was provided as to the direct role of PTP1B in IR dephosphorylation and insulin sensitivity in vivo [43, 44]. PTP1B$^{-/-}$ mice were born normal and healthy with increased insulin sensitivity and glucose homeostasis, as well as increased IR phosphorylation in liver and muscle, but not WAT. These mice were resistant to diet-induced obesity (DIO), had reduced adiposity, decreased leptin/body fat ratios, and increased energy expenditure on HFD, whilst its insulin sensitivity was tissue-specific, as glucose uptake was elevated in muscle, but not in WAT [44]. A set of elegant experiments using PTP1B$^{-/-}$ mice treated with gold thioglucose, which resulted in ablation of hypothalamic neurons, demonstrated that this partially blocked resistance to obesity, suggesting that some, but perhaps not all, of the effects of resistance to DIO were due to the role of PTP1B in hypothalamic leptin-sensitive neurons [45].

To investigate tissue types responsible for the improvements in glucose homeostasis and insulin sensitivity as well as resistance to DIO and decrease in adiposity, tissue-specific PTP1B$^{-/-}$ mice were generated [46]. Neuronal-PTP1B$^{-/-}$ mice exhibited reduced body mass and adiposity due to leptin hypersensitivity, despite increased circulating and mRNA leptin levels, resulting in decreased food intake and increased energy expenditure in these mice [46]. These mice were also insulin-sensitive, but it is unclear if this is due to the direct role of PTP1B on the IR in the brain, or an indirect effect of reduced body mass and adiposity. Furthermore, mice lacking PTP1B specifically in pro-opiomelanocortin (POMC) neurons also exhibited decreased adiposity, increased leptin sensitivity and energy expenditure, and improved glucose homeostasis on an HFD compared with wild-type mice [47]. These mice were found to also exhibit alterations in their homeostatic response to cold exposure, as their plasma triiodothyronine (T3) and brown adipose tissue (BAT) weight were increased in response to a 4-day cold challenge (4 °C) in comparison to controls [48]. More recently, the direct role of PTP1B in leptin receptor-expressing neurons was determined by analyzing LepRb-PTP1B$^{-/-}$ mice [49]. These mice strikingly recapitulated the phenotype of global as well as neuronal-PTP1B$^{-/-}$ mice in regard to their resistance to DIO, reduced adiposity, and leptin hypersensitivity, thereby establishing that PTP1B indeed regulates body weight and adiposity due to its role in leptin receptor-expressing neurons [49]. However, the effects of PTP1B in regulation of insulin sensitivity independently of its effects on body mass and adiposity came from studies investigating its role in peripheral insulin-sensitive tissues [5, 6, 50]. The direct effects of PTP1B deficiency on insulin sensitivity and IR phosphorylation, independent of body mass, were evident in muscle-PTP1B$^{-/-}$ mice, which had the same weight and adiposity as their littermate controls on normal diet and HFD, but exhibited almost complete protection against HFD-induced insulin resistance [51]. There was also basal hyperphosphorylation of the IR in the muscles from muscle-PTP1B$^{-/-}$ mice, accompanied by increased glucose uptake into the muscle and improvements in whole-body glucose homeostasis. These studies suggested that at least in the muscle, IR is the direct substrate for PTP1B. In addition, treatment of muscle-PTP1B$^{-/-}$ mice with an antidiabetic drug Rosiglitazone (Rosi) resulted in complete protection against HFD-induced insulin resistance and glucose intolerance. Liver-PTP1B$^{-/-}$ mice also weighed the same as their littermate controls on chow and HFD, with comparable adiposity; however, again, these mice were protected against HFD-induced insulin resistance as well as exhibited a degree of protection against HFD-induced serum and liver lipid accumulation [38]. Conversely, transgenic over-expression of PTP1B in muscle and/or liver resulted in development of insulin resistance in mice [52, 53]. Adipocyte-PTP1B$^{-/-}$ mice, on the other hand, exhibited mild glucose intolerance, increase in leptin secretion and adipocyte cell size as well as an increase in basal lipogenesis [2]. These mice also had threefold higher expression of the hypoxia marker hypoxia-induced factor-1-alpha (*Hif-1*α), which was equivalent to levels measured in HFD-fed control mice; in addition, HFD-feeding increased *Hif-1*α expression in adiponectin-PTP1B$^{-/-}$ mice even further [2]. Strikingly, there were no changes in adipocyte IR phosphorylation in adipocyte-PTP1B$^{-/-}$ mice in vivo or ex vivo, suggesting that IR is not a direct substrate for PTP1B in adipocytes and that PTP1B may have a different role to play in this tissue [2].

Role of PTP1B in Insulin Sensitivity and ER Stress

The first direct evidence that PTP1B plays an important role in ER stress response signal transduction came from studies utilizing PTP1B$^{-/-}$ mouse embryonic fibroblasts [40] (Table 5.1). In the absence of PTP1B, IRE-1-dependent JNK and p38 activation, XBP-1 splicing, and EDEM transcription were decreased in response to pharmacological ER stress inducers, in addition to attenuation of stress-induced apoptosis [40]. These studies suggested that PTP1B was somehow involved in IRE1 signaling and that a functional tyrosine phosphatase domain was required for PTP1B to exert these effects. Other studies have shown that PTP1B is over-expressed in multiple insulin- and leptin-responsive tissues in mice under chronic ER stress conditions such as diet-induced or genetic obesity or in cells with pro-inflammatory cytokine or free fatty acid treatment [54–56]. Utilization of liver-PTP1B$^{-/-}$ mice demonstrated for the first time in vivo that PTP1B indeed is an important physiological player in ER stress response signal transduction [38]. Liver-PTP1B$^{-/-}$ mice exhibited reduced ER stress response compared to WT controls, protection against the induction of HFD-induced ER stress response, and phosphorylation of p38 was significantly lower in livers from these mice compared with controls, as was phosphorylation of JNK, PERK, and eIF2α, suggesting that liver PTP1B indeed plays an

Table 5.1 Tissue-divergent role of PTP1B in regulation of whole-body glucose/lipid/energy homeostasis and ER stress

Tissue/cell type examined	Overall phenotype	ER stress response genes	Pathways involved	ER stress inducers used	References
Mouse embryonic fibroblasts from $PTP1B^{-/-}$ global mice	Improvement in insulin sensitivity and glucose homeostasis; resistance to obesity; increased energy expenditure [43–45]	↓ JNK phosphorylation, p38 phosphorylation, XBP1 splicing, EDEM (ER degradation-enhancing-mannosidase-like protein), apoptosis	IRE1 signaling	Azetidine-2-carboxylic acid (Azc) (10 mM) and Tunicamycin (Tun) (10 μg/mL)	[40]
Liver-$PTP1B^{-/-}$ and HepG2 liver cell line	Improvement in insulin sensitivity and glucose homeostasis; lower liver and plasma lipids [38]	↓ JNK phosphorylation, p38 phosphorylation, PERK phosphorylation, eIF2α phosphorylation, XBP1 splicing, CHOP, ATF4, GRP78 (BIP), GADD34, GRP94, ERDJ4, ATF6 cleavage, EDEM	IRE1, PERK, and ATF6 signaling	High-fat diet (55 % fat; Harlan Teklad), thapsigargin (300 nM), Tun (5 μg/mL) or in vivo Tun injection (1 μg/g body weight)	[25, 38, 75]
Min6 pancreatic cell line PTP1B knockdown	Role of pancreatic PTP1B not determined yet; however high-fat diet increases PTP1B levels in pancreas	↑ PERK phosphorylation, eIF2α phosphorylation, XBP1 splicing, GRP78 (BIP), ER-stress-induced apoptosis	IRE1, PERK	Palmitate (1 mM), Tun (2 ng/mL)	[65]
Brown adipocytes from adipose-$PTP1B^{-/-}$ mice	Mild glucose intolerance, leptin resistance, increase in circulating leptin and hypoxic markers [2]	↑ PERK phosphorylation (Thr 980 and Tyr615), eIF2α phosphorylation, XBP1 splicing, CHOP, BIP, ATF6 cleavage, ATF4 mRNA translation	IRE1, PERK, ATF6	Tun (2 ng/mL), thapsigargin (1 μM)	[66]

important role in ER stress-induced IRE1 signaling. CHOP and XBP1s levels also were decreased in HFD-fed liver-PTP1B$^{-/-}$ mice, consistent with impaired ER stress-induced PERK signaling. These in vivo studies demonstrated that PTP1B deficiency affects multiple "arms" of the ER stress response pathway [38]. Further in depth analysis of the role of hepatic PTP1B in the involvement of ER stress response induction and insulin resistance revealed that chronic HFD-feeding, as well as induction of ER stress response using acute pharmacological inducers (e.g. tunicamycin, thapsigargin), led to an elevation of PTP1B mRNA and protein levels in mouse livers and hepatic cell lines [25]. HFD-feeding itself also increased processing of ATF6 into its active, cleaved form in control mice whilst this was found to be decreased to normal physiological levels in mice with a liver-specific PTP1B deletion [25]. GRP78/BIP was coordinately up-regulated in control HFD-fed mice, but completely reduced to chow diet levels in liver-PTP1B$^{-/-}$ HFD-fed mice. This is consistent with ATF6α being solely responsible for transcriptional induction of ER chaperones, e.g., GRP78/BIP [32]. Foufelle and co-workers also found increased levels of cleaved ATF6 in livers of *ob/ob* mice [57]. However in other studies, cleaved ATF6 levels were found to be reduced in *ob/ob*, *db/db,* and DIO mice and p85α-deficient cell lines with reduced nuclear XBP1 [58, 59]. Altogether, these studies suggest that PTP1B appears to contribute to the ER stress response induction directly since liver-specific PTP1B deletion in mice and/or siRNA knockdown in hepatic cells blunts the full activation of all three arms of the ER stress pathway, namely IRE1-α/XBP-1, PERK/eIF2-α, and ATF6 [25, 38, 40]. Indeed, studies examining the mechanisms of hypoglycemic effects of *Astragalus* polysaccharide (APS) found that APS exerts its insulin sensitizing effects in the liver through decreasing ER stress and inhibiting PTP1B activity [60] and that ATF6 is involved in mediating the expression of PTP1B upon induction of ER stress [59].

Interestingly, the IR has also been reported to be directly affected by ER stress inducers [61, 62]. Tunicamycin treatment of 3T3-L1 adipocytes leads to accumulation in the ER of a 180 kDa IR pro-peptide devoid of glycosylation, which is not further processed and does not reach the cell surface. Thus, glycosylation of the insulin pro-receptor is crucial for proper processing and formation of functional receptor. IR is also downregulated at the mRNA level in obesity/insulin resistance and up-regulated in rodent models of type 1 diabetes, inversely correlating with circulating insulin levels. Thus, it appears that both PTP1B and IR are regulated by multiple stimulants of the metabolic and ER stress-related variety. However, it is not clear if these regulatory pathways interact physiologically such as with the proposed interactions of XBP-1 with p85 and FOXO (Fig. 5.1) [58, 63, 64].

Consistent with other data, analysis of PTP1B function in ER stress response in pancreas, and pancreatic MIN6 cells, revealed that HFD-feeding and chemical treatment of cells with palmitate increased PTP1B expression levels [65]. Overexpression of PTP1B in MIN6 cells, however, led to a significant decrease in palmitate- and tunicamycin-induced PERK and eIF2a phosphorylation; conversely, PTP1B knockdown using siRNA potentiated ER stress as assessed by PERK and eIF2a phosphorylation as well as BIP, CHOP, and XBP1s expression [65]. Examining cell death in these cells, using p53 as a readout of apoptosis, PTP1B knockdown led to increased p53 expression as well as Bax/Bcl-xl ratio; in addition, expression of active caspases-8, -9, -3, and -7 was increased in these knockdown cells.

In addition, PTP1B deficiency in brown adipocytes was also found to potentiate PERK/eIF2a signaling by the same group [66]. Brown adipocytes from adipose-specific PTP1B$^{-/-}$ mice exhibited increased PERK and eIF1α phosphorylation, as well as increased expression of BIP, XBP1s, and CHOP; the same was the case in subcutaneous WAT of these mice as well as brown preadipocyte cell lines [66]. Whilst these findings may appear paradoxical to the findings from immortalized fibroblasts from global-PTP1B$^{-/-}$ mice [40] and liver-PTP1B$^{-/-}$ mice [25, 38], they are in line with the published phenotype of adipose-specific PTP1B$^{-/-}$ mice [2], which were found to exhibit mild glucose intolerance, increased cell size, and basal lipogenesis.

It is unclear how PTP1B expression or activity can directly affect the induction and maintenance of ER stress and why these appear to differ between different tissues and/or cell types; however, these may depend on pathophysiological circumstances (for example, under obesogenic HFD-feeding, inflammatory conditions, PTP1B deficiency may be protective against ER stress induction and maintenance whilst it may play a less active role under normal physiological conditions) as well as whether ER stress induction is measured in vivo vs. under isolated and controlled tissue culture conditions.

Potential Mechanistic Link Between PTP1B and PERK^{Tyr615}

Interestingly, adipose PTP1B deficiency increased PERK tyrosine phosphorylation on Tyr615, suggesting a direct interaction between PTP1B and PERK [66]. Indeed, a direct enzyme-substrate interaction between PTP1B and pTyr619-PERK has been demonstrated in human embryonic kidney (HEK) 293T cells (Fig. 5.1) [67]. This study examined the effects of hydrogen sulfide (H_2S), a gaseous signaling molecule implicated in regulation of many metabolic processes including its potential as an anti-inflammatory target, on PTP1B activity and found that H_2S reversibly inactivated PTP1B both in vitro and in vivo, via sulfhydration of the active Cys residue on PTP1B upon induction of ER stress. Sulfhydration inhibited the ability of PTP1B to dephosphorylate PERK. This process required the presence of cystathionine-γ-lyase (CSE), a critical enzyme in H_2S production, which inhibits PTP1B phosphatase activity; suppression of CSE, on the other hand, led to decreased phosphorylation and activation of PERK at the Tyr619 site. Furthermore, mutation of Tyr619 in PERK abrogated its interaction with the PTP1B-D181A substrate-trapping mutant, suggesting that this site is critical to mediating the interaction of PTP1B with PERK. Importantly, all the effects of suppressing H_2S production in response to ER stress were found to be abolished in the presence of a small molecule inhibitor of PTP1B [67].

ER Stress and Other Tyrosine Phosphatases

T cell protein tyrosine phosphatase (TCPTP) shares ~72 % catalytic domain sequence identity with PTP1B and has also been implicated in IR regulation [68]. There are two variants of TCPTP; a 48 kDa variant which is targeted to the ER via its hydrophobic C-terminus and a 45 kDa variant targeted to the nucleus. In contrast to PTP1B, ER stress-induced by HFD-feeding of mice for 22 weeks led to a decrease in pancreatic TCPTP protein levels [65]. Knockdown of TCPTP in MIN6 cells led to decreased phosphorylation of PERK and eIF2a in response to pharmacological inducers of ER stress response, palmitate and tunicamycin. Thus, in the context of these studies it appears that PTP1B and TCPTP function coordinately to regulate ER stress signaling in β cells [65].

Conclusion

Overall, all the data so far suggest that PTP1B is a key player in the regulation of body mass, insulin sensitivity, and lipid homeostasis, through regulation of diverse signaling pathways, some of which are briefly discussed here and elsewhere in this book. The importance of PTP1B involvement in the ER stress response pathway is becoming readily apparent but is yet to be fully understood; however, considering that the ER stress response pathway has been found to be involved in the etiology of insulin resistance [19, 34–36], leptin resistance [37], aging [69–72], neurodegenerative disorders [73], cancer [74], and other diseases, PTP1B inhibitors may indeed be a viable therapeutic target in the treatment and/or prevention of a number of these diseases.

Acknowledgments MD is supported by RCUK, British Heart Foundation, Diabetes UK, Tenovus Scotland, European Foundation for the Study of Diabetes (EFSD)/Lilly. NM is the recipient of a British Heart Foundation intermediate basic research fellowship.

References

1. Boute N, Boubekeur S, Lacasa D, Issad T (2003) Dynamics of the interaction between the insulin receptor and protein tyrosine-phosphatase 1B in living cells. EMBO Rep 4(3): 313–319
2. Owen C, Czopek A, Agouni A, Grant L, Judson R, Lees EK et al (2012) Adipocyte-specific protein tyrosine phosphatase 1B deletion increases lipogenesis, adipocyte cell size and is a minor regulator of glucose homeostasis. PLoS One 7(2):e32700
3. Lizcano JM, Alessi DR (2002) The insulin signalling pathway. Curr Biol 12(7):R236–R238
4. King MJ, Sale GJ (1990) Dephosphorylation of insulin-receptor autophosphorylation sites by particulate and soluble phosphotyrosyl-protein phosphatases. Biochem J 266(1): 251–259
5. Delibegovic M, Mody N (2009) Protein tyrosine phosphatase 1B (PTP1B) in obesity and Type 2 diabetes. Acta Med Sal 38(1):2–7

6. Bence KK (2010) Hepatic PTP1B deficiency: the promise of a treatment for metabolic syndrome? J Clin Metab Diabetes 1(1):27–33
7. Stuible M, Tremblay ML (2010) In control at the ER: PTP1B and the down-regulation of RTKs by dephosphorylation and endocytosis. Trends Cell Biol 20(11):672–679
8. Frangioni JV, Beahm PH, Shifrin V, Jost CA, Neel BG (1992) The nontransmembrane tyrosine phosphatase PTP-1B localizes to the endoplasmic reticulum via its 35 amino acid C-terminal sequence. Cell 68(3):545–560
9. Frangioni JV, Oda A, Smith M, Salzman EW, Neel BG (1993) Calpain-catalyzed cleavage and subcellular relocation of protein phosphotyrosine phosphatase 1B (PTP-1B) in human platelets. EMBO J 12(12):4843–4856
10. Kuchay SM, Kim N, Grunz EA, Fay WP, Chishti AH (2007) Double knockouts reveal that protein tyrosine phosphatase 1B is a physiological target of calpain-1 in platelets. Mol Cell Biol 27(17):6038–6052
11. Yip SC, Cotteret S, Chernoff J (2012) Sumoylated protein tyrosine phosphatase 1B localizes to the inner nuclear membrane and regulates the tyrosine phosphorylation of emerin. J Cell Sci 125(Pt 2):310–316
12. Haj FG, Verveer PJ, Squire A, Neel BG, Bastiaens PI (2002) Imaging sites of receptor dephosphorylation by PTP1B on the surface of the endoplasmic reticulum. Science 295(5560): 1708–1711
13. Eden ER, White IJ, Tsapara A, Futter CE (2010) Membrane contacts between endosomes and ER provide sites for PTP1B-epidermal growth factor receptor interaction. Nat Cell Biol 12(3):267–272
14. Shi K, Egawa K, Maegawa H, Nakamura T, Ugi S, Nishio Y et al (2004) Protein-tyrosine phosphatase 1B associates with insulin receptor and negatively regulates insulin signaling without receptor internalization. J Biochem 136(1):89–96
15. Romsicki Y, Reece M, Gauthier JY, Asante-Appiah E, Kennedy BP (2004) Protein tyrosine phosphatase-1B dephosphorylation of the insulin receptor occurs in a perinuclear endosome compartment in human embryonic kidney 293 cells. J Biol Chem 279(13):12868–12875
16. Dadke S, Cotteret S, Yip SC, Jaffer ZM, Haj F, Ivanov A et al (2007) Regulation of protein tyrosine phosphatase 1B by sumoylation. Nat Cell Biol 9(1):80–85
17. Zhang K, Kaufman RJ (2008) From endoplasmic-reticulum stress to the inflammatory response. Nature 454(7203):455–462
18. Ron D, Walter P (2007) Signal integration in the endoplasmic reticulum unfolded protein response. Nat Rev Mol Cell Biol 8(7):519–529
19. Gregor MG, Hotamisligil GS (2007) Adipocyte stress: the endoplasmic reticulum and metabolic disease. J Lipid Res 48:1905–1914
20. Harding HP, Zhang Y, Bertolotti A, Zeng H, Ron D (2000) Perk is essential for translational regulation and cell survival during the unfolded protein response. Mol Cell 5(5):897–904
21. Jiang HY, Wek RC (2005) Phosphorylation of the alpha-subunit of the eukaryotic initiation factor-2 (eIF2alpha) reduces protein synthesis and enhances apoptosis in response to proteasome inhibition. J Biol Chem 280(14):14189–14202
22. Harding HP, Zhang Y, Zeng H, Novoa I, Lu PD, Calfon M et al (2003) An integrated stress response regulates amino acid metabolism and resistance to oxidative stress. Mol Cell 11(3):619–633
23. Malhotra JD, Kaufman RJ (2007) The endoplasmic reticulum and the unfolded protein response. Semin Cell Dev Biol 18(6):716–731
24. Malhotra JD, Kaufman RJ (2007) Endoplasmic reticulum stress and oxidative stress: a vicious cycle or a double-edged sword? Antioxid Redox Signal 9(12):2277–2293
25. Agouni A, Mody N, Owen C, Czopek AJ, Zimmer D, Bentires-Alj M et al (2011) Liver-specific deletion of protein tyrosine phosphatase (PTP) 1B improves obesity- and pharmacologically-induced endoplasmic reticulum stress. Biochem J 438(2):369–378
26. Calfon M, Zeng H, Urano F, Till JH, Hubbard SR, Harding HP et al (2002) IRE1 couples endoplasmic reticulum load to secretory capacity by processing the XBP-1 mRNA. Nature 415(6867):92–96

27. Lee AH, Iwakoshi NN, Glimcher LH (2003) XBP-1 regulates a subset of endoplasmic reticulum resident chaperone genes in the unfolded protein response. Mol Cell Biol 23(21): 7448–7459
28. Mori K (2009) Signalling pathways in the unfolded protein response: development from yeast to mammals. J Biochem 146(6):743–750
29. Yoshida H, Matsui T, Hosokawa N, Kaufman RJ, Nagata K, Mori K (2003) A time-dependent phase shift in the mammalian unfolded protein response. Dev Cell 4(2):265–271
30. Haze K, Yoshida H, Yanagi H, Yura T, Mori K (1999) Mammalian transcription factor ATF6 is synthesized as a transmembrane protein and activated by proteolysis in response to endoplasmic reticulum stress. Mol Biol Cell 10(11):3787–3799
31. Wu J, Rutkowski DT, Dubois M, Swathirajan J, Saunders T, Wang J et al (2007) ATF6alpha optimizes long-term endoplasmic reticulum function to protect cells from chronic stress. Dev Cell 13(3):351–364
32. Yamamoto K, Sato T, Matsui T, Sato M, Okada T, Yoshida H et al (2007) Transcriptional induction of mammalian ER quality control proteins is mediated by single or combined action of ATF6alpha and XBP1. Dev Cell 13(3):365–376
33. Ji C, Kaplowitz N (2006) ER stress: can the liver cope? J Hepatol 45(2):321–333
34. Hotamisligil GS (2006) Inflammation and metabolic disorders. Nature 444(7121):860–867
35. Ozcan U, Cao Q, Yilmaz E, Lee AH, Iwakoshi NN, Ozdelen E et al (2004) Endoplasmic reticulum stress links obesity, insulin action, and type 2 diabetes. Science 306(5695):457–461
36. Ozcan U, Yilmaz E, Ozcan L, Furuhashi M, Vaillancourt E, Smith RO et al (2006) Chemical chaperones reduce ER stress and restore glucose homeostasis in a mouse model of type 2 diabetes. Science 313(5790):1137–1140
37. Ozcan L, Ergin AS, Lu A, Chung J, Sarkar S, Nie D et al (2009) Endoplasmic reticulum stress plays a central role in development of leptin resistance. Cell Metab 9(1):35–51
38. Delibegovic M, Zimmer D, Kauffman C, Rak K, Hong EG, Cho YR et al (2009) Liver-specific deletion of protein-tyrosine phosphatase 1B (PTP1B) improves metabolic syndrome and attenuates diet-induced endoplasmic reticulum stress. Diabetes 58(3):590–599
39. Hoggard N, Agouni A, Mody N, Delibegovic M (2012) Serum levels of RBP4 and adipose tissue levels of PTP1B are increased in obese men resident in northeast Scotland without associated changes in ER stress response genes. Int J Gen Med 5:403–411
40. Gu F, Nguyen DT, Stuible M, Dube N, Tremblay ML, Chevet E (2004) Protein-tyrosine phosphatase 1B potentiates IRE1 signaling during endoplasmic reticulum stress. J Biol Chem 279(48):49689–49693
41. Salmeen A, Andersen JN, Myers MP, Tonks NK, Barford D (2000) Molecular basis for the dephosphorylation of the activation segment of the insulin receptor by protein tyrosine phosphatase 1B. Mol Cell 6(6):1401–1412
42. Kusari J, Kenner KA, Suh KI, Hill DE, Henry RR (1994) Skeletal muscle protein tyrosine phosphatase activity and tyrosine phosphatase 1B protein content are associated with insulin action and resistance. J Clin Invest 93(3):1156–1162
43. Elchebly M, Payette P, Michaliszyn E, Cromlish W, Collins S, Loy AL et al (1999) Increased insulin sensitivity and obesity resistance in mice lacking the protein tyrosine phosphatase-1B gene. Science 283(5407):1544–1548
44. Klaman LD, Boss O, Peroni OD, Kim JK, Martino JL, Zabolotny JM et al (2000) Increased energy expenditure, decreased adiposity, and tissue-specific insulin sensitivity in protein-tyrosine phosphatase 1B-deficient mice. Mol Cell Biol 20(15):5479–5489
45. Zabolotny JM, Bence-Hanulec KK, Stricker-Krongrad A, Haj F, Wang Y, Minokoshi Y et al (2002) PTP1B regulates leptin signal transduction in vivo. Dev Cell 2(4):489–495
46. Bence KK, Delibegovic M, Xue B, Gorgun CZ, Hotamisligil GS, Neel BG et al (2006) Neuronal PTP1B regulates body weight, adiposity and leptin action. Nat Med 12(8):917–924
47. Banno R, Zimmer D, De Jonghe BC, Atienza M, Rak K, Yang W et al (2010) PTP1B and SHP2 in POMC neurons reciprocally regulate energy balance in mice. J Clin Invest 120(3):720–734
48. De Jonghe BC, Hayes MR, Kanoski SE, Zimmer DJ, Grill HJ, Bence KK (2012) Food intake reductions and increases in energetic responses by hindbrain leptin and melanotan II are

enhanced in mice with POMC-specific PTP1B deficiency. Am J Physiol Endocrinol Metab 303(5):E644–E651
49. Tsou RC, Zimmer DJ, De Jonghe BC, Bence KK (2012) Deficiency of PTP1B in leptin receptor-expressing neurons leads to decreased body weight and adiposity in mice. Endocrinology 153(9):4227–4237
50. Tsou RC, Bence KK (2012) The genetics of PTPN1 and obesity: insights from mouse models of tissue-specific PTP1B deficiency. J Obes 2012:926857
51. Delibegovic M, Bence KK, Mody N, Hong EG, Ko HJ, Kim JK et al (2007) Improved glucose homeostasis in mice with muscle-specific deletion of protein-tyrosine phosphatase 1B. Mol Cell Biol 27(21):7727–7734
52. Zabolotny JM, Haj FG, Kim YB, Kim HJ, Shulman GI, Kim JK et al (2004) Transgenic overexpression of protein-tyrosine phosphatase 1B in muscle causes insulin resistance, but overexpression with leukocyte antigen-related phosphatase does not additively impair insulin action. J Biol Chem 279(23):24844–24851
53. Haj FG, Zabolotny JM, Kim YB, Kahn BB, Neel BG (2005) Liver-specific protein-tyrosine phosphatase 1B (PTP1B) re-expression alters glucose homeostasis of PTP1B–/– mice. J Biol Chem 280(15):15038–15046
54. Dube N, Tremblay ML (2005) Involvement of the small protein tyrosine phosphatases TC-PTP and PTP1B in signal transduction and diseases: from diabetes, obesity to cell cycle, and cancer. Biochim Biophys Acta 1754(1–2):108–117
55. Zabolotny JM, Kim YB, Welsh LA, Kershaw EE, Neel BG, Kahn BB (2008) Protein-tyrosine phosphatase 1B expression is induced by inflammation in vivo. J Biol Chem 283(21): 14230–14241
56. Parvaneh L, Meshkani R, Bakhtiyari S, Mohammadtaghvaie N, Gorganifiruzjaee S, Taheripak G et al (2010) Palmitate and inflammatory state additively induce the expression of PTP1B in muscle cells. Biochem Biophys Res Commun 396(2):467–471
57. Kammoun HL, Chabanon H, Hainault I, Luquet S, Magnan C, Koike T et al (2009) GRP78 expression inhibits insulin and ER stress-induced SREBP-1c activation and reduces hepatic steatosis in mice. J Clin Invest 119(5):1201–1215
58. Winnay JN, Boucher J, Mori MA, Ueki K, Kahn CR (2010) A regulatory subunit of phosphoinositide 3-kinase increases the nuclear accumulation of X-box-binding protein-1 to modulate the unfolded protein response. Nat Med 16(4):438–445
59. Wang N, Zhang D, Mao X, Zou F, Jin H, Ouyang J (2009) Astragalus polysaccharides decreased the expression of PTP1B through relieving ER stress induced activation of ATF6 in a rat model of type 2 diabetes. Mol Cell Endocrinol 307(1–2):89–98
60. Mao XQ, Yu F, Wang N, Wu Y, Zou F, Wu K et al (2009) Hypoglycemic effect of polysaccharide enriched extract of Astragalus membranaceus in diet induced insulin resistant C57BL/6J mice and its potential mechanism. Phytomedicine 16(5):416–425
61. Lane MD, Ronnett G, Slieker LJ, Kohanski RA, Olson TL (1985) Post-translational processing and activation of insulin and EGF proreceptors. Biochimie 67(10–11):1069–1080
62. Zhou L, Zhang J, Fang Q, Liu M, Liu X, Jia W et al (2009) Autophagy-mediated insulin receptor down-regulation contributes to endoplasmic reticulum stress-induced insulin resistance. Mol Pharmacol 76(3):596–603
63. Park SW, Zhou Y, Lee J, Lu A, Sun C, Chung J et al (2010) The regulatory subunits of PI3K, p85alpha and p85beta, interact with XBP-1 and increase its nuclear translocation. Nat Med 16(4):429–437
64. Zhou Y, Lee J, Reno CM, Sun C, Park SW, Chung J et al (2011) Regulation of glucose homeostasis through a XBP-1-FoxO1 interaction. Nat Med 17(3):356–365
65. Bettaieb A, Liu S, Xi Y, Nagata N, Matsuo K, Matsuo I et al (2011) Differential regulation of endoplasmic reticulum stress by protein tyrosine phosphatase 1B and T cell protein tyrosine phosphatase. J Biol Chem 286(11):9225–9235
66. Bettaieb A, Matsuo K, Matsuo I, Wang S, Melhem R, Koromilas AE et al (2012) Protein tyrosine phosphatase 1B deficiency potentiates PERK/eIF2alpha signaling in brown adipocytes. PLoS One 7(4):e34412

67. Krishnan N, Fu C, Pappin DJ, Tonks NK (2011) H2S-Induced sulfhydration of the phosphatase PTP1B and its role in the endoplasmic reticulum stress response. Sci Signal 4(203):ra86
68. Tiganis T (2013) PTP1B and TCPTP—nonredundant phosphatases in insulin signaling and glucose homeostasis. FEBS J 280(2):445–458
69. Naidoo N, Ferber M, Master M, Zhu Y, Pack AI (2008) Aging impairs the unfolded protein response to sleep deprivation and leads to proapoptotic signaling. J Neurosci 28(26): 6539–6548
70. Naidoo N (2009) The endoplasmic reticulum stress response and aging. Rev Neurosci 20(1):23–37
71. Paz Gavilan M, Vela J, Castano A, Ramos B, del Rio JC, Vitorica J et al (2006) Cellular environment facilitates protein accumulation in aged rat hippocampus. Neurobiol Aging 27(7):973–982
72. Hussain SG, Ramaiah KV (2007) Reduced eIF2alpha phosphorylation and increased proapoptotic proteins in aging. Biochem Biophys Res Commun 355(2):365–370
73. Doyle KM, Kennedy D, Gorman AM, Gupta S, Healy SJ, Samali A (2011) Unfolded proteins and endoplasmic reticulum stress in neurodegenerative disorders. J Cell Mol Med 15(10):2025–2039
74. Kriss CL, Pinilla-Ibarz JA, Mailloux AW, Powers JJ, Tang CH, Kang CW et al (2012) Overexpression of TCL1 activates the endoplasmic reticulum stress response: a novel mechanism of leukemic progression in mice. Blood 120(5):1027–1038
75. Agouni A, Owen C, Czopek A, Mody N, Delibegovic M (2010) In vivo differential effects of fasting, re-feeding, insulin and insulin stimulation time course on insulin signaling pathway components in peripheral tissues. Biochem Biophys Res Commun 401(1):104–111

Chapter 6
Role of Protein Tyrosine Phosphatase 1B in Hepatocyte-Specific Insulin and Growth Factor Signaling

Águeda González-Rodríguez and Ángela M. Valverde

Abstract Insulin resistance in major insulin target tissues such as liver, adipose tissues, and skeletal muscle is a key pathogenic feature of type 2 diabetes mellitus (T2DM). Among them, the liver plays a major role in controlling blood glucose homeostasis through the balance between glucose utilization and storage (Int J Biochem Cell Biol 36:753–758, 2004; Diabetologia 44:983–991, 2001). The molecular mechanism underlying hepatic insulin resistance is not completely understood; however, it is believed to involve impairment of the insulin receptor (IR) signaling network. A number of epidemiologic and clinical studies have shown a close association between nonalcoholic fatty liver disease (NAFLD) and chronic hepatitis C virus (HCV) infection and insulin resistance (N Engl J Med 346:1221–1231, 2002; Diabetologia 48:634–642, 2005; J Hepatol 44:253–261, 2006; N Engl J Med 345:41–52, 2001). Therefore, one of the challenges facing researchers in the field is the selection of therapeutic targets against hepatic insulin resistance among components of the insulin signaling cascade. In this chapter, we will focus on the regulation of hepatic insulin signaling by protein tyrosine phosphatase 1B (PTP1B) since many studies during the last decade have revealed that PTP1B is a critical node of the insulin signaling cascade and, therefore, its inhibition is of promise for alleviating insulin resistance and T2DM. On the other hand, since receptor tyrosine kinase (RTK)-mediated signaling is also negatively modulated by PTP1B, particularly in the liver, we will also review the impact of the inhibition of PTP1B in the context of cell proliferation and survival.

Á. González-Rodríguez, Ph.D. • Á.M. Valverde, Ph.D. (✉)
Institute of Biomedicine Alberto Sols (CSIC/UAM), Consejo Superior de Investigaciones Científicas, Centro de Investigación Biomédica en Red de Diabetes y Enfermedades, Metabólicas Asociadas (CIBERDEM, ISCIII), Madrid 28029, Spain
e-mail: avalverde@iib.uam.es

K.K. Bence (ed.), *Protein Tyrosine Phosphatase Control of Metabolism*,
DOI 10.1007/978-1-4614-7855-3_6, © Springer Science+Business Media New York 2013

Abbreviations

2′5′OAS	2′5′ oligoadenylate synthase
bFGF	Basic fibroblast growth factor
CREB	cAMP response element-binding protein
ER	Endoplasmic reticulum
GH	Growth hormone
GSK	Glycogen synthase kinase
HCC	Hepatocellular carcinoma
HCSs	Stellate cells
HCV	Hepatitis C virus
HGP	Hepatic glucose production
HIF	Hypoxia-inducible factor
HOMA-IR	Homeostasis model assessment of insulin resistance
IFNα	Interferon α
IL6	Interleukin 6
IR	Insulin receptor
IRS2	Insulin receptor substrate 2
JAK2	Janus kinase 2
Mx	Myxovirus-resistance
NAFLD	Nonalcoholic fatty liver disease
NOX	NADPH oxidase
PCNA	Proliferating cell nuclear antigen
PDK-1	Phosphoinositide-dependent protein kinase-1
PH	Partial hepatectomy
PI	Phosphatidylinositol
PKC	Protein kinase C
PP2A	Protein phosphatase 2A
PTP1B	Protein tyrosine phosphatase 1B
RTK	Receptor tyrosine kinase
SH2	Src homology 2
Sirt	Sirtuin
SOCS	Suppressor of cytokine signaling
SREBP	Sterol regulatory element-binding protein
STAT3	Signal transducer and activator of transcription 3
T2DM	Type 2 diabetes mellitus
TCPTP	T-cell protein tyrosine phosphatase
TGF	Transforming growth factor
TNFα	Tumor necrosis factor α

Dysregulation of Hepatic PTP1B Expression in Metabolic Diseases

Increased expression of protein tyrosine phosphatase 1B (PTP1B) in peripheral tissues can attenuate the insulin signal and contribute to insulin resistance in rodents and humans. Hepatic PTP1B expression and/or activity are widely modulated in response to metabolic stressors in insulin-resistant states and, particularly, in obesity. Related to insulin resistance and Type 2 diabetes mellitus (T2DM), mRNA, protein levels, and enzymatic activity of PTP1B are increased in the livers from diabetic insulin receptor substrate 2 (IRS2)-deficient mice [37]. At the molecular level, in vitro experiments in cultured hepatocytes in high glucose medium have shown enhanced transactivation of PTP1B gene and its expression through Sp1 activation by protein kinase C (PKC) [44], suggesting that this up-regulated expression may contribute to glucose toxicity in diabetes. Likewise, it has been shown that hepatic expression of PTP1B is increased in fructose-fed models of insulin resistance in hamster [15] and rats [54]. Regarding obesity, elevated PTP1B expression in the liver induced by high-fat diet coincides with increased levels of tumor necrosis factor α (TNFα) and CD68, two markers of hepatic inflammation associated with steatosis [93], as well as with induction of endoplasmic reticulum (ER)-stress response signaling [1]. Again, in vitro experiments have demonstrated that TNFα treatment itself induces an increase in PTP1B expression [93] and enzymatic activity [32] in hepatocyte cell lines. Likewise, ER stressors directly increased PTP1B mRNA and protein levels in HepG2 cells [1]. During aging-induced obesity, another situation that concurs with metabolic damage, increased mRNA levels, protein content, and enzymatic activity of PTP1B in both liver and skeletal muscle paralleled adipose tissue inflammation [36]. Regarding non-parenchymal hepatic cells, in stellate cells (HCSs) leptin signals via Janus kinase 2 (JAK2)/signal transducer and activator of transcription 3 (STAT3) pathway producing effects that enhance extracellular matrix deposition. On the other hand, adiponectin-mediated signaling induced PTP1B expression and activity via receptor AdipoR1, but not AdipoR2 in HCSs cells [42]. In this cellular model, adiponectin inhibits leptin signaling at least partially by enhancing PTP1B activity, thus promoting JAK2 desphosphorylation and preventing leptin receptor (Ob-Rb) activation. Such a mechanism would represent a novel molecular link accounting for the hepatoprotective effect of adiponectin in the context of hepatic fibrosis. Of note, in contrast to the metabolic effects of adiponectin in hepatocytes, its effects on PTP1B expression described in this study are cell-specific, occurring only in activated HSCs, which are the primary mediator of extracellular matrix metabolism and are present normally only during hepatic wound-healing.

PTP1B Is a Critical Negative Modulator of Hepatic Insulin Signaling

Given the importance of insulin signaling in maintaining hepatic glucose homeostasis, it must be regulated carefully [48, 70]. The cascade begins when activation of the IR by autophosphorylation on several tyrosine residues leads to the recruitment and tyrosine phosphorylation of IRS proteins 1 and 2 [74, 88]. Then, phosphorylated IRSs bind proteins containing Src homology 2 (SH2) domains such as phosphatidylinositol (PI) 3-kinase [6], which has a central role in the metabolic actions elicited by insulin. Over the last years the role of insulin signaling pathways in the liver has also been studied in mice lacking hepatic IR (LIRKO) [28, 58], IRS2 [22, 50, 84, 85, 89, 90], phosphoinositide-dependent protein kinase-1 (PDK-1) [62], protein kinase B (PKB/Akt) [19], and also in mice over-expressing a dominant negative mutant of PI 3-kinase [60]. Various lines of genetic evidence support the concept that insulin negatively regulates hepatic glucose production (HGP) through a main pathway involving IR, IRS2, PI 3-kinase, and Akt.

On the other hand, PTP1B is a major negative regulator of insulin and leptin sensitivity, acting to dephosphorylate the IR and JAK2 [9, 24]. PTP1B may also dephosphorylate more distal components of these signaling pathways, such as IRS1 [33]. In humans, *PTPN1* polymorphisms are associated with insulin resistance, obesity, and other characteristics of metabolic syndrome in some populations [8, 18, 46, 83].

The importance of PTP1B in hepatic insulin signaling emerged from the initial in vitro studies performed in rat KRC-7 hepatoma cells by Ahmad et al. [2] who showed that osmotic loading of PTP1B antibodies decreased insulin-induced IRS1 tyrosine phosphorylation and PI 3-kinase activity. Moreover, over-expression of PTP1B in Fao hepatoma cells impairs insulin-stimulated glucose metabolism [25]. Conversely, reduction of PTP1B with a specific PTP1B antisense oligonucleotide (PTP1B ASO) in these cells increased insulin signaling [16].

The generation of PTP1B-deficient (PTP1B$^{-/-}$) mice by different strategies in two independent laboratories revealed a tissue-specific increase in insulin sensitivity at 10–14 week of age owing to enhanced phosphorylation of IR in liver and skeletal muscle, resistance to weight gain on high-fat diet, and increased basal metabolic rate [27, 47]. The key role of the liver as a target tissue of PTP1B was demonstrated by the specific recovery of PTP1B expression in the liver of PTP1B$^{-/-}$ mice that led to a marked attenuation of their enhanced insulin sensitivity [41]. By contrast, liver-specific deletion of PTP1B improves glucose tolerance and metabolic syndrome and attenuates diet-induced ER stress [1, 7, 20]. In the light of these data, hepatocytes from adult PTP1B$^{-/-}$ mice have increased insulin sensitivity that specifically affects the insulin-mediated Akt/Foxo1 signaling pathway that controls the inhibition of gluconeogenic genes *Pck1* and *G6pc* [34].

Unexpectedly, PTP1B-deficient neonatal hepatocytes (3–5-day-old) did not display increased sensitivity to insulin in the early steps of insulin signaling. Conversely, these cells showed a prolonged insulin signaling and, as a consequence, prolonged inhibitory effect of insulin on *Pck1* and *G6pc* mRNAs. In the light of these data, reduction of PTP1B by two different siRNA oligos in wild-type neonatal hepatocytes prolonged insulin signaling similar to the observations in PTP1B-deficient cell lines. These data suggest that in the liver of PTP1B$^{-/-}$ mice there is a switch from prolonged insulin action to enhanced insulin sensitivity in a postnatal-developmental manner [34]. Besides the lack of enhancement of insulin signaling in PTP1B$^{-/-}$ neonatal hepatocytes, these cells showed increased basal glucose uptake as compared to hepatocytes from adult PTP1B$^{-/-}$ mice [38]. This occurs without changes in hexokinase, glucokinase, and glucose 6-phosphatase enzymatic activities. Interestingly, the glucose transporter GLUT2 was up-regulated at the protein level in neonatal hepatocytes and livers from PTP1B$^{-/-}$ neonates. These results were accompanied by a significant increase in the net free intrahepatic glucose levels in the livers of PTP1B$^{-/-}$ neonates indicating positive correlation of the in vitro and in vivo data. The excess of glucose that reaches the liver in the absence of PTP1B is not metabolized through glycolysis or glycogen synthesis since glycogen, lactate, and pyruvate levels are decreased as compared to the wild-type controls [29]. These results suggest that alternative pathways that metabolize glucose are increased in the absence of PTP1B at suckling. In this regard, glucose-6-phosphate dehydrogenase activity, a rate-limiting enzyme of the pentose phosphate cycle, is significantly augmented in the liver of PTP1B$^{-/-}$ suckling mice as compared with the wild-type controls. These data indicate an enhanced activation of the pentose phosphate cycle with the consequent increase in the generation of ribose phosphates that are required for DNA synthesis and cell proliferation, as reported in fetal hepatocytes from rats [61]. Therefore, the physiological relevance of increased glucose uptake and the net free intrahepatic glucose in the liver of PTP1B$^{-/-}$ mice during suckling is correlated with an increase in liver growth which will be discussed below in the section on cellular proliferation.

PTP1B Deficiency in Mouse Models of Insulin Resistance Recovers Hepatic Insulin Signaling

Since inhibition of PTP1B has been demonstrated to enhance PI 3-kinase/Akt-mediated signaling in the liver [27, 47], recent studies have investigated whether deletion of PTP1B could restore insulin signaling and hepatic insulin sensitivity in different models of insulin resistance in mice. The first in vivo strategy was the reduction of PTP1B protein levels by 60 % in the ob/ob mice by using the PTP1B ASO [39, 73, 95]. In the liver of PTP1B ASO-treated mice insulin signaling proteins IRS1 and IRS2 and the PI3-kinase regulatory subunit p50α were increased and, conversely, the expression of the p85α regulatory subunit was decreased. These changes in protein expression correlated with increased insulin-stimulated Akt phosphorylation. PTP1B ASO down-regulated gluconeogenic genes (*Pck1* and *Fbp1*) and also decreased TNFα protein levels and phosphorylation of the transcription factor cAMP response element-binding protein (CREB). These initial findings suggested that PTP1B inhibition in the liver may have clinical benefit in T2DM.

Mice heterozygous for IR and IRS1 deficiency (DHet) provide a model of polygenic T2DM in which early-onset, genetically programmed insulin resistance leads to diabetes, similar to the human disease [10]. Absence of PTP1B in DHet mice markedly improved glucose tolerance and insulin sensitivity at 10–11 weeks of age and reduced the incidence of diabetes and hyperplastic pancreatic islets at 6 months of age [91]. In the liver, insulin-stimulated tyrosine phosphorylation of IR, IRS1, and IRS2 and serine phosphorylation of Akt, glycogen synthase kinase (GSK) 3β, and S6K1 were impaired in DHet and recovered by PTP1B deficiency. In addition, increased *Pck1* in DHet mouse liver was attenuated by PTP1B deficiency. Thus, this model has revealed that even in the setting of high genetic risk for diabetes, reducing PTP1B is partially protective, further demonstrating its attractiveness as a target for prevention and treatment of T2DM.

Given that the phenotype of PTP1B-deficient mice contrasts with that of mice lacking IRS2 ($IRS2^{-/-}$), which develop progressive deterioration of glucose homeostasis because of insulin resistance in the liver and a lack of beta-cell compensation [90], it has been investigated whether deletion of PTP1B expression by genetic or pharmacological approaches could restore hepatic sensitivity to insulin and beta-cell function in systemic $IRS2^{-/-}$ mice that displayed fasting hyperglycemia (>300 mg/dL) at 12–16 weeks of age [37, 51]. Two independent studies have demonstrated that PTP1B disruption in systemic $IRS2^{-/-}$ mice ($IRS2^{-/-}/PTP1B^{-/-}$) normalized peripheral insulin sensitivity and glucose tolerance [37, 51]. In the liver, Foxo1 is the molecular link between Akt phosphorylation and gluconeogenic gene expression [37, 65, 69]. The lack of effect of insulin to phosphorylate Foxo1 through activation of PI (3,4,5) tris phosphate (PIP3)/Akt-mediated signaling in mice lacking $IRS2^{-/-}$ is reversed by PTP1B deletion in the liver of double $IRS2^{-/-}/PTP1B^{-/-}$ mice [37]. This is a critical finding since in this organ both tyrosine phosphorylated IRS1 and IRS2 can efficiently activate the PI 3-kinase pathway to mediate the physiological responses to insulin in regulating carbohydrate and lipid metabolism, as has been reported recently in mice with liver-specific deletions of both IRS1 and IRS2 [23, 37, 49]. However, the initial studies in livers of systemic $IRS2^{-/-}$ mice had already demonstrated the inability of IRS1 to activate PI 3-kinase in response to insulin in the absence of IRS2 [84, 90]. Remarkably, as stated above, mRNA, protein levels, and the phosphatase activity of PTP1B were elevated in the livers of systemic $IRS2^{-/-}$ mice as compared to wild-type mice [37]. More importantly, up-regulation of PTP1B in liver of $IRS2^{-/-}$ mice enhanced its association with the IR and, consequently, the increased dephosphorylation of IR is likely to be responsible for the inability of IRS1 to mediate insulin signaling in the liver of diabetic $IRS2^{-/-}$ mice. Therefore, these studies have unravelled the molecular mechanism (depicted in Fig. 6.1) by which IRS1 was unable to compensate IRS2 deficiency in diabetic mice. The absence of PTP1B in $IRS2^{-/-}/PTP1B^{-/-}$ mice promoted insulin sensitivity in the liver by enabling the activation of an IR/IRS1-mediated compensatory mechanism, culminating in the inhibition of *Pck1* and *G6pc* in hepatocytes [37].

Very recently, the role of PTP1B in the development of aging-associated obesity, inflammation, and peripheral insulin resistance has been evaluated by assessing metabolic parameters and insulin signaling at 3 and 16 months in wild-type and PTP1B-deficient mice [36]. Whereas fat mass and adipocyte size were increased in wild-type control mice at 16 months, these parameters did not change with aging in $PTP1B^{-/-}$ mice. Increased levels of pro-inflammatory cytokines, crown-like structures, and hypoxia-inducible factor (HIF)-1α were observed only in adipose tissue from 16-month-old wild-type mice. In the liver, hepatic steatosis and increased lipogenic gene expression that occurred with aging-associated obesity were not observed in $PTP1B^{-/-}$ mice. Similarly, islet hyperplasia and hyperinsulinemia were observed

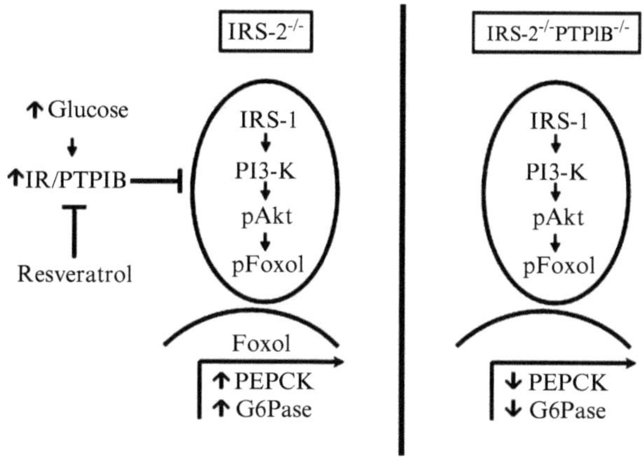

Fig. 6.1 Schematic representation of the mechanism by which inhibition of PTP1B promotes insulin sensitivity in the liver of IRS2-deficient mice. The lack of PTP1B promotes insulin sensitivity in the liver of IRS2$^{-/-}$ mice through the restoration of IRS1-mediated Akt/Foxo1 phosphorylation and the inhibition of gluconeogenic enzymes. This molecular mechanism can be mimicked by resveratrol through down-regulation of PTP1B expression and activity in the liver of IRS2$^{-/-}$ mice

in wild-type mice with aging-associated obesity, but not in PTP1B$^{-/-}$ animals. Leanness in 16-month-old PTP1B$^{-/-}$ mice was associated with increased energy expenditure. Whole body insulin sensitivity decreased in 16-month-old wild-type mice; however, hyperinsulinemic-euglycemic clamp studies revealed that PTP1B deficiency prevented this obesity-related decreased peripheral insulin sensitivity. At a molecular level, PTP1B expression and enzymatic activity were up-regulated in liver and muscle of 16-month-old wild-type mice as were the activation of stress kinases JNK and p38 MAPK and the expression of the senescence marker p53. As a result, IR-mediated Akt/Foxo1 signaling was attenuated in both tissues of aged control mice. In the liver, insulin was unable to decrease *Pck1* and *G6pc* in primary hepatocytes from 16-month-old wild-type mice. Activation of stress kinases and increased p53 expression were not observed in liver and muscle of PTP1B$^{-/-}$ mice at 16 months that preserved insulin-mediated signaling and the inhibition of gluconeogenic genes as in the young stage. Collectively, these data implicate a role of PTP1B in the development of chronic low grade inflammation and insulin resistance associated with obesity during aging (Fig. 6.2).

Insulin resistance is a risk factor for nonresponse to interferon (IFN)/ribavirin therapy in patients with chronic hepatitis C [72]. In a recent study, the role of PTP1B in the interferon resistance associated with insulin resistance was evaluated in HepG2 cells and obese mice [32]. Silencing IRS2 mRNA in HepG2 hepatic cells induced insulin resistance measured by elevated PTP1B expression and enzymatic activity and by the inability of insulin to activate Akt. This insulin-resistant state inhibited the effect of interferon α (IFNα) in inducing STAT1 tyrosine phosphorylation and 2′5′ oligoadenylate synthase (2′5′OAS) and myxovirus-resistance (Mx) gene expression. Likewise, TNFα-induced insulin resistance suppressed the IFNα

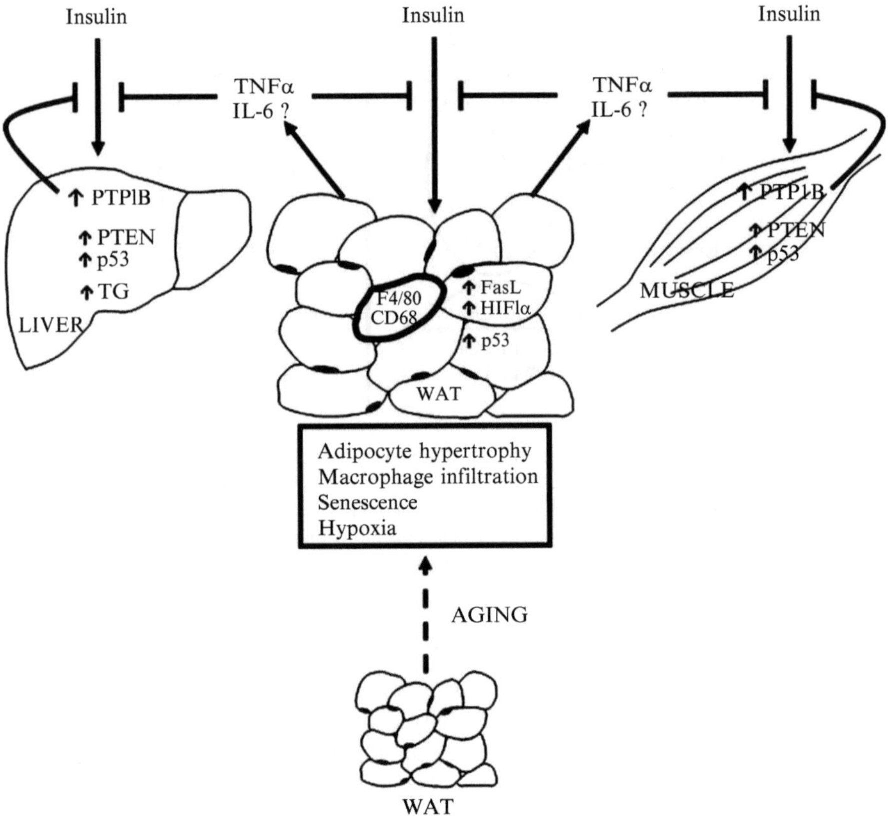

Fig. 6.2 Schematic representation of the role of PTP1B in age-related obesity, inflammation, and insulin resistance. During aging, increased adiposity leads to inflammation, hypoxia as well as an increase in p53, a senescence marker in WAT. As a result, circulating pro-inflammatory cytokines reach the liver and muscle increasing PTP1B expression. This, in turn, negatively interferes with insulin signaling leading to peripheral insulin resistance. PTP1B deficiency protects against adiposity with aging preventing the alterations in the endocrine milieu of WAT and insulin resistance in peripheral tissues

response. Treatment of HepG2 cells with pervanadate and, more interestingly, knocking down PTP1B specifically restored insulin and IFNα response. Alternatively, treatment of insulin-resistant HepG2 cells with metformin inhibited PTP1B activity and improved IFNα response. The beneficial effect of PTP1B inhibition in the response to IFNα was also evaluated in insulin-resistant *ob/ob* mice with increased PTP1B gene expression and activity in the liver. These mice did not respond to IFNα administration, but in agreement with the in vitro experiments, treatment with metformin decreased PTP1B and restored the response to IFNα. These data highlight the use of drugs that lead to PTP1B inhibition among other targets such as metformin, to improve the response of chronic hepatitis C patients to combined therapy with IFNα/ribavirin.

Role of PTP1B in Hepatic Lipogenesis

The insulin resistance/metabolic syndrome is characterized by the variable coexistence of hyperinsulinemia, hyperlipidemia, obesity, and hypertension. Many individuals with obesity and insulin resistance suffer from hepatic steatosis, or "fatty liver." Several lines of evidence indicate that fatty liver in insulin-resistant states is caused by activation of the sterol regulatory element-binding protein (SREBP)-1c, which is elevated in response to high insulin levels [43, 77]. Although insulin resistance is present in peripheral tissues, compensatory hyperinsulinemia is postulated to persistently activate SREBP-1c transcription and cleavage, resulting in increased lipogenic gene expression, enhanced fatty acid synthesis, and accelerated triglyceride accumulation in the liver in obese insulin-resistant mice [79].

Shimizu et al. [78] have found a strong correlation between SREBP-1c and PTP1B expression in insulin-resistant rats fed a high fructose diet. These results were confirmed by over-expression of PTP1B in rat hepatocytes which led to increased mRNA content and promoter activity of SREBP-1a and SREBP-1c, resulting in increased mRNA expression of *Fasn* (fatty acid synthase). Moreover, inhibition of protein phosphatase 2A (PP2A) activity normalized the PTP1B-enhanced SREBP-1a and SREBP-1c expression. Thus, the proposed molecular mechanism involves the activation of PP2A by tyrosine 307 dephosphorylation by PTP1B resulting in enhanced activity of the SREBP-1 promoter by modulating its transcriptional activity via Sp1 sites. These sites are located at about the −90 bp region that is thought to be important in the regulation of SREBP-1 gene expression. On the basis of this study, PTP1B represents a novel therapeutic target for the amelioration of postprandial (diet-induced) hypertriglyceridemia through a serial activation of the PTP1B-PP2A axis. In the light of these data, it has been demonstrated that treatment of *ob/ob* mice with PTP1B ASO results in a decrease in genes involved in hepatic lipogenesis, including SREBP-1c [87]. Furthermore, rats fed a high fructose diet developed insulin resistance coincident with increased PTP1B and SREBP-1c gene expression in the liver [54]. Consistent with the PTP1B ASO study, hepatic SREBP-1a and SREBP-1c gene expression levels were significantly lower in Alb-Cre-PTP1B$^{-/-}$ mice than in wild-type controls, which is counterintuitive to what would be expected in the context of enhanced insulin sensitivity seen in these mice. This was accompanied with decreased triglyceride and cholesterol levels together with diminished lipogenic gene expression [20].

PTP1B Is a Drug Target Against Insulin Resistance

Many treatments are available for T2D, including sulfonylureas, thiazolidinediones, biguanides, and α-glycosidase inhibitors. These current treatment options are often effective over the short and medium term, but do not alter the underlying progression of the disease. In addition, weight gain, hypoglycemia, gastrointestinal events, and peripheral edema are important adverse events associated with these agents which limit compliance. The need for new treatments is widely recognized and the search is on [17]. In this regard, pharmacological agents which are capable of inhibiting PTP1B are expected to potentiate the action of insulin and leptin and, therefore, be beneficial in the treatment of both T2DM and obesity.

Several groups have earlier demonstrated that by employing the knowledge derived from both structural and medicinal chemistry, the use of PTP1B inhibitors with the requisite potency and selectivity is now a viable option to be considered. This notion is supported by a recent study showing the beneficial effects of the dietary supplement curcumin in the prevention of hepatic steatosis [54]. This compound reduced PTP1B expression in the liver of fructose-fed rats and, as a result, tyrosine phosphorylation of IR and IRS1 and activation of Akt were enhanced. Moreover, curcumin down-regulated suppressor of cytokine signaling (SOCS) 3 expression, exhibiting improvement of hepatic leptin signaling transduction and reduction of liver VLDL overproduction and triglyceride hypersynthesis in the leptin-resistant state of fructose-fed rats. In light of these data, treatment of HFD-fed diabetic rats with flavonoids of *Litsea coreana* results in a significant increase in insulin sensitivity, marked decrease in body weight, serum-free fatty acids, and triglyceride content together with decreased PTP1B expression [55]. Recently, a novel proteoglycan PTP1B inhibitor, named Fudan-Yueyang-Ganoderma lucidum (FYGL), from *Ganoderma lucidum* was orally administered to diabetic *db/db* mice for 4 weeks [86]. This treatment decreased fasting plasma glucose level, serum insulin concentration, and the homeostasis model assessment of insulin resistance (HOMA-IR). Pharmacology research suggests that FYGL decreases the plasma glucose level by inhibiting PTP1B expression and activity, consequently regulating tyrosine phosphorylation of the IRβ-subunit and the level of hepatic glycogen, resulting in improved insulin sensitivity. Therefore, FYGL is also promising as an insulin sensitizer for the therapy of T2DM and accompanied dyslipidemias.

Resveratrol, a plant-derived polyphenol that is a potent activator of sirtuin (SIRT)1, has been reported to repress PTP1B expression and attenuate high-fat diet-induced insulin resistance in the liver [81]. Resveratrol also exerts a protective effect against hepatic insulin resistance in $IRS2^{-/-}$ mice by reducing both PTP1B expression and phosphatase activity. Consequently, the ability of insulin to stimulate IRS1-mediated PI 3-kinase signaling is restored to normal in IRS2-deficient mice treated with resveratrol. Hepatocytes from resveratrol-treated $IRS2^{-/-}$ mice respond to insulin by inhibiting *Pck1* and *G6pc* genes. These data are consistent with those obtained by the genetic ablation of PTB1B in the IRS2-deficient model and suggest that resveratrol is a suitable candidate for treatment of hepatic insulin resistance [37] (Fig. 6.1).

Regarding synthetic PTP1B inhibitors, the effects of compound CCF06240 on insulin sensitivity and lipid metabolic abnormalities in vivo and in vitro have been evaluated [57]. In lipid accumulated HepG2 cells, CCF06240 was found to reverse the increased PTP1B activity, enhance the insulin-induced tyrosine phosphorylation in insulin signaling pathway, attenuate free fatty acid-insulin-induced cellular lipid accumulation, and down-regulate the expression of genes related to fatty acid synthesis. In high-fat diet-fed mice, oral administration of CCF06240 reduced body weight and the levels of serum triglycerides.

Although the development of specific PTP1B inhibitors is of potential clinical interest in therapies against insulin resistance and T2DM, the success of targeting this phosphatase has been limited. Poor oral availability, as well as lack of specificity without simultaneous inhibition of the closely related T-cell protein tyrosine phosphatase (TCPTP), will have to be overcome to generate effective and specific PTP1B inhibitors for therapeutic use.

Role of PTP1B in Hepatocyte Proliferation and Liver Growth

PTP1B is also a key regulator of cell growth through its ability to dephosphorylate receptors of the tyrosine kinase superfamily (receptor tyrosine kinase [RTK]) and modulate the duration and intensity of the signals emerging from these activated receptors. It has been shown by genetic and biochemical approaches that PTP1B dephosphorylates and inactivates the EGF receptor (EGFR), the PDGF receptor (PDGFR) [40], the HGF receptor (HGFR/Met) [75], and the IGF-I receptor (IGF-IR) [11], all of which have been implicated in the control of proliferation, survival and/or metabolic functions of hepatocytes, and many other cellular models.

In mammals, the regulation of physiological functions such as growth and glucose metabolism varies in a postnatal-developmental manner. The liver undergoes an accelerated growth during suckling mainly based on cell proliferation [63]. In this regard, glucose utilization through the pentose phosphate pathway is elevated in immature liver in coincidence with suckling [61]. This pathway provides ribose-5-phosphate intermediates required for DNA synthesis, thus supporting mitogenesis. In order to keep the hepatic metabolism tightly adapted to diet, a physiological condition maintaining low insulin action in the liver is established during suckling [45]. At the same time, rapid hepatic growth is essentially driven by growth hormone (GH), whose effects are mediated through the IGF-I/IGF-IR signaling cascade [56]. Thus, it must be emphasized that IGF-I actions constitute important age-related conditions for the proper development of the immature liver. Given that IGF-IR is a PTP1B substrate [11], differences in PTP1B expression might have consequences during the early stages of postnatal development of the liver as compared to adulthood. Interestingly, IRS1 expression, which is critical in mediating the mitogenic pathway of both insulin and IGF-I [5], was up-regulated in hepatocytes and liver extracts from neonatal PTP1B$^{-/-}$ mice as compared to the wild-type littermates in parallel with an increase in the mitogenic capacity. Moreover, the liver-to-body mass ratio, DNA content, and proliferating cell nuclear antigen (PCNA) levels were increased in PTP1B$^{-/-}$ neonates as compared to the corresponding wild-type controls [29]. These effects are specific of the neonatal period since they are absent in adult mice. Enhanced proliferation concurred with increased levels of IR_A isoform in neonatal hepatocytes lacking PTP1B [38]. It is noteworthy that IR_A isoform is mainly expressed in highly proliferating fetal and cancer cells whereas IR_B isoform is expressed mainly in adult insulin-responsive tissues such as the liver to maintain glucose homeostasis [31]. In the light of these data, hepatic STAT5B tyrosine phosphorylation, which is activated by GH, was increased in the livers of

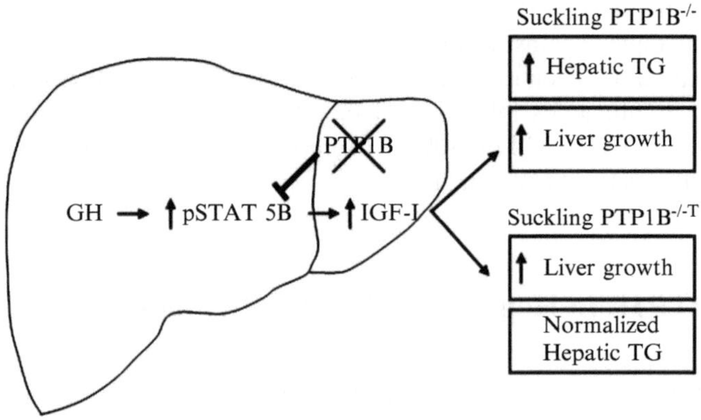

Fig. 6.3 Proposed model for the effects of PTP1B deficiency in the liver of suckling mice. PTP1B deficiency increased STAT5B phosphorylation, IGF-I levels, and IGF-IR tyrosine phosphorylation in the livers of suckling mice leading to increased liver growth. These effects did not change by embryonic gene transfer (PTP1B$^{-/-T}$ means transferred embryos). On the other hand, PTP1B deficiency increased lipid content in lactating females, thereby hepatic TG accumulated in their offspring. These effects were abolished by embryonic transfer

PTP1B$^{-/-}$ suckling mice and, consequently, IGF-I mRNA and protein levels were up-regulated. Moreover, IGF-IR tyrosine phosphorylation was also enhanced in livers of PTP1B$^{-/-}$ suckling mice and this might favor its autocrine/paracrine effects in liver cells, thereby exacerbating liver growth. Since both STAT5B and IGF-IR are PTP1B substrates [4, 11], these results reinforce the unique role of PTP1B in the control of the GH/IGF-I axis during the neonatal stage. It is important to highlight that the effects on liver growth are due to PTP1B deficiency and not females. PTP1B$^{-/-}$ embryos that have been transferred into wild-type females also showed elevated IGF-I expression and enhanced liver growth after birth [29]. Collectively, these studies highlight the role of PTP1B in the control of physiological liver development after birth (Fig. 6.3).

In the adult stage, hepatic regeneration after partial hepatectomy (PH) is a unique process that requires the reentry of quiescent hepatocytes into the cell cycle in order to proliferate and, therefore, restore the original liver mass after surgery [59, 82]. A recent study has demonstrated that PTP1B deficiency accelerates hepatic regeneration in mice after PH [71]. The expression of cell proliferation markers was enhanced in PTP1B$^{-/-}$ regenerating livers in parallel with BrdU incorporation (Fig. 6.4). Among the complexity of the molecular events involved in hepatic regeneration after liver resection, phosphorylation of JNK and STAT3, early triggers of hepatic regeneration in response to elevations in TNFα and interleukin 6 (IL6), was increased in PTP1B$^{-/-}$ regenerating livers and in PTP1B$^{-/-}$ hepatocytes stimulated with these cytokines. As demonstrated previously for JAK2 [92], recombinant PTP1B dephosphorylates tyrosine 185 residue of JNK. In parallel, enhanced EGFR and HGFR-mediated signaling was observed in regenerating livers lacking PTP1B and in PTP1B$^{-/-}$ hepatocytes stimulated with EGF or HGF. PTP1B deficiency also induced additional benefits on hepatic regeneration such as a more rapid increase in intrahepatic lipid accumulation and attenuation of stress-mediated signaling at latter

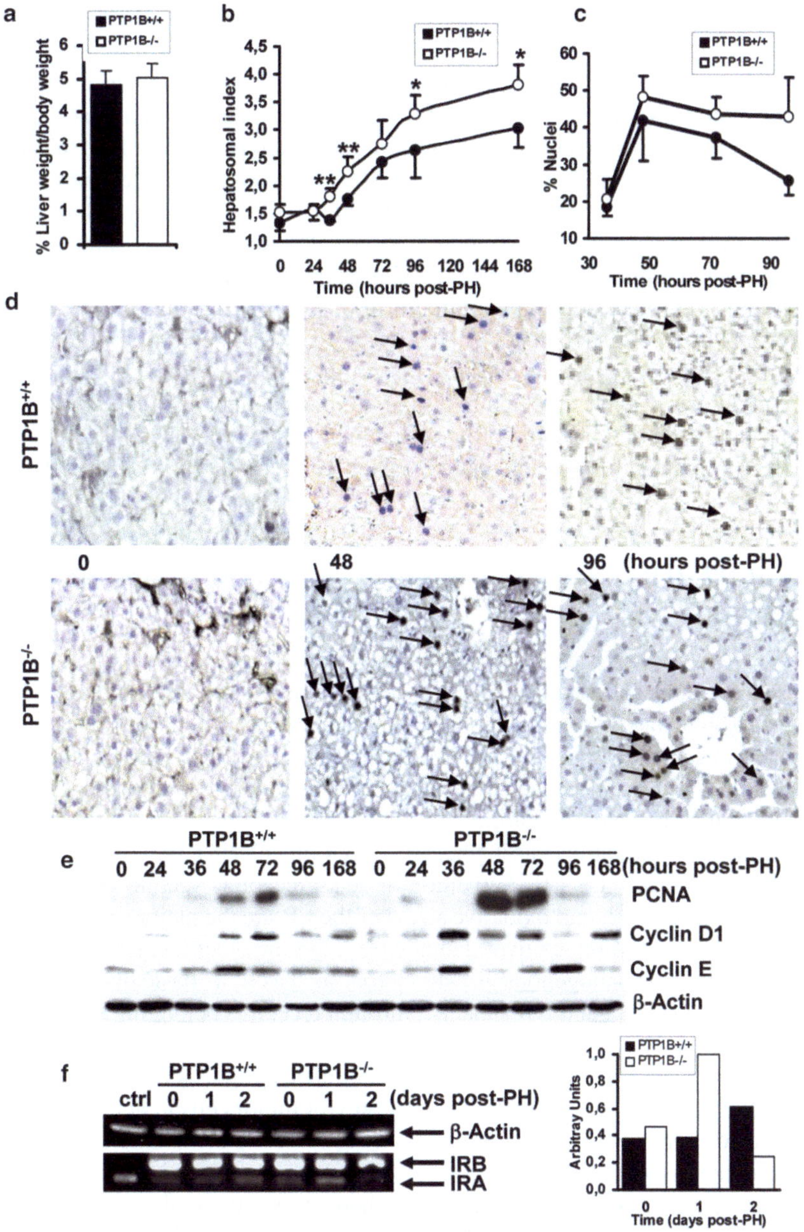

Fig. 6.4 Increased liver/body weight ratio and hepatocyte proliferation after partial hepatectomy (PH) in PTP1B-deficient mice. PH was performed in wild-type and PTP1B$^{-/-}$ mice and regenerating livers were removed at the indicated time-periods. (**a**) Comparison of liver-to-body weight ratio in wild-type (PTP1B$^{+/+}$) and PTP1B$^{-/-}$ mice. (**b**) Liver regeneration index (liver weight/body weight) at the indicated time-periods after PH. (**c**) Quantification of BrdU positive nuclei in regenerating livers from wild-type and PTP1B$^{-/-}$ mice at 24–96 h post-PH. (**d**) Representative microscopy images of BrdU incorporation at 48 and 96 h post-PH. BrdU positive nuclei are indicated by *arrows*. Original magnification: ×10. (**e**) PCNA, cyclin D1, cyclin E levels at the indicated time-periods after PH. (**f**) IR isoforms at 24 and 48 h after PH (reprinted from Am J Pathol, 178, Revuelta-Cervantes, J., Mayoral, R., Miranda, S., Gonzalez-Rodriguez, A., Fernandez, M., Martin-Sanz, P., and Valverde, A.M., Protein Tyrosine Phosphatase 1B (PTP1B) deficiency accelerates hepatic regeneration in mice, 1591–1604, Copyright 2011, with permission from Elsevier)

time-periods. Finally, it was demonstrated that PTP1B deficiency up-regulated the transforming growth factor (TGF)-β inhibitor SnoN and delayed termination of liver regeneration. The most relevant conclusion of this study was the fact that overall process of liver regeneration in the absence of PTP1B was accelerated but not deregulated. Therefore, it was proposed that PTP1B could serve as a novel drug target to improve liver regeneration in patients with acute or chronic liver injury. Of note, despite potential-enhanced oncogenic signaling, PTP1B-deficient mice do not overtly develop tumors during adulthood. In this aspect, a mouse model of cancer combined with PTP1B deficiency will enable us to determine how PTP1B deficiency potentiates or reduces tumorigenesis. This complex issue needs to be addressed. In this regard it has been described that loss of inhibitory effect of PTP1B may contribute to progression and invasion of hepatocellular carcinoma (HCC) through activation of Wnt/β-Catenin signaling and expansion of liver tumor initiating cells [94].

Role of PTP1B in Hepatocyte Survival Against Apoptotic Triggers

Cell survival and cell death are governed by stimulatory and inhibitory signals. Whereas trophic factors simultaneously stimulate mitosis and inhibit cell death, negative growth signals regulate the opposite of these biological effects. Deregulation of apoptosis represents an underlying cause or contributor to many diseases. In the liver, a balance between apoptotic and anti-apoptotic signals must be tightly controlled to prevent alterations during hepatic development and pathological conditions such as liver disease [13, 14]. Activation of death receptors such as Fas and TNFα, mitochondrial damage, and oxidative stress of the ER are the major triggers of apoptosis that ultimately produce liver damage [66]. On the other hand, in the liver, trophic factors include endogenous growth factors such as EGF, basic fibroblast growth factor (bFGF), TGF-β, and IGFs that act through receptors belonging to the tyrosine kinase superfamily [76]. Consistent with this notion, inhibitors of tyrosine kinases and protein tyrosine phosphatases can also modulate apoptosis in the liver. The fact that PTP1B has been related with the sensitivity of tumor cells to apoptosis induced by TNFα [68] prompted several groups to investigate the involvement of this phosphatase in the susceptibility of hepatocytes to undergo apoptosis induced by different stimuli in cellular systems and in models of hepatotoxicity in mice.

In immortalized hepatocytes, PTP1B deficiency concurs with the attenuation of the cellular apoptotic machinery in response to growth factor withdrawal, whereas PTP1B over-expression accelerates cell death [35]. Early activation of caspase-3 occurred in hepatocytes that over-express PTP1B, but was nearly abolished in PTP1B$^{-/-}$ cells. At the molecular level, PTP1B over-expression/deficiency altered the balance of pro-(Bim) and anti-(Bcl-x_L) apoptotic members of the Bcl-2 family upon serum withdrawal. Likewise, cytosolic cytochrome C increased rapidly in hepatocytes with increased PTP1B expression whereas it was retained in the mitochondria of PTP1B$^{-/-}$ cells. Similar modulation was observed in the analysis of DNA laddering and hypodiploid cells. Regarding signaling mediators of cell death and survival signals, transcription factors of the Foxo family have been implicated in the regulation of several pro-apoptotic genes, including mitochondria-associated proteins such as Bim and members of the death receptor pathway such as FasL [21, 80]. Growth factor stimulation induces phosphorylation of Foxo1 by Akt, thereby precluding its entry into the nucleus [64]. In serum-deprived hepatocytes with PTP1B over-expression, a rapid entry of Foxo1 into the nucleus and an earlier activation of caspase-8 were observed. However, both events were suppressed in PTP1B$^{-/-}$ hepatocytes. Moreover, PTP1B deficiency conferred resistance to apoptosis directly induced by constitutively active Foxo1 suggesting additional actions of PTP1B downstream of Foxo1 nuclear translocation. This in vitro study in hepatocytes has revealed the involvement of PTP1B in the signaling network that ultimately modulates cell fate in response to apoptotic stimuli.

TGF-β plays a dual role in hepatocytes, mediating both tumor suppressor and promoter effects [30]. The suppressor effects of this cytokine can be negatively regulated by activation of survival signals, mostly dependent on tyrosine kinase activity. Using immortalized hepatocytes, Ortiz et al. [67] have found that PTP1B deficiency conferred resistance against TGF-β-mediated apoptosis and growth inhibition. At the molecular level, this effect correlated with lower Smad2/Smad3 phosphorylation and nuclear translocation, the lack of up-regulation of the inhibitory Smad7 and sustained activation of Akt and ERK1/2. Interestingly, in the presence of the general tyrosine kinase inhibitor genistein both responses were recovered. Moreover, PTP1B$^{-/-}$ hepatocytes stimulated with TGF-β showed elevated NF-κB activation. In the light of these results, knockdown of the NF-κB p65 subunit increased the response to TGF-β of PTP1B$^{-/-}$ hepatocytes in terms of Smad2/3 phosphorylation and apoptosis. The lack of PTP1B also promoted an altered NADPH oxidase (NOX) expression pattern induced by TGF-β, strongly increasing the NOX1/NOX4 ratio, which was reverted by genistein and NF-κB p65 subunit knockdown. Alternatively, NOX1 knockdown in PTP1B$^{-/-}$ hepatocytes recovered the apoptotic response to TGF-β through the inhibition of nuclear translocation of NF-κB p65 subunit, increase of Smad2 phosphorylation, and decrease of levels of the inhibitory Smad7. Altogether, these results highlight the role of PTP1B in the signaling cascades activated by TGF-β that involve Smads2/3, Smad7, NF-κB, and NOX1/4 family members. Further work will be necessary to study the potential role of this novel pathway in pathological conditions characterized by kinase hyperactivation, such as liver cancer.

The role of PTP1B in response to liver damage has been studied in vivo in mice injected with a lethal dose of the Fas/CD95 agonist antibody Jo2, a model of fulminant hepatitis [75]. In agreement with the in vitro studies in hepatocytes, PTP1B$^{-/-}$ mice are resistant to Fas-induced liver damage and lethality. This is due to reduced hepatic apoptosis and reduced levels of circulating liver enzymes compared to the lethal effects in the wild-type mice including tachypnea, shallow breathing, and prostration indicative of severe liver failure. Histological analysis of livers from the Jo-2-injected wild-type mice that showed distress revealed parenchymal necrosis, hemorrhage, and hepatocyte apoptosis assayed by TUNEL, whereas livers from the majority of PTP1B$^{-/-}$ mice showed no significant histologic pathological features. Consistent with the histological data, cleavage of caspases-8, -9, -3, and -6, normally activated following Fas receptor oligomerization, was detected in liver extracts from wild-type mice 6 h post-Jo-2 treatment, but was significantly abrogated in PTP1B$^{-/-}$ mice. Because activation of caspase-8 is one of the first events following activation of the Fas receptor, these data demonstrate that one of the earliest events in Fas-induced apoptosis is abrogated in PTP1B$^{-/-}$ mice. Protection against Fas-mediated apoptosis in the liver of PTP1B$^{-/-}$ mice correlated with an elevation and/or activation of numerous anti-apoptotic signaling proteins including FLIP$_L$, ERK1/2, and NF-κB. In addition, the resistance of PTPIB$^{-/-}$ hepatocytes to Fas-mediated apoptosis also required tyrosine kinase activity, because as occurred with TGF-β [67], PTP1B$^{-/-}$ hepatocytes were rendered susceptible to Fas-mediated apoptosis by pretreatment with genistein. Upon analysis of different PTP1B substrates, it was found that HGFR/Met showed sustained tyrosine phosphorylation and increased activation of its downstream mediator ERK1/2 in both livers and primary hepatocytes lacking PTP1B after Jo-2 treatment. The observation that the majority of PTP1B$^{-/-}$ mice are protected against Fas-mediated liver damage suggests that pharmacological manipulation of PTP1B activity may constitute a viable therapeutic modality for treatment against hepatotoxins and liver damage. A similar inhibition of caspase induction and death receptor-induced liver damage was demonstrated for the compound suramin [26]. Notably, suramin is an inhibitor of PTP1B at concentrations shown to provide protection to death receptor-mediated liver damage. This effect is promising since the efficacy of PTP1B inhibitors in hepatoprotection remains unexplored.

Perspectives

The close association between hepatic insulin resistance and progression to nonalcoholic fatty liver disease (NAFLD) and T2DM is a severe health problem [3, 12, 52, 53]. Therefore, the identification of the factors contributing to the development of hepatic insulin resistance including critical nodes of the insulin signaling pathway such as PTP1B will be of benefit in the discovery of new drug targets. The present chapter provides evidence of the complexity of PTP1B cellular actions in the liver that involve regulation of metabolism, proliferation, and survival. Since PTP1B inhibitors are potential antidiabetic drugs and, in addition, might be used to alleviate hepatotoxicity, their tissue-specific effects in these processes should be carefully evaluated.

References

1. Agouni A, Mody N, Owen C, Czopek A, Zimmer D, Bentires-Alj M, Bence KK, Delibegovic M (2011) Liver-specific deletion of protein tyrosine phosphatase (PTP) 1B improves obesity- and pharmacologically induced endoplasmic reticulum stress. Biochem J 438:369–378
2. Ahmad F, Li PM, Meyerovitch J, Goldstein BJ (1995) Osmotic loading of neutralizing antibodies demonstrates a role for protein-tyrosine phosphatase 1B in negative regulation of the insulin action pathway. J Biol Chem 270:20503–20508

3. Angulo P (2002) Nonalcoholic fatty liver disease. N Engl J Med 346:1221–1231
4. Aoki N, Matsuda T (2000) A cytosolic protein-tyrosine phosphatase PTP1B specifically dephosphorylates and deactivates prolactin-activated STAT5a and STAT5b. J Biol Chem 275:39718–39726
5. Araki E, Lipes MA, Patti ME, Bruning JC, Haag B III, Johnson RS, Kahn CR (1994) Alternative pathway of insulin signalling in mice with targeted disruption of the IRS-1 gene. Nature 372:186–190
6. Backer JM, Myers MG Jr, Shoelson SE, Chin DJ, Sun XJ, Miralpeix M, Hu P, Margolis B, Skolnik EY, Schlessinger J et al (1992) Phosphatidylinositol 3′-kinase is activated by association with IRS-1 during insulin stimulation. EMBO J 11:3469–3479
7. Bence KK (2010) Hepatic PTP1B deficiency: the promise of a treatment for metabolic syndrome? J Clin Metab Diabetes 1:27–33
8. Bento JL, Palmer ND, Mychaleckyj JC, Lange LA, Langefeld CD, Rich SS, Freedman BI, Bowden DW (2004) Association of protein tyrosine phosphatase 1B gene polymorphisms with type 2 diabetes. Diabetes 53:3007–3012
9. Bourdeau A, Dube N, Tremblay ML (2005) Cytoplasmic protein tyrosine phosphatases, regulation and function: the roles of PTP1B and TC-PTP. Curr Opin Cell Biol 17:203–209
10. Bruning JC, Winnay J, Bonner-Weir S, Taylor SI, Accili D, Kahn CR (1997) Development of a novel polygenic model of NIDDM in mice heterozygous for IR and IRS-1 null alleles. Cell 88:561–572
11. Buckley DA, Cheng A, Kiely PA, Tremblay ML, O'Connor R (2002) Regulation of insulin-like growth factor type I (IGF-I) receptor kinase activity by protein tyrosine phosphatase 1B (PTP-1B) and enhanced IGF-I-mediated suppression of apoptosis and motility in PTP-1B-deficient fibroblasts. Mol Cell Biol 22:1998–2010
12. Bugianesi E, Gastaldelli A, Vanni E, Gambino R, Cassader M, Baldi S, Ponti V, Pagano G, Ferrannini E, Rizzetto M (2005) Insulin resistance in non-diabetic patients with non-alcoholic fatty liver disease: sites and mechanisms. Diabetologia 48:634–642
13. Bursch W, Wastl U, Hufnagl K, Schulte-Hermann R (2005) No increase of apoptosis in regressing mouse liver after withdrawal of growth stimuli or food restriction. Toxicol Sci 85:507–514
14. Canbay A, Friedman S, Gores GJ (2004) Apoptosis: the nexus of liver injury and fibrosis. Hepatology 39:273–278
15. Carpentier A, Taghibiglou C, Leung N, Szeto L, Van Iderstine SC, Uffelman KD, Buckingham R, Adeli K, Lewis GF (2002) Ameliorated hepatic insulin resistance is associated with normalization of microsomal triglyceride transfer protein expression and reduction in very low density lipoprotein assembly and secretion in the fructose-fed hamster. J Biol Chem 277: 28795–28802
16. Clampit JE, Meuth JL, Smith HT, Reilly RM, Jirousek MR, Trevillyan JM, Rondinone CM (2003) Reduction of protein-tyrosine phosphatase-1B increases insulin signaling in FAO hepatoma cells. Biochem Biophys Res Commun 300:261–267
17. Cole P (2008) Insights into inflammatory mechanisms and new therapeutic strategies. Drug News Perspect 21:573–579
18. Cheyssac C, Lecoeur C, Dechaume A, Bibi A, Charpentier G, Balkau B, Marre M, Froguel P, Gibson F, Vaxillaire M (2006) Analysis of common PTPN1 gene variants in type 2 diabetes, obesity and associated phenotypes in the French population. BMC Med Genet 7:44
19. Cho H, Mu J, Kim JK, Thorvaldsen JL, Chu Q, Crenshaw EB III, Kaestner KH, Bartolomei MS, Shulman GI, Birnbaum MJ (2001) Insulin resistance and a diabetes mellitus-like syndrome in mice lacking the protein kinase Akt2 (PKB beta). Science 292:1728–1731
20. Delibegovic M, Zimmer D, Kauffman C, Rak K, Hong EG, Cho YR, Kim JK, Kahn BB, Neel BG, Bence KK (2009) Liver-specific deletion of protein-tyrosine phosphatase 1B (PTP1B) improves metabolic syndrome and attenuates diet-induced endoplasmic reticulum stress. Diabetes 58:590–599
21. Dijkers PF, Medema RH, Pals C, Banerji L, Thomas NS, Lam EW, Burgering BM, Raaijmakers JA, Lammers JW, Koenderman L et al (2000) Forkhead transcription factor FKHR-L1 modulates cytokine-dependent transcriptional regulation of p27(KIP1). Mol Cell Biol 20:9138–9148

22. Dong X, Park S, Lin X, Copps K, Yi X, White MF (2006) Irs1 and Irs2 signaling is essential for hepatic glucose homeostasis and systemic growth. J Clin Invest 116:101–114
23. Dong XC, Copps KD, Guo S, Li Y, Kollipara R, DePinho RA, White MF (2008) Inactivation of hepatic Foxo1 by insulin signaling is required for adaptive nutrient homeostasis and endocrine growth regulation. Cell Metab 8:65–76
24. Dube N, Bourdeau A, Heinonen KM, Cheng A, Loy AL, Tremblay ML (2005) Genetic ablation of protein tyrosine phosphatase 1B accelerates lymphomagenesis of p53-null mice through the regulation of B-cell development. Cancer Res 65:10088–10095
25. Egawa K, Maegawa H, Shimizu S, Morino K, Nishio Y, Bryer-Ash M, Cheung AT, Kolls JK, Kikkawa R, Kashiwagi A (2001) Protein-tyrosine phosphatase-1B negatively regulates insulin signaling in l6 myocytes and Fao hepatoma cells. J Biol Chem 276:10207–10211
26. Eichhorst ST, Krueger A, Muerkoster S, Fas SC, Golks A, Gruetzner U, Schubert L, Opelz C, Bilzer M, Gerbes AL et al (2004) Suramin inhibits death receptor-induced apoptosis in vitro and fulminant apoptotic liver damage in mice. Nat Med 10:602–609
27. Elchebly M, Payette P, Michaliszyn E, Cromlish W, Collins S, Loy AL, Normandin D, Cheng A, Himms-Hagen J, Chan CC et al (1999) Increased insulin sensitivity and obesity resistance in mice lacking the protein tyrosine phosphatase-1B gene. Science 283:1544–1548
28. Escribano O, Guillen C, Nevado C, Gomez-Hernandez A, Kahn CR, Benito M (2009) Beta-Cell hyperplasia induced by hepatic insulin resistance: role of a liver-pancreas endocrine axis through insulin receptor A isoform. Diabetes 58:820–828
29. Escriva F, Gonzalez-Rodriguez A, Fernandez-Millan E, Rondinone CM, Alvarez C, Valverde AM (2010) PTP1B deficiency enhances liver growth during suckling by increasing the expression of insulin-like growth factor-I. J Cell Physiol 225:214–222
30. Fabregat I, Roncero C, Fernandez M (2007) Survival and apoptosis: a dysregulated balance in liver cancer. Liver Int 27:155–162
31. Frasca F, Pandini G, Scalia P, Sciacca L, Mineo R, Costantino A, Goldfine ID, Belfiore A, Vigneri R (1999) Insulin receptor isoform A, a newly recognized, high-affinity insulin-like growth factor II receptor in fetal and cancer cells. Mol Cell Biol 19:3278–3288
32. Garcia-Ruiz I, Solis-Munoz P, Gomez-Izquierdo E, Munoz-Yague MT, Valverde AM, Solis-Herruzo JA (2012) Protein tyrosine phosphatases are involved in the interferon resistance associated with insulin resistance in HepG2 cells and obese mice. J Biol Chem 287(23):19564–19573
33. Goldstein BJ (2001) Protein-tyrosine phosphatase 1B (PTP1B): a novel therapeutic target for type 2 diabetes mellitus, obesity and related states of insulin resistance. Curr Drug Targets Immune Endocr Metabol Disord 1:265–275
34. Gonzalez-Rodriguez A, Clampit JE, Escribano O, Benito M, Rondinone CM, Valverde AM (2007) Developmental switch from prolonged insulin action to increased insulin sensitivity in protein tyrosine phosphatase 1B-deficient hepatocytes. Endocrinology 148:594–608
35. Gonzalez-Rodriguez A, Escribano O, Alba J, Rondinone CM, Benito M, Valverde AM (2007) Levels of protein tyrosine phosphatase 1B determine susceptibility to apoptosis in serum-deprived hepatocytes. J Cell Physiol 212:76–88
36. Gonzalez-Rodriguez A, Mas-Gutierrez JA, Mirasierra M, Fernandez-Perez A, Lee YJ, Ko HJ, Kim JK, Romanos E, Carrascosa JM, Ros M et al (2012) Essential role of protein tyrosine phosphatase 1B in obesity-induced inflammation and peripheral insulin resistance during aging. Aging Cell 11:284–296
37. Gonzalez-Rodriguez A, Mas Gutierrez JA, Sanz-Gonzalez S, Ros M, Burks DJ, Valverde AM (2010) Inhibition of PTP1B restores IRS1-mediated hepatic insulin signaling in IRS2-deficient mice. Diabetes 59:588–599
38. Gonzalez-Rodriguez A, Nevado C, Escriva F, Sesti G, Rondinone CM, Benito M, Valverde AM (2008) PTP1B deficiency increases glucose uptake in neonatal hepatocytes: involvement of IRA/GLUT2 complexes. Am J Physiol Gastrointest Liver Physiol 295:G338–G347
39. Gum RJ, Gaede LL, Koterski SL, Heindel M, Clampit JE, Zinker BA, Trevillyan JM, Ulrich RG, Jirousek MR, Rondinone CM (2003) Reduction of protein tyrosine phosphatase 1B increases insulin-dependent signaling in ob/ob mice. Diabetes 52:21–28

40. Haj FG, Markova B, Klaman LD, Bohmer FD, Neel BG (2003) Regulation of receptor tyrosine kinase signaling by protein tyrosine phosphatase-1B. J Biol Chem 278:739–744
41. Haj FG, Zabolotny JM, Kim YB, Kahn BB, Neel BG (2005) Liver-specific protein-tyrosine phosphatase 1B (PTP1B) re-expression alters glucose homeostasis of PTP1B-/-mice. J Biol Chem 280:15038–15046
42. Handy JA, Fu PP, Kumar P, Mells JE, Sharma S, Saxena NK, Anania FA (2011) Adiponectin inhibits leptin signalling via multiple mechanisms to exert protective effects against hepatic fibrosis. Biochem J 440:385–395
43. Horton JD, Goldstein JL, Brown MS (2002) SREBPs: activators of the complete program of cholesterol and fatty acid synthesis in the liver. J Clin Invest 109:1125–1131
44. Inada S, Ikeda Y, Suehiro T, Takata H, Osaki F, Arii K, Kumon Y, Hashimoto K (2007) Glucose enhances protein tyrosine phosphatase 1B gene transcription in hepatocytes. Mol Cell Endocrinol 271:64–70
45. Issad T, Coupe C, Ferre P, Girard J (1987) Insulin resistance during suckling period in rats. Am J Physiol 253:E142–E148
46. Kipfer-Coudreau S, Eberle D, Sahbatou M, Bonhomme A, Guy-Grand B, Froguel P, Galan P, Basdevant A, Clement K (2004) Single nucleotide polymorphisms of protein tyrosine phosphatase 1B gene are associated with obesity in morbidly obese French subjects. Diabetologia 47:1278–1284
47. Klaman LD, Boss O, Peroni OD, Kim JK, Martino JL, Zabolotny JM, Moghal N, Lubkin M, Kim YB, Sharpe AH et al (2000) Increased energy expenditure, decreased adiposity, and tissue-specific insulin sensitivity in protein-tyrosine phosphatase 1B-deficient mice. Mol Cell Biol 20:5479–5489
48. Klover PJ, Mooney RA (2004) Hepatocytes: critical for glucose homeostasis. Int J Biochem Cell Biol 36:753–758
49. Kubota N, Kubota T, Itoh S, Kumagai H, Kozono H, Takamoto I, Mineyama T, Ogata H, Tokuyama K, Ohsugi M et al (2008) Dynamic functional relay between insulin receptor substrate 1 and 2 in hepatic insulin signaling during fasting and feeding. Cell Metab 8:49–64
50. Kubota N, Tobe K, Terauchi Y, Eto K, Yamauchi T, Suzuki R, Tsubamoto Y, Komeda K, Nakano R, Miki H et al (2000) Disruption of insulin receptor substrate 2 causes type 2 diabetes because of liver insulin resistance and lack of compensatory beta-cell hyperplasia. Diabetes 49:1880–1889
51. Kushner JA, Haj FG, Klaman LD, Dow MA, Kahn BB, Neel BG, White MF (2004) Islet-sparing effects of protein tyrosine phosphatase-1b deficiency delays onset of diabetes in IRS2 knockout mice. Diabetes 53:61–66
52. Larter CZ, Farrell GC (2006) Insulin resistance, adiponectin, cytokines in NASH: Which is the best target to treat? J Hepatol 44:253–261
53. Lauer GM, Walker BD (2001) Hepatitis C virus infection. N Engl J Med 345:41–52
54. Li JM, Li YC, Kong LD, Hu QH (2010) Curcumin inhibits hepatic protein-tyrosine phosphatase 1B and prevents hypertriglyceridemia and hepatic steatosis in fructose-fed rats. Hepatology 51:1555–1566
55. Lu YX, Zhang Q, Li J, Sun YX, Wang LY, Cheng WM, Hu XY (2010) Antidiabetic effects of total flavonoids from Litsea Coreana leve on fat-fed, streptozotocin-induced type 2 diabetic rats. Am J Chin Med 38:713–725
56. Lupu F, Terwilliger JD, Lee K, Segre GV, Efstratiadis A (2001) Roles of growth hormone and insulin-like growth factor 1 in mouse postnatal growth. Dev Biol 229:141–162
57. Ma YM, Tao RY, Liu Q, Li J, Tian JY, Zhang XL, Xiao ZY, Ye F (2011) PTP1B inhibitor improves both insulin resistance and lipid abnormalities in vivo and in vitro. Mol Cell Biochem 357:65–72
58. Michael MD, Kulkarni RN, Postic C, Previs SF, Shulman GI, Magnuson MA, Kahn CR (2000) Loss of insulin signaling in hepatocytes leads to severe insulin resistance and progressive hepatic dysfunction. Mol Cell 6:87–97
59. Michalopoulos GK (2007) Liver regeneration. J Cell Physiol 213:286–300

60. Miyake K, Ogawa W, Matsumoto M, Nakamura T, Sakaue H, Kasuga M (2002) Hyperinsulinemia, glucose intolerance, and dyslipidemia induced by acute inhibition of phosphoinositide 3-kinase signaling in the liver. J Clin Invest 110:1483–1491
61. Molero C, Valverde AM, Benito M, Lorenzo M (1992) Phosphoenolpyruvate carboxykinase and glucose-6-phosphate dehydrogenase expression in fetal hepatocyte primary cultures under proliferative conditions. Exp Cell Res 200:295–300
62. Mora A, Lipina C, Tronche F, Sutherland C, Alessi DR (2005) Deficiency of PDK1 in liver results in glucose intolerance, impairment of insulin-regulated gene expression and liver failure. Biochem J 385:639–648
63. Nadal E, Dogliotti PL, Rodriguez JC, Zuccaro G (2000) Craniofacial distraction osteogenesis en bloc. J Craniofac Surg 11:246–251, discussion 252–243
64. Nakae J, Barr V, Accili D (2000) Differential regulation of gene expression by insulin and IGF-1 receptors correlates with phosphorylation of a single amino acid residue in the forkhead transcription factor FKHR. EMBO J 19:989–996
65. Nakae J, Kitamura T, Silver DL, Accili D (2001) The forkhead transcription factor Foxo1 (Fkhr) confers insulin sensitivity onto glucose-6-phosphatase expression. J Clin Invest 108:1359–1367
66. Ockner RK (2001) Apoptosis and liver diseases: recent concepts of mechanism and significance. J Gastroenterol Hepatol 16:248–260
67. Ortiz C, Caja L, Bertran E, Gonzalez-Rodriguez A, Valverde AM, Fabregat I, Sancho P (2012) Protein-tyrosine phosphatase 1B (PTP1B) deficiency confers resistance to transforming growth factor-beta (TGF-beta)-induced suppressor effects in hepatocytes. J Biol Chem 287:15263–15274
68. Perez M, Haschke B, Donato NJ (1999) Differential expression and translocation of protein tyrosine phosphatase 1B-related proteins in ME-180 tumor cells expressing apoptotic sensitivity and resistance to tumor necrosis factor: potential interaction with epidermal growth factor receptor. Oncogene 18:967–978
69. Puigserver P, Rhee J, Donovan J, Walkey CJ, Yoon JC, Oriente F, Kitamura Y, Altomonte J, Dong H, Accili D et al (2003) Insulin-regulated hepatic gluconeogenesis through FOXO1-PGC-1alpha interaction. Nature 423:550–555
70. Radziuk J, Pye S (2001) Production and metabolic clearance of glucose under basal conditions in Type II (non-insulin-dependent) diabetes mellitus. Diabetologia 44:983–991
71. Revuelta-Cervantes J, Mayoral R, Miranda S, Gonzalez-Rodriguez A, Fernandez M, Martin-Sanz P, Valverde AM (2011) Protein tyrosine phosphatase 1B (PTP1B) deficiency accelerates hepatic regeneration in mice. Am J Pathol 178:1591–1604
72. Romero-Gomez M, Del Mar Viloria M, Andrade RJ, Salmeron J, Diago M, Fernandez-Rodriguez CM, Corpas R, Cruz M, Grande L, Vazquez L et al (2005) Insulin resistance impairs sustained response rate to peginterferon plus ribavirin in chronic hepatitis C patients. Gastroenterology 128:636–641
73. Rondinone CM, Trevillyan JM, Clampit J, Gum RJ, Berg C, Kroeger P, Frost L, Zinker BA, Reilly R, Ulrich R et al (2002) Protein tyrosine phosphatase 1B reduction regulates adiposity and expression of genes involved in lipogenesis. Diabetes 51:2405–2411
74. Saltiel AR, Kahn CR (2001) Insulin signalling and the regulation of glucose and lipid metabolism. Nature 414:799–806
75. Sangwan V, Paliouras GN, Cheng A, Dube N, Tremblay ML, Park M (2006) Protein-tyrosine phosphatase 1B deficiency protects against Fas-induced hepatic failure. J Biol Chem 281: 221–228
76. Schulte-Hermann R, Bursch W, Grasl-Kraupp B (1995) Active cell death (apoptosis) in liver biology and disease. Prog Liver Dis 13:1–35
77. Shimano H (2001) Sterol regulatory element-binding proteins (SREBPs): transcriptional regulators of lipid synthetic genes. Prog Lipid Res 40:439–452
78. Shimizu S, Ugi S, Maegawa H, Egawa K, Nishio Y, Yoshizaki T, Shi K, Nagai Y, Morino K, Nemoto K et al (2003) Protein-tyrosine phosphatase 1B as new activator for hepatic lipogenesis via sterol regulatory element-binding protein-1 gene expression. J Biol Chem 278: 43095–43101

79. Shimomura I, Matsuda M, Hammer RE, Bashmakov Y, Brown MS, Goldstein JL (2000) Decreased IRS-2 and increased SREBP-1c lead to mixed insulin resistance and sensitivity in livers of lipodystrophic and ob/ob mice. Mol Cell 6:77–86
80. Suhara T, Kim HS, Kirshenbaum LA, Walsh K (2002) Suppression of Akt signaling induces Fas ligand expression: involvement of caspase and Jun kinase activation in Akt-mediated Fas ligand regulation. Mol Cell Biol 22:680–691
81. Sun C, Zhang F, Ge X, Yan T, Chen X, Shi X, Zhai Q (2007) SIRT1 improves insulin sensitivity under insulin-resistant conditions by repressing PTP1B. Cell Metab 6:307–319
82. Taub R (2004) Liver regeneration: from myth to mechanism. Nat Rev Mol Cell Biol 5:836–847
83. Tsou RC, Bence KK (2012) The genetics of PTPN1 and obesity: insights from mouse models of tissue-specific PTP1B deficiency. J Obes 2012:926857
84. Valverde AM, Burks DJ, Fabregat I, Fisher TL, Carretero J, White MF, Benito M (2003) Molecular mechanisms of insulin resistance in IRS-2-deficient hepatocytes. Diabetes 52:2239–2248
85. Valverde AM, Fabregat I, Burks DJ, White MF, Benito M (2004) IRS-2 mediates the antiapoptotic effect of insulin in neonatal hepatocytes. Hepatology 40:1285–1294
86. Wang CD, Teng BS, He YM, Wu JS, Pan D, Pan LF, Zhang D, Fan ZH, Yang HJ, Zhou P (2012) Effect of a novel proteoglycan PTP1B inhibitor from Ganoderma lucidum on the amelioration of hyperglycaemia and dyslipidaemia in db/db mice. Br J Nutr 108(11):2014–2025
87. Waring JF, Ciurlionis R, Clampit JE, Morgan S, Gum RJ, Jolly RA, Kroeger P, Frost L, Trevillyan J, Zinker BA et al (2003) PTP1B antisense-treated mice show regulation of genes involved in lipogenesis in liver and fat. Mol Cell Endocrinol 203:155–168
88. White MF (2002) IRS proteins and the common path to diabetes. Am J Physiol Endocrinol Metab 283:E413–E422
89. Withers DJ, Burks DJ, Towery HH, Altamuro SL, Flint CL, White MF (1999) Irs-2 coordinates Igf-1 receptor-mediated beta-cell development and peripheral insulin signalling. Nat Genet 23:32–40
90. Withers DJ, Gutierrez JS, Towery H, Burks DJ, Ren JM, Previs S, Zhang Y, Bernal D, Pons S, Shulman GI et al (1998) Disruption of IRS-2 causes type 2 diabetes in mice. Nature 391:900–904
91. Xue B, Kim YB, Lee A, Toschi E, Bonner-Weir S, Kahn CR, Neel BG, Kahn BB (2007) Protein-tyrosine phosphatase 1B deficiency reduces insulin resistance and the diabetic phenotype in mice with polygenic insulin resistance. J Biol Chem 282:23829–23840
92. Zabolotny JM, Bence-Hanulec KK, Stricker-Krongrad A, Haj F, Wang Y, Minokoshi Y, Kim YB, Elmquist JK, Tartaglia LA, Kahn BB et al (2002) PTP1B regulates leptin signal transduction in vivo. Dev Cell 2:489–495
93. Zabolotny JM, Kim YB, Welsh LA, Kershaw EE, Neel BG, Kahn BB (2008) Protein-tyrosine phosphatase 1B expression is induced by inflammation in vivo. J Biol Chem 283:14230–14241
94. Zheng LY, Zhou DX, Lu J, Zhang WJ, Zou DJ (2012) Down-regulated expression of the protein-tyrosine phosphatase 1B (PTP1B) is associated with aggressive clinicopathologic features and poor prognosis in hepatocellular carcinoma. Biochem Biophys Res Commun 420:680–684
95. Zinker BA, Rondinone CM, Trevillyan JM, Gum RJ, Clampit JE, Waring JF, Xie N, Wilcox D, Jacobson P, Frost L et al (2002) PTP1B antisense oligonucleotide lowers PTP1B protein, normalizes blood glucose, and improves insulin sensitivity in diabetic mice. Proc Natl Acad Sci U S A 99:11357–11362

Chapter 7
PTP1B in Obesity-Related Cardiovascular Function

Pimonrat Ketsawatsomkron, David W. Stepp, and Eric J. Belin de Chantemèle

Abstract Protein tyrosine phosphatase 1B (PTP1B), a molecular brake on leptin and insulin signaling pathways, is an important regulator of metabolic functions. Due to the close proximity between the signaling pathways regulating the metabolic and the cardiovascular systems, it has been hypothesized that PTP1B might have a role in the control of cardiovascular functions in the context of metabolic disorders. Evidence from population-based studies demonstrates that polymorphisms of the human PTP1B gene, *PTPN1*, predispose to both type II diabetes and cardiovascular disease, including hypertension and atherosclerosis. Consistent with these findings, genetic deletion of PTP1B in lean and obese animal models has proven to be protective against obesity-induced hypertension and endothelial dysfunction. While both human and animals studies mostly support indirect effects of PTP1B on the cardiovascular system via its control of metabolic function, evidence from in vitro studies suggests direct effects of PTP1B on endothelial and vascular smooth muscle cell integrity. Taken together these data identify PTP1B as a major regulator of cardiovascular function and suggest that manipulation of PTP1B expression and activity could provide a novel therapeutic approach to prevent obesity-related cardiovascular disease.

P. Ketsawatsomkron, Ph.D.
Department of Pharmacology, University of Iowa, Iowa City, IA 52242, USA

D.W. Stepp, Ph.D.
Vascular Biology Center, Department of Physiology, Georgia Health Sciences University, Augusta, GA 30912, USA

E.J.B. de Chantemèle, Ph.D. (✉)
Department of Physiology, Georgia Health Sciences University, Augusta, GA 30912, USA
e-mail: ebelindechanteme@gru.edu

K.K. Bence (ed.), *Protein Tyrosine Phosphatase Control of Metabolism*,
DOI 10.1007/978-1-4614-7855-3_7,

Abbreviations

Akt	Protein kinase B
Ang II	Angiotensin II
BAEC	Bovine aortic endothelial cell
bFGF	Basic fibroblast growth factor
BMO	Bis(maltolato)-oxovanadium
EGF	Epidermal growth factor
eNOS	Endothelial nitric oxide synthase
ET-1	Endothelin-1
FAK	Focal adhesion kinase
HbA1c	Glycosylated hemoglobin
IR	Insulin receptor
IRS1	Insulin receptor substrate 1
JAK2	Janus kinase 2
KO	Knockout
LZR	Lean Zucker rat
MAP	Mean arterial pressure
MAPK	Mitogen-activated protein kinase
NFκB	Nuclear factor κB
NO	Nitric oxide
NOX1	NADPH oxidase 1
NOXA1	NADPH oxidase activator 1
Noxo1	NADPH oxidase organizer 1
OZR	Obese Zucker rat
PDGF	Platelet-derived growth factor
PI3K	Phosphatidylinositide 3-kinases
PKA	Protein kinase A
PTP1B	Protein tyrosine phosphatase 1B
PTPN1	Protein tyrosine phosphatase non-receptor type 1
ROS	Reactive oxygen species
SMC	Smooth muscle cell
STAT	Signal transducer and activator of transcription
TNFalpha	Tumor necrosis factor alpha
VCAM	Vascular cell adhesion molecule
VEGF	Vascular endothelial growth factor
VEGFR	Vascular endothelial growth factor receptor
VSMC	Vascular smooth muscle cell

Introduction

In the early nineties, a growing interest emerged on the use of insulin sensitizers to prevent insulin resistance and type II diabetes. The group of John McNeill in Vancouver, Canada, asked the question of whether improving insulin sensitivity with compounds such as vanadate or bis(maltolato)-oxovanadium (BMO) would prevent the cardiovascular dysfunction associated with insulin resistance and obesity. Using spontaneously hypertensive rats [1], a model of insulin resistance, or obese Zucker rats (OZRs), a model of insulin resistance associated with obesity [2], McNeill's group demonstrated that chronic treatment with BMO for a duration of 8–14 weeks ameliorated age-dependent increases in systolic blood pressure. Using a similar approach, this group also reported that a 25-day sodium orthovanadate treatment blunted obesity-induced increases in aortic constriction in response to sympathetic stimulation, endothelin 1, or depolarization [3]. However, improving insulin sensitivity with sodium orthovanadate did not restore aortic endothelial function nor did it prevent obesity-induced changes in mesenteric artery reactivity [3]. Although the cardio-protective effects of insulin sensitizers were substantial, their mechanisms of action were, at that time, completely unknown. Concomitantly to these studies, a growing interest was developing in protein tyrosine phosphatases (PTPs) and their crucial role in signaling pathways was rapidly being unraveled. Vanadate, used in the previous studies as an insulin sensitizer, was soon identified as an inhibitor of PTPs [4, 5] and notably as a nonspecific inhibitor of protein tyrosine phosphatase 1B (PTP1B) [6], arguing for a role for PTPs and more specifically for PTP1B in the regulation of cardiovascular function.

PTP1B is an ubiquitously expressed PTP which regulates insulin signaling by dephosphorylating the insulin receptor tyrosine kinase, reducing its kinase activity and thus inhibiting insulin signaling. Deletion of PTP1B in mice results not only in an increase in insulin sensitivity [7] but also in resistance to diet-induced obesity, most likely due to the role played by PTP1B in the control of leptin signaling [8, 9]. Indeed, PTP1B is also a molecular break on leptin signaling pathway [8, 9]. As described in more detail in the other chapters of this book, activation of the leptin signaling pathway requires leptin binding to its long form receptor. Upon binding, leptin triggers the autophosphorylation of its receptor-associated tyrosine kinase, janus kinase 2 (JAK2), which leads to the phosphorylation of tyrosine residues on the leptin receptor. Tyrosine phosphorylation facilitates the binding of signal transducer and activator of transcription (STAT) proteins and activates downstream transcriptional targets [10]. Consequently, by dephosphorylating JAK2, PTP1B plays a pivotal role in regulating the leptin signaling pathway.

Obesity, a complex metabolic disorder associated with insulin resistance, hyperleptinemia, and hyperlipidemia, is a major risk factor for the development of cardiovascular disease including hypertension [11], endothelial dysfunction, coronary heart disease, and heart failure. Insulin and leptin tightly control not only metabolic function but also regulate cardiovascular function through their ability to directly modulate sympathetic tone and relax blood vessels via the stimulation of nitric oxide (NO) released from vascular endothelial cells. This complex interaction of leptin and insulin in the control of metabolic and cardiovascular functions positions PTP1B, which regulates both signaling pathways, as a major regulator of metabolic and cardiovascular function. In this chapter, our goal is to highlight studies ranging from human genetics to animal models and cell culture, in order to decipher the role of PTP1B in obesity-related cardiovascular diseases.

PTP1B Gene Polymorphisms and Cardiovascular Function

As mentioned earlier, due to its crucial role in the regulation of insulin and leptin signaling pathways, PTP1B has been suggested to be involved in the polygenic pathogenesis of obesity and type 2 diabetes. Hence, the PTP1B gene, *PTPN1*, has been explored as a candidate gene for insulin resistance, metabolic syndrome, and obesity-related cardiovascular dysfunction. *PTPN1* is located on chromosome 20q13 in humans and on the distal arm of the chromosome 2 in mice. In both species, these two chromosomes have been linked to obesity, fasting serum insulin levels, and type 2 diabetes [12, 13]. To determine whether *PTPN1* is involved in the occurrence of metabolic syndrome and its related cardiovascular diseases, searches for polymorphisms in both the regulatory and the coding region of the human PTP1B gene have been conducted which led to the identification of several single nucleotide polymorphisms (SNPs). Because of the low allele frequency (<2 %) and/or the nature of the sequence variation (silent or intronic), only one single SNP localized in the 3′ UTR region of the PTP1B gene (184insG) was considered for association studies with markers of insulin resistance and cardiovascular dysfunction in two Italian populations [14]. Within these two different populations, the 1484insG variation was the only SNP associated with several features of insulin resistance (higher values of the insulin resistance $HOMA_{IR}$ index, serum triglycerides, and total/HDL cholesterol ratio), but also with a higher blood pressure in females [14]. Similar associations between *PTPN1* polymorphisms, indexes of insulin resistance, and hypertension have also been reported in other ethnicities. While studying a French cohort, Cheyssac *et al.* identified a risk allele homozygous for SNP rs2426159 for increased systolic blood pressure [15]. Olivier *et al.*, in a study focused on Americans of Chinese and Japanese descent, found an association of individual SNPs and the resulting haplotype, with alterations in lipid profiles and hypertension. A strong association of common risk haplotypes with hypertension was found in the two populations studied [16]. Furthermore, within the 14 *PTPN1* SNPs identified and the six genotyped, the strongest association with body mass index, a common risk factor for hypertension, metabolic syndrome, and cardiovascular complications, was seen with SNP g.54281T>A. Consistent with this latter study, Gu *et al.* demonstrated that T allele carriers of g.54281T>A and A allele carriers of I5/37C>A had higher risk of hypertension in a Chinese population living in China [17]. Santaniemi *et al.* identified different polymorphisms and concluded that the *PTPN1* IVS6+G82A polymorphism is associated with BMI, albuminuria, HbA1c, and hypertension in Finnish type 2 diabetic patients [18]. Although no unique SNP was identified in the above studies, they all concluded that there is indeed an association between *PTPN1* polymorphisms and markers of metabolic dysfunction and hypertension. Reasons for non-replication of association studies or identification of different polymorphisms between studies are numerous and might include population heterogeneity, ethnic stratification, and population-specific linkage disequilibrium between markers and causal variants, as well as sample size.

Studies associating PTP1B gene variation with metabolic and cardiovascular dysfunctions raised the question of whether cardiovascular disease and type 2 diabetes occur together because they share common predisposing causes or whether the metabolic dysregulation leads to cardiovascular disease. Using data from the Diabetes Heart Study, designed to identify genes contributing to the cardiovascular complications of type 2 diabetes, Donald Bowden's group confirmed that *PTPN1* polymorphisms are associated with type 2 diabetes [19] and assessed whether subclinical arteriosclerosis in type 2 diabetic patients is associated with *PTPN1* polymorphisms. They tested the hypothesis that the common haplotypes known to increase the risk of type 2 diabetes and insulin resistance would also be associated with an increase in cardiovascular disease [20]. By genotyping 23 SNPs in 590 Caucasian Americans with type 2 diabetes, these authors reported that the *PTPN1* 184insG SNP, mentioned earlier and associated with hypertension in Italian females [14], was indeed associated with carotid artery-calcified plaques. However, the rarity of this SNP in the Diabetes Heart Study cohort (as well as in the Italian population [14]) suggests that these results should be viewed with caution, as mean genotypic values were highly influenced by a few individuals. Nevertheless, the haplotype analysis performed revealed that the nine-SNP haplotype GACTTCAGO, previously linked to type 2 diabetes, insulin resistance, and fasting blood glucose, is significantly associated with increased coronary artery-calcified plaques, suggesting an influence of the PTP1B gene on the phenotype of coronary artery calcification in type 2 diabetic patients. Interestingly, correlation studies revealed that coronary artery-calcified plaques and body mass index were not correlated and adjustment for BMI did not alter the outcome of the study. This suggests that the association of PTP1B SNPs with coronary artery-calcified plaques occurs independently of obesity. All together, these studies implied that *PTPN1* contributes to genetic risk of type 2 diabetes, obesity, hypertension, and subclinical cardiovascular diseases and suggested a pleiotropic role for *PTPN1* in the determination of these traits.

While human genetic studies provide an appropriate tool to identify *PTPN1* polymorphisms and their association with markers of metabolic and cardiovascular impairments, a strictly genetic approach has the limitation of being purely descriptive and of not allowing for an assessment of protein function. The question of whether PTP1B may play a direct role in the control of cardiovascular function can be addressed in animals and cell culture models via the use of pharmacological and genetic approaches.

PTP1B, Insulin Sensitivity, and Cardiovascular Function

Vascular tone and more specifically vascular relaxation is under the control of a thin layer of endothelial cells lining the interior surface of the vascular lumen. Blood flow (shear stress) stimulates the endothelial cells and triggers the secretion of the vasorelaxant agent NO by the endothelium NO synthase (eNOS). Flow-mediated eNOS stimulation involves the activation of the PI3 Kinase/Akt signaling pathway [21, 22], which requires tyrosine phosphorylation. This led to the hypothesis that PTP1B may control flow (shear stress)-mediated vascular relaxation. Using three structurally different PTP1B inhibitors, the group of Vincent Richard, in Rouen,

France reported that PTP1B inhibition restored chronic heart failure-induced vascular dysfunction [23]. Consistent with the beneficial effects of these inhibitors, this group also demonstrated that despite reducing baseline flow-mediated relaxation, PTP1B deficiency in mice prevents chronic heart failure-induced endothelial and cardiac dysfunction [24]. However, although it was demonstrated that PTP1B is expressed in endothelial cells and that the beneficial effects of PTP1B inhibition involved the PI3 Kinase/Akt signaling pathway [23], the specific mechanisms by which PTP1B inhibition restored the impaired flow-mediated dilation in chronic heart failure are still unknown. As mentioned earlier, the insulin receptor (IR) is one of the main targets of PTP1B. Consequently, the likelihood that PTP1B inhibition restores endothelial function through an increase in insulin sensitivity in the model of chronic heart failure persists.

Contrary to the role of PTP1B in flow-mediated dilation, the role of PTP1B in the control of insulin sensitivity is well described. Given that PTP1B knockout (KO) mice demonstrate increased insulin sensitivity [7, 25], our group hypothesized that improving insulin sensitivity via PTP1B deletion in obese mice would prevent obesity-induced cardiovascular dysfunction. To test this hypothesis, PTP1B-deficient mice (on a mixed C57Bl/6-Balb/C background) were crossed with obese leptin receptor-deficient (*db/db*) mice. The resultant double KO mouse model, despite being obese, had improved peripheral insulin sensitivity as reflected by a reduced HbA1c index, improved glucose tolerance, and a higher level of IR phosphorylation in muscle and fat compared to the obese leptin receptor-deficient mice [26]. Surprisingly, hepatic insulin signaling was unimproved, suggesting that PTP1B's role in insulin resistance may be tissue-specific. To determine whether increasing insulin sensitivity in obese mice affects vascular function, we isolated mesenteric arteries and analyzed their reactivity on a pressurized myograph. We reported that correction of peripheral insulin resistance in obese mice prevents obesity-induced impaired vascular relaxation (Fig. 7.1). This improvement of endothelial function was associated with a reduction in the level of reactive oxygen species (ROS) produced at the level of the vasculature, as well as with a reduction in the expression of the ROS-generating enzyme NADPH oxidase 1 (NOX1) and its two regulatory enzymes Noxa1 and Noxo1 [26]. Whether this improvement is the reflection of enhanced insulin sensitivity or whether it involves a direct effect of PTP1B on ROS generation at the vascular level was not determined in this study. However, lean mice deficient in PTP1B did not exhibit improved endothelial function nor did they present a reduced expression level in ROS-generating enzymes. These latter data support an indirect effect of PTP1B deletion via increased insulin sensitivity. Consistent with this notion, liver-specific deletion of PTP1B does not protect against diet-induced obesity, but maintains glucose tolerance, improves lipid homeostasis, and prevents mice from obesity-induced endothelial dysfunction [27], further supporting the theory that PTP1B deficiency does not improve cardiovascular function through direct mechanisms but rather via enhanced insulin sensitivity and improvement of the associated plasma milieu.

Another piece of data generated by our group further supports the theory that PTP1B-deficiency prevents obesity-induced cardiovascular dysfunction. The double KO model described previously [26] and generated by crossing obese leptin receptor-deficient mice with PTP1B KO mice was used to demonstrate whether deletion

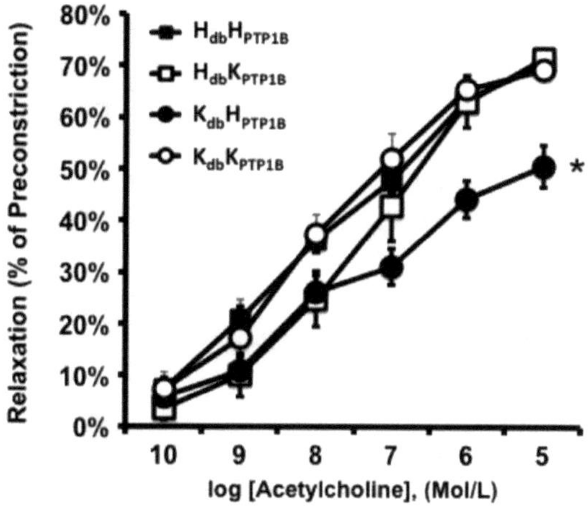

Fig. 7.1 Increasing insulin sensitivity via PTP1B deletion, in obese mice, prevents obesity-induced endothelial dysfunction. Endothelium-dependent relaxation to acetylcholine in lean control ($H_{db}H_{PTP1B}$), lean PTP1B KO ($H_{db}K_{PTP1B}$), obese ($K_{db}H_{PTP1B}$), and obese PTP1B KO mice ($K_{db}K_{PTP1B}$). Data are mean ± sem, $*p<0.05$ vs. $H_{db}H_{PTP1B}$ (adapted from Ali MI, Ketsawatsomkron P, Belin de Chantemele EJ, Mintz JD, Muta K, Salet C, et al. Deletion of Protein Tyrosine Phosphatase 1b Improves Peripheral Insulin Resistance and Vascular Function in Obese, Leptin-Resistant Mice via Reduced Oxidant Tone. Circ Res. 2009;105:1013–22)

of PTP1B in obese animals would blunt the increase in blood pressure associated with obesity [28]. Our study demonstrated that deletion of PTP1B in obese leptin receptor-deficient mice abolished obesity-induced hypertension. However, PTP1B deletion did not affect the blood pressure of lean animals on the mixed C57Bl/6-Balb/C background. This suggests that PTP1B modulates blood pressure by indirect means rather than through direct effects of PTP1B on mechanisms of control of arterial pressure. However, the hypothesis of the indirect effects of PTP1B on the cardiovascular function was recently challenged by the group of Debra Diz. Indeed, these authors observed that an acute bilateral injection of an allosteric PTP1B inhibitor in the nucleus tractus solitarius resulted in a transient decrease in resting blood pressure and heart rate in Sprague-Dawley rats. These direct effects of PTP1B inhibition on blood pressure and heart rate were associated with an impaired baroreflex sensitivity characterized by an reduced bradycardic response to phenylephrine [29]. Control data demonstrating specific inhibition of PTP1B in the nucleus tractus solitarius would nevertheless strongly support the observations of this study.

PTP1B, Leptin Sensitivity, and Cardiovascular Function

As described earlier, besides being a major regulator of the insulin signaling pathway, PTP1B is also a molecular brake on the leptin signaling pathway [8, 9]. Well-recognized for its role in the control of appetite and energy expenditure, leptin is

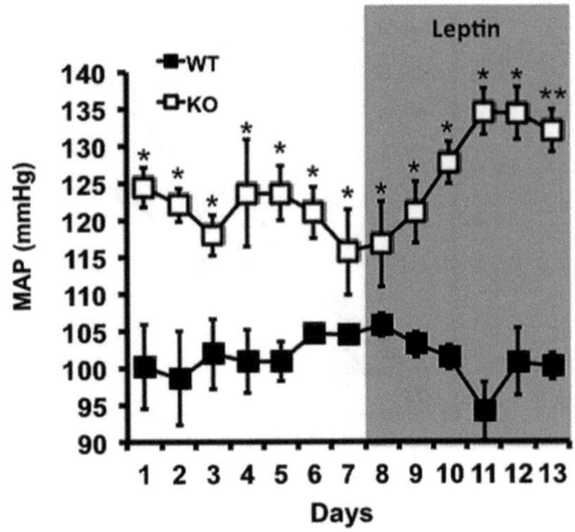

Fig. 7.2 PTP1B deletion increases baseline blood pressure and hemodynamic response to leptin infusion: Baseline and mean arterial pressure (MAP) response to leptin infusion (10 μg/day, 0.5 μL/h), determined by telemetry, in wild-type and PTP1B knockout mice. Data are mean ± sem, *$p < 0.05$ vs. WT, **$p < 0.001$ vs. WT (adapted from Belin de Chantemele EJ, Muta K, Mintz J, Tremblay ML, Marrero MB, Fulton DJ, et al. Protein Tyrosine Phosphatase 1B, a Major Regulator of Leptin-Mediated Control of Cardiovascular Function. Circulation. 2009;120:753–63)

also involved in the control of cardiovascular functions through its ability to increase sympathetic activation of the kidneys, the vasculature, and the heart [30–33]. Because obese patients present with an increased sympathetic tone [34–36], it has been postulated that leptin could be the link between the metabolic and cardiovascular dysfunction associated with obesity. The key role of PTP1B in the control of leptin signaling led us to hypothesize that PTP1B modulates leptin-mediated cardiovascular action. The role of PTP1B in leptin-mediated control of cardiovascular function was analyzed by determining the cardiovascular phenotype of PTP1B KO mice on the Balb/C background. Analyzing the cardiovascular function of conscious animals via radiotelemetry, we observed an increase in baseline blood pressure and a trend towards an increase in heart rate in the KO animals (Fig. 7.2) [37]. The increase in blood pressure and heart rate was also associated with an increased sympathetic tone as reflected by an exaggerated blood pressure response to ganglionic blockade. Since PTP1B controls both leptin and insulin sensitivity, and since both leptin and insulin control sympathetic tone, we questioned whether the increase in blood pressure was the reflection of increased leptin and/or insulin sensitivity. To distinguish between these possibilities, the animals were challenged with a 7-day infusion of either leptin or insulin. Leptin but not insulin treatment further raised the elevated baseline blood pressure of the KO animals, suggesting that PTP1B modulates leptin-mediated control of blood pressure. Although these data clearly suggest that PTP1B influences the blood pressure response to exogenous leptin, no mechanism responsible for the elevated baseline blood pressure has been identified yet.

Whether PTP1B deletion directly influences blood pressure or only sensitizes the cardiovascular response to endogenous leptin still remains to be determined. Interestingly, as mentioned above, the PTP1B KO mice on a mixed C57Bl/6 – Balb/C background, used to generate the double KO mice described previously, are normotensive [28]. The aforementioned data suggest an influence of the genetic background on the blood pressure response to PTP1B deletion.

Notably, while studying the vascular function of the PTP1B KO mice on the Balb/C background, an impaired vascular adrenergic tone was also identified in the PTP1B-deficient mice. PTP1B KO mice presented a reduced constriction to the α_1-adrenergic receptor agonist phenylephrine both in vivo and ex vivo [37], compared to wild-type controls. The constrictor response to serotonin or potassium chloride (depolarization) as well as endothelium-dependent relaxation was not affected in these mice. The reduction in vascular adrenergic tone observed with PTP1B deletion has been reproduced in wild-type mice with chronic leptin infusion [38]. In both cases, the reduction was blocked by pretreatment with the α_1-adrenergic receptor antagonist prazosin. These studies suggest that PTP1B-deficiency, similar to leptin infusion, modulates vascular adrenergic tone through increased sympathetic tone and likely rules out a direct effect of PTP1B on the adrenergic signaling pathway.

Evidence for a Role of PTP1B in the Vasculature

Insulin resistance syndromes generally are accompanied by both macrovascular and microvascular diseases. Multiple lines of evidence suggest that an imbalance of insulin signaling in the vascular tissues, namely in endothelial cells and smooth muscle cells (SMCs), plays a role in determining vascular complications. PTP1B has been shown to be a key modulator of the insulin signaling pathway in various insulin-sensitive tissues [39]. It is important to note that the major components of the insulin signaling pathway in endothelium and smooth muscle appear to be similar to those in liver, skeletal muscle, and adipose tissues, raising the possibility that PTP1B regulates insulin signaling in the vasculature. The physiological effects of insulin in the vasculature, however, extend beyond the regulation of cellular metabolism. Highlights of studies indicating a role for PTP1B in vascular cell types, both in an insulin-dependent and insulin-independent manner, are discussed below.

Endothelial Cell

PTP1B mRNA and protein have been shown to be expressed in endothelial cells in both isolated cultures and in intact vessels [23, 40, 41]. Overexpression of PTP1B in bovine aortic endothelial cells (BAECs) was sufficient to reduce insulin-mediated activation of Akt and NO production [42], suggesting that PTP1B might be important in the regulation of endothelial function. Dysfunction of the vascular endothelium, a condition frequently associated with insulin resistance, is tightly associated with cardiovascular complications [43]. Insulin not only regulates glucose homeostasis in the vasculature, but it also is involved in the production of NO [44], a critical regulator of vascular tone and growth. Through activation of IRS-1/PI3-K/Akt, insulin stimulation in endothelial cells results in NO release [45–48]. In contrast, activation of mitogen-activated protein kinase (MAPK) is required for insulin-induced secretion of endothelin-1 (ET-1) [49] as shown previously in BAECs. Pretreatment with the PI3-K inhibitor wortmannin had no effect on insulin-induced activation of MAPK and ET-1 release, while pretreatment with the MAPK inhibitor, PD98059, completely blocked this effect of insulin [50]. In addition, PD98059 had no effect on Akt activation induced by insulin [50], suggesting that insulin-induced activation of Ras/MAPK pathways and PI3-K/Akt is totally independent, at least in vascular endothelial cells. The impairment of insulin-induced activation of the IRS-1/PI3-K/Akt pathway, with simultaneous sustained activation of the Ras/MAPK, has been shown in the vasculature of different models of insulin resistance [49, 51–54]. Insulin-induced activation of IR/IRS1/PI3-K/Akt was impaired significantly in the vasculature of obese compared to lean rats [51, 52, 54]. In contrast, the activation of p42/44 MAPK stimulated by insulin in the vasculature was comparable between the groups, suggesting a selective resistance to the PI3-K pathway but not the MAPK pathway in an insulin-resistant state [51, 52]. Consistently, the response to acetylcholine-induced vasodilation was reduced significantly in the aorta of high fat-fed rats and coronary artery of OZRs compared to control rats [54, 55]. These data suggest that endothelial dysfunction occurs both in conduit and resistant vessels of obese animal models. In addition, in high fat-fed rats, insulin-induced activation of IRS-1/PI3-K is decreased, leading to reduced aortic eNOS protein expression and activity, while p42/44 MAPK is hyperactivated [54]. Although impairment of insulin-induced eNOS gene expression has been reported in obese animals [53], this finding appears inconsistent. Fulton *et al.* reported that protein expression of eNOS in aorta did not differ between OZR and lean Zucker rat (LZR) [42]. Alternative mechanisms have been proposed for endothelial dysfunction during insulin resistance rather than decreased eNOS expression. For example, Molnar *et al.* reported that eNOS protein expression was not changed in the femoral artery from high fat-fed mice, but disruption of eNOS dimers, leading to a reduced NO production, was found instead [56]. Therefore, insulin resistance in endothelium might result in imbalance of NO and ET-1 production. It remains unclear whether PTP1B participates in this process.

On the other hand, PTP activity including PTP1B can be inhibited by NO through *S*-nitrosylation of the active site cysteine residue. Gain of eNOS function, accompanied by increased NO production, leads to enhanced insulin signaling by inhibition of PTP activity *in vitro*. Interestingly, ablation of eNOS by siRNA suppressed insulin-induced activation of IR, insulin receptor substrate (IRS), and Akt in murine endothelial cells [57]. Taken together, these findings suggest a model whereby enhanced insulin signaling is achieved indirectly via NO inhibition of PTPs. It remains, however, to be determined whether this mechanism is evoked in states of insulin resistance, when NO is diminished.

In addition to the IR, other PTP1B substrates include the epidermal growth factor (EGF), insulin-like growth factor 1, platelet-derived growth factor (PDGF) receptors, and cadherin proteins [58], suggesting that alteration of PTP1B activity could affect the physiological functions conferred by these signaling molecules. For example, angiogenesis has been shown to be regulated by PTP1B, at least partly through regulation of vascular endothelial growth factor receptor (VEGFR). Activation of VEGFR by vascular endothelial growth factor (VEGF) is important in initiating angiogenesis. Transduction of catalytically inactive PTP1B enhanced, while overexpression of wild-type PTP1B inhibited autophosphorylation of VEGFR-stimulated by VEGF in human umbilical vein endothelial cells. The inhibition of VEGFR signaling by PTP1B is associated with inhibition of VEGF-stimulated endothelial cell proliferation and stabilizes VE-cadherin-mediated cell–cell adhesion. All of these findings *in vitro* indicate that PTP1B acts to negatively counteract VEGFR activation [59]. Moreover, the angiogenic response to hindlimb ischemia involves a dramatic increase in PTP1B expression and activity, in a mouse model of femoral artery resection [59]. It remains unclear whether inhibition of PTP1B will promote angiogenesis during ischemia in vivo. Notably, PTP1B expression has also been found in circulating angiogenic cells (CACs), progenitor cell types critical for neovascularization, and endothelial repairs [60]. Elevated expression of PTP1B was observed in CACs isolated from obese, hyperleptinemic individuals. Pharmacological inhibition of PTP1B restored the sensitivity of CACs to the angiogenic effects of leptin, consistent with other studies that leptin signaling is a target of PTP1B.

PTP1B has also been shown to participate in regulation of pulmonary endothelial barrier during acute lung injury. Overexpression of dominant negative PTP1B in cultured pulmonary endothelial cells, as well as in the vasculature of intact murine lungs resulted in decreased pulmonary endothelial barrier and the formation of edema. Conversely, increased expression of wild-type PTP1B enhanced endothelial barrier function. Increase in PTP1B oxidation accompanied by reduced activity was observed following LPS treatment in lung microvascular endothelial cells. Overexpression of the oxidant resistant form of PTP1B significantly inhibited LPS-mediated increases in endothelial monolayer permeability and pulmonary edema in mice [61]. Alternatively, activation of endothelial PTP1B is required for vascular adhesion molecule, VCAM-mediated leukocyte transendothelial migration, a critical process for inflammatory responses [62].

Vascular Smooth Muscle Cells

Studies in vascular smooth muscle cell (VSMC) have demonstrated significant roles for PTP1B, particularly in regulation of growth signaling. Studies by Hassid *et al.* reported that knockdown of PTP1B using antisense oligonucleotides resulted in increased VSMC mobility *via* enhanced tyrosine phosphorylation of several focal adhesion proteins, including paxillin, focal adhesion kinase (FAK), and p130^{cas} [63].

Additionally, treatment of VSMC with a NO donor leads to an upregulation of PTP1B and a decrease in VSMC mobility and phosphorylation of adhesion proteins [64, 65]. These effects of NO were blunted by PTP1B antisense oligos, suggesting that upregulation of PTP1B is essential for anti-mobility effect of NO in VSMC [64]. Consistently, insulin-mediated receptor phosphorylation and p44/42 MAPK signaling were impaired in the presence of NO. Overexpression of mutant PTP1B was able to inhibit the NO-mediated antagonism of insulin activation, particularly inhibition of p44/42 MAPK signaling [66], suggesting that PTP1B counteracts growth signaling in VSMC. These reports appear to be counterintuitive, given that insulin-mediated effects through MAPK pathway remain sensitive, despite the fact that PTP1B was upregulated during insulin resistance [9]. If PTP1B is required to antagonize VSMC growth, upregulation of PTP1B in states of insulin resistance might prevent VSMC proliferation. Another possibility is that enhanced PTP1B expression during insulin resistance might be a part of compensation mechanism. Consistent with this contention, two potent growth factors of neointima formation, PDGF and basic fibroblast growth factor (bFGF), were shown to increase PTP1B expression [67, 68]. It was also reported that PTP1B counteracts PDGF-induced motility and proliferation in rat aortic SMC cultures [65, 69]. Caution must be taken when considering these effects in the context of a disease state due to the complexity of signaling interactions. As shown by Zhuang *et al.*, prolonged exposure of VSMC to insulin (as occurs during hyperinsulinemia) enhanced PDGF-induced mitogenicity by suppressing PTP1B mRNA and protein expression [70]. Clearly, these findings need to be tested *in vivo*. Confounding effects from plasticity of smooth muscle cultures may not represent what actually occurs in the intact vessel.

Studies in rat carotid following balloon catheter injury showed that PTP1B mRNA and protein levels were highly upregulated in proliferating and migrating SMCs during neointima formation [69, 71]. Given that multiple growth factors and cytokines are upregulated in the injured blood vessel [69], it is possible that induction of PTP1B expression was attributable to PDGF and bFGF during neointima formation. However, it remains unclear what role PTP1B might play during vascular injury. As mentioned earlier, an upregulation of PTP1B perhaps serves as a compensatory mechanism to turn off growth signaling during vascular injury. To address this question, subsequent studies by the same group demonstrated that transduction of dominant negative PTP1B in injured rat carotid artery was sufficient to increase neointima formation [72], showing that the counter regulatory effect of PTP1B in growth signaling also exists *in vivo*.

The aforementioned findings provide evidence that PTP1B might act as a negative regulator of growth signaling in smooth muscle. However, this notion has been challenged by others. Haj *et al.* reported that deletion of PTP1B in fibroblasts results in sustained and increased tyrosyl phosphorylation of the EGF and PDGF receptors, but had a little effect, if any, on cascades downstream of EGF and PDGF receptors [58]. In contrast to this, another study has shown that loss of PTP1B in fibroblasts leads to a selectively diminished activation of Ras and p44/42 MAPK, whereas activation of Akt is increased in response to PDGF [73]. These data indicate that PTP1B potentially acts as a positive regulator for the mitogenic (Ras/MAPK) pathway. Studies in rat aortic SMC cultures by our group also support this hypothesis. We reported that angiotensin (Ang) II-mediated SMC insulin resistance was dependent upon PTP1B activation. Insulin-induced phosphorylation of IR/IRS-1/Akt was significantly suppressed by pre-incubation of Ang II [74], whereas activation of growth signaling (insulin-induced activation of p44/42 MAPK) was enhanced by Ang II. PTP1B activity was significantly increased following Ang II stimulation. Knocking down of PTP1B with antisense oligonucleotides reversed the inhibitory effect of Ang II on insulin-induced activation of IRS-1 and Akt. In contrast, the combined effect of Ang II and insulin on VSMC growth was completely blunted in PTP1B antisense transfected group. Additionally, we also observed that Ang II-induced activation of PTP1B was dependent upon activation of protein kinase A (PKA) and JAK2 in VSMC [74]. Our data provides further evidence that PTP1B selectively interferes with IR/IRS-1/Akt, while having a positive impact on MAPK signaling.

Insulin maintains VSMC quiescence and counteracts PDGF-stimulated dedifferentiation *via* a PI3K-dependent mechanism but promotes VSMC migration through MAPK [75]. Our data are also in line with the observation that during insulin resistance insulin-mediated IR/IRS-1/Akt is usually impaired whereas growth signaling remains intact. Moreover, PTP1B protein expression in human media artery SMCs was upregulated following long-term exposure to high glucose [76], resulting in VSMC insulin resistance. Other inflammatory signaling such as TNF alpha through activation of nuclear factor κB (NFκB) has been shown to increase PTP1B expression in other tissues [9]. It is interesting to speculate that these inflammatory signals attribute to upregulation of vascular PTP1B, hence, leading to imbalance of insulin signaling and vascular complications in type 2 diabetes.

Conclusion

Taken together, the literature reviewed above indicates that PTP1B, besides being a key regulator of metabolic function, is a major player in the control of cardiovascular function in the context of obesity and type 2 diabetes. Whether PTP1B modulates cardiovascular function directly or indirectly through the control of metabolic function remains an active avenue of research. Indeed, whereas *in vitro* studies clearly demonstrate that PTP1B regulates vascular function by interfering with various receptor tyrosine kinase signaling pathways, both in endothelial and SMCs, human genetic studies and animals studies argue favorably for indirect effects *via* the control of the metabolic function. Although further studies are required to determine precisely how PTP1B intervenes in the control of cardiovascular function, current evidence clearly implicates this PTP in the pathogenesis of cardiovascular dysfunction and suggests that manipulations of PTP1B expression and activity could provide a novel therapeutic avenue for the treatment of obesity-related cardiovascular dysfunctions.

References

1. Bhanot S, Bryer-Ash M, Cheung A, McNeill JH (1994) Bis(maltolato)oxovanadium(IV) attenuates hyperinsulinemia and hypertension in spontaneously hypertensive rats. Diabetes 43: 857–861
2. Yuen VG, Pederson RA, Dai S, Orvig C, McNeill JH (1996) Effects of low and high dose administration of bis(maltolato)oxovanadium(IV) on fa/fa Zucker rats. Can J Physiol Pharmacol 74:1001–1009
3. Hopfner RL, McNeill JR, Gopalakrishnan V (1998) Vanadate treatment normalizes exaggerated vascular smooth muscle responses in the obese Zucker rat. Eur J Pharmacol 357:61–65
4. Swarup G, Cohen S, Garbers DL (1982) Inhibition of membrane phosphotyrosyl-protein phosphatase activity by vanadate. Biochem Biophys Res Commun 107:1104–1109
5. Huyer G, Liu S, Kelly J, Moffat J, Payette P, Kennedy B et al (1997) Mechanism of inhibition of protein-tyrosine phosphatases by vanadate and pervanadate. J Biol Chem 272:843–851
6. Mohammad A, Wang J, McNeill JH (2002) Bis(maltolato)oxovanadium(IV) inhibits the activity of PTP1B in Zucker rat skeletal muscle in vivo. Mol Cell Biochem 229:125–128
7. Elchebly M, Payette P, Michaliszyn E, Cromlish W, Collins S, Loy AL et al (1999) Increased insulin sensitivity and obesity resistance in mice lacking the protein tyrosine phosphatase-1B gene. Science 283:1544–1548
8. Cheng A, Uetani N, Simoncic PD, Chaubey VP, Lee-Loy A, McGlade CJ et al (2002) Attenuation of leptin action and regulation of obesity by protein tyrosine phosphatase 1B. Dev Cell 2:497–503
9. Zabolotny JM, Bence-Hanulec KK, Stricker-Krongrad A, Haj F, Wang Y, Minokoshi Y et al (2002) PTP1B regulates leptin signal transduction in vivo. Dev Cell 2:489–495
10. Hegyi K, Fülöp K, Kovács K, Tóth S, Falus A (2004) Leptin-induced signal transduction pathways. Cell Biol Int 28:159–169
11. Kannel W, Brand M, Skinner J, Dawber T, McNamara P (1967) The relation of adiposity to blood pressure and development of hypertension: the Framingham study. Ann Intern Med 67:48–59
12. Lembertas AV, Perusse L, Chagnon YC, Fisler JS, Warden CH, Purcell DAH et al (1997) Identification of an obesity quantitative trait locus on mouse chromosome 2 and evidence of linkage to body fat and insulin on the human homologous region 20q. J Clin Invest 100:1240–1247
13. Tsou RC, Bence KK (2012) The genetics of PTPN1 and obesity: insights from mouse models of tissue-specific PTP1B deficiency. J Obes 2012:926857
14. Di Paola R, Frittitta L, Miscio G, Bozzali M, Baratta R, Centra M et al (2002) A variation in 3′ UTR of hPTP1B increases specific gene expression and associates with insulin resistance. Am J Hum Genet 70:806–812
15. Cheyssac C, Lecoeur C, Dechaume A, Bibi A, Charpentier G, Balkau B et al (2006) Analysis of common PTPN1 gene variants in type 2 diabetes, obesity and associated phenotypes in the French population. BMC Med Genet 7:44
16. Olivier M, Hsiung CA, Chuang L-M, Ho L-T, Ting C-T, Bustos VI et al (2004) Single nucleotide polymorphisms in protein tyrosine phosphatase 1β (PTPN1) are associated with essential hypertension and obesity. Hum Mol Genet 13:1885–1892
17. Gu P, Jiang W, Du H, Shao J, Lu B, Wang J et al (2010) Protein tyrosine phosphatase 1B gene polymorphisms and essential hypertension: a case-control study in Chinese population. J Endocrinol Invest 33:483–488
18. Santaniemi M, Ukkola O, Kesaniemi YA (2004) Tyrosine phosphatase 1B and leptin receptor genes and their interaction in type 2 diabetes. J Intern Med 256:48–55
19. Bento JL, Palmer ND, Mychaleckyj JC, Lange LA, Langefeld CD, Rich SS et al (2004) Association of protein tyrosine phosphatase 1B gene polymorphisms with type 2 diabetes. Diabetes 53:3007–3012
20. Burdon KP, Bento JL, Langefeld CD, Campbell JK, Carr JJ, Wagenknecht LM et al (2006) Association of protein tyrosine phosphatase-N1 polymorphisms with coronary calcified plaque in the Diabetes Heart Study. Diabetes 55:651–658

21. Dimmeler S, Fleming I, Fisslthaler B, Hermann C, Busse R, Zeiher AM (1999) Activation of nitric oxide synthase in endothelial cells by Akt-dependent phosphorylation. Nature 399: 601–605
22. Fulton D, Gratton J-P, McCabe TJ, Fontana J, Fujio Y, Walsh K et al (1999) Regulation of endothelium-derived nitric oxide production by the protein kinase Akt. Nature 399:597–601
23. Vercauteren M, Remy E, Devaux C, Dautreaux B, Henry J-P, Bauer F et al (2006) Improvement of peripheral endothelial dysfunction by protein tyrosine phosphatase inhibitors in heart failure. Circulation 114:2498–2507
24. Gomez E, Vercauteren M, Kurtz B, Ouvrard-Pascaud A, Mulder P, Henry JP et al (2012) Reduction of heart failure by pharmacological inhibition or gene deletion of protein tyrosine phosphatase 1B. J Mol Cell Cardiol 52:1257–1264
25. Klaman LD, Boss O, Peroni OD, Kim JK, Martino JL, Zabolotny JM et al (2000) Increased energy expenditure, decreased adiposity, and tissue-specific insulin sensitivity in protein-tyrosine phosphatase 1B-deficient mice. Mol Cell Biol 20:5479–5489
26. Ali MI, Ketsawatsomkron P, Belin de Chantemele EJ, Mintz JD, Muta K, Salet C et al (2009) Deletion of protein tyrosine phosphatase 1b improves peripheral insulin resistance and vascular function in obese, leptin-resistant mice via reduced oxidant tone. Circ Res 105:1013–1022
27. Agouni A, Tual-Chalot S, Duluc L, Martinez CM, Mody N, Andriantsitohaina R et al (2012) Liver-specific deletion of protein tyrosine phosphatase (PTP) 1B improves endothelial dysfunction and cardiovascular alterations associated with obesity in mice. FASEB J 26:526.5
28. Belin De Chantemele EJ, Ali MI, Mintz J, Rainey WE, Tremblay ML, Fulton DJ et al (2012) Increasing peripheral insulin sensitivity by PTP1B deletion improves control of blood pressure in obesity. Hypertension 60:1273–1279
29. Arnold AC, Nautiyal M, Diz DI (2012) Protein phosphatase 1b in the solitary tract nucleus is necessary for normal baroreflex function. J Cardiovasc Pharmacol 59:472–478
30. Dunbar JC, Hu Y, Lu H (1997) Intracerebroventricular leptin increases lumbar and renal sympathetic nerve activity and blood pressure in normal rats. Diabetes 46:2040–2043
31. Mark AL, Sivitz WI (2002) Uncoupling metabolism and coupling leptin to cardiovascular disease. Arterioscler Thromb Vasc Biol 22:881–883
32. Rahmouni K, Morgan DA, Morgan GM, Mark AL, Haynes WG (2005) Role of selective leptin resistance in diet-induced obesity hypertension. Diabetes 54:2012–2018
33. Haynes WG, Morgan DA, Walsh SA, Mark AL, Sivitz WI (1997) Receptor-mediated regional sympathetic nerve activation by leptin. J Clin Invest 100:270–278
34. Grassi G, Seravalle G, Cattaneo BM, Bolla GB, Lanfranchi A, Colombo M et al (1995) Sympathetic activation in obese normotensive subjects. Hypertension 25:560–563
35. Grassi G, Seravalle G, Dell'Oro R, Turri C, Bolla GB, Mancia G (2000) Adrenergic and reflex abnormalities in obesity-related hypertension. Hypertension 36:538–542
36. Vaz M, Jennings G, Turner A, Cox H, Lambert G, Esler M (1997) Regional sympathetic nervous activity and oxygen consumption in obese normotensive human subjects. Circulation 96:3423–3429
37. Belin de Chantemele EJ, Muta K, Mintz J, Tremblay ML, Marrero MB, Fulton DJ et al (2009) Protein tyrosine phosphatase 1B, a major regulator of leptin-mediated control of cardiovascular function. Circulation 120:753–763
38. Belin de Chantemele EJ, Mintz JD, Rainey WE, Stepp DW (2011) Impact of leptin-mediated sympatho-activation on cardiovascular function in obese mice. Hypertension 58:271–279
39. Asante-Appiah E, Kennedy BP (2003) Protein tyrosine phosphatases: the quest for negative regulators of insulin action. Am J Physiol Endocrinol Metab 284:E663–E670
40. Deem TL, Abdala-Valencia H, Cook-Mills JM (2007) VCAM-1 activation of endothelial cell protein tyrosine phosphatase 1B. J Immunol 178:3865–3873
41. Ishino Y, Zhu C, Harris DL, Joyce NC (2008) Protein tyrosine phosphatase-1B (PTP1B) helps regulate EGF-induced stimulation of S-phase entry in human corneal endothelial cells. Mol Vis 14:61–70
42. Fulton D, Harris MB, Kemp BE, Venema RC, Marrero MB, Stepp DW (2004) Insulin resistance does not diminish eNOS expression, phosphorylation, or binding to HSP-90. Am J Physiol Heart Circ Physiol 287:H2384–H2393

43. Zhang H, Dellsperger KC, Zhang C (2012) The link between metabolic abnormalities and endothelial dysfunction in type 2 diabetes: an update. Basic Res Cardiol 107:237
44. Baron AD, Clark MG (1997) Role of blood flow in the regulation of muscle glucose uptake. Annu Rev Nutr 17:487–499
45. Montagnani M, Chen H, Barr VA, Quon MJ (2001) Insulin-stimulated activation of eNOS is independent of Ca2+ but requires phosphorylation by Akt at Ser(1179). J Biol Chem 276:30392–30398
46. Montagnani M, Ravichandran LV, Chen H, Esposito DL, Quon MJ (2002) Insulin receptor substrate-1 and phosphoinositide-dependent kinase-1 are required for insulin-stimulated production of nitric oxide in endothelial cells. Mol Endocrinol 16:1931–1942
47. Zeng G, Nystrom FH, Ravichandran LV, Cong LN, Kirby M, Mostowski H et al (2000) Roles for insulin receptor, PI3-kinase, and Akt in insulin-signaling pathways related to production of nitric oxide in human vascular endothelial cells. Circulation 101:1539–1545
48. Zeng G, Quon MJ (1996) Insulin-stimulated production of nitric oxide is inhibited by wortmannin. Direct measurement in vascular endothelial cells. J Clin Invest 98:894–898
49. Potenza MA, Marasciulo FL, Chieppa DM, Brigiani GS, Formoso G, Quon MJ et al (2005) Insulin resistance in spontaneously hypertensive rats is associated with endothelial dysfunction characterized by imbalance between NO and ET-1 production. Am J Physiol Heart Circ Physiol 289(2):H813–H822
50. Kim JA, Montagnani M, Koh KK, Quon MJ (2006) Reciprocal relationships between insulin resistance and endothelial dysfunction: molecular and pathophysiological mechanisms. Circulation 113:1888–1904
51. Eringa EC, Stehouwer CDA, Roos MH, Westerhof N, Sipkema P (2007) Selective resistance to vasoactive effects of insulin in muscle resistance arteries of obese Zucker (fa/fa) rats. Am J Physiol Endocrinol Metab 293(5):E1134–E1139
52. Jiang ZY, Lin YW, Clemont A, Feener EP, Hein KD, Igarashi M et al (1999) Characterization of selective resistance to insulin signaling in the vasculature of obese Zucker (fa/fa) rats. J Clin Invest 104:447–457
53. Kuboki K, Jiang ZY, Takahara N, Ha SW, Igarashi M, Yamauchi T et al (2000) Regulation of endothelial constitutive nitric oxide synthase gene expression in endothelial cells and in vivo: a specific vascular action of insulin. Circulation 101:676–681
54. Zecchin HG, Priviero FB, Souza CT, Zecchin KG, Prada PO, Carvalheira JB et al (2007) Defective insulin and acetylcholine induction of endothelial cell-nitric oxide synthase through insulin receptor substrate/Akt signaling pathway in aorta of obese rats. Diabetes 56:1014–1024
55. Picchi A, Gao X, Belmadani S, Potter BJ, Focardi M, Chilian WM et al (2006) Tumor necrosis factor-alpha induces endothelial dysfunction in the prediabetic metabolic syndrome. Circ Res 99:69–77
56. Molnar J, Yu S, Mzhavia N, Pau C, Chereshnev I, Dansky HM (2005) Diabetes induces endothelial dysfunction but does not increase neointimal formation in high-fat diet fed C57BL/6J mice. Circ Res 96(11):1178–1184
57. Hsu MF, Meng TC (2010) Enhancement of insulin responsiveness by nitric oxide-mediated inactivation of protein-tyrosine phosphatases. J Biol Chem 285:7919–7928
58. Haj FG, Markova B, Klaman LD, Bohmer FD, Neel BG (2003) Regulation of receptor tyrosine kinase signaling by protein tyrosine phosphatase-1B. J Biol Chem 278:739–744
59. Nakamura Y, Patrushev N, Inomata H, Mehta D, Urao N, Kim HW et al (2008) Role of protein tyrosine phosphatase 1B in vascular endothelial growth factor signaling and cel–cell adhesions in endothelial cells. Circ Res 102:1182–1191
60. Heida NM, Leifheit-Nestler M, Schroeter MR, Muller JP, Cheng IF, Henkel S et al (2010) Leptin enhances the potency of circulating angiogenic cells via src kinase and integrin (alpha) vbeta5: implications for angiogenesis in human obesity. Arterioscler Thromb Vasc Biol 30: 200–206
61. Grinnell KL, Chichger H, Braza J, Duong H, Harrington EO (2012) Protection against LPS-induced pulmonary edema through the attenuation of protein tyrosine phosphatase-1B oxidation. Am J Respir Cell Mol Biol 46:623–632

62. Abdala-Valencia H, Berdnikovs S, Cook-Mills JM (2011) Mechanisms for vascular cell adhesion molecule-1 activation of ERK1/2 during leukocyte transendothelial migration. PLoS One 6:e26706
63. Hassid A, Huang S, Yao J (1999) Role of PTP-1B in aortic smooth muscle cell motility and tyrosine phosphorylation of focal adhesion proteins. Am J Physiol 277:H192–H198
64. Hassid A, Yao J, Huang S (1999) NO alters cell shape and motility in aortic smooth muscle cells via protein tyrosine phosphatase 1B activation. Am J Physiol Heart Circ Physiol 277:H1014–H1026
65. Zhuang D, Balani P, Pu Q, Thakran S, Hassid A (2011) Suppression of PKG by PDGF or nitric oxide in differentiated aortic smooth muscle cells: obligatory role of protein tyrosine phosphatase 1B. Am J Physiol Heart Circ Physiol 300:H57–H63
66. Sreejayan N, Lin Y, Hassid A (2002) NO attenuates insulin signaling and motility in aortic smooth muscle cells via protein tyrosine phosphatase 1B-mediated mechanism. Arterioscler Thromb Vasc Biol 22:1086–1092
67. Ferns GA, Raines EW, Sprugel KH, Motani AS, Reidy MA, Ross R (1991) Inhibition of neointimal smooth muscle accumulation after angioplasty by an antibody to PDGF. Science 253: 1129–1132
68. Jackson CL, Reidy MA (1993) Basic fibroblast growth factor: its role in the control of smooth muscle cell migration. Am J Pathol 143:1024–1031
69. Chang Y, Zhuang D, Zhang C, Hassid A (2004) Increase of PTP levels in vascular injury and in cultured aortic smooth muscle cells treated with specific growth factors. Am J Physiol Heart Circ Physiol 287:H2201–H2208
70. Zhuang D, Pu Q, Ceacareanu B, Chang Y, Dixit M, Hassid A (2008) Chronic insulin treatment amplifies PDGF-induced motility in differentiated aortic smooth muscle cells by suppressing the expression and function of PTP1B. Am J Physiol Heart Circ Physiol 295:H163–H173
71. Wright MB, Seifert RA, Bowen-Pope DF (2000) Protein-tyrosine phosphatases in the vessel wall: differential expression after acute arterial injury. Arterioscler Thromb Vasc Biol 20: 1189–1198
72. Chang Y, Ceacareanu B, Zhuang D, Zhang C, Pu Q, Ceacareanu AC et al (2006) Counterregulatory function of protein tyrosine phosphatase 1B in platelet-derived growth factor- or fibroblast growth factor-induced motility and proliferation of cultured smooth muscle cells and in neointima formation. Arterioscler Thromb Vasc Biol 26:501–507
73. Dube N, Cheng A, Tremblay ML (2004) The role of protein tyrosine phosphatase 1B in Ras signaling. Proc Natl Acad Sci U S A 101:1834–1839
74. Ketsawatsomkron P, Stepp DW, Fulton DJ, Marrero MB (2010) Molecular mechanism of angiotensin II-induced insulin resistance in aortic vascular smooth muscle cells: roles of protein tyrosine phosphatase-1B. Vascul Pharmacol 53:160–168
75. Wang CC, Gurevich I, Draznin B (2003) Insulin affects vascular smooth muscle cell phenotype and migration via distinct signaling pathways. Diabetes 52:2562–2569
76. Popov D, Nemecz M, Dumitrescu M, Georgescu A, Bohmer FD (2009) Long-term high glucose concentration influences Akt, ERK1/2, and PTP1B protein expression in human aortic smooth muscle cells. Biochem Biophys Res Commun 388:51–55

Chapter 8
Role of the SHP2 Protein Tyrosine Phosphatase in Cardiac Metabolism

Maria I. Kontaridis, Eleni V. Geladari, and Charalampia V. Geladari

Abstract The heart relies heavily on oxidation and requires an integrally responsive metabolic function to maintain cardiac homeostasis. There is increasing evidence that it is this loss of metabolic flexibility that ultimately leads to cardiac dysfunction in disease conditions such as diabetes, ischemic heart disease, hypertrophic cardiomyopathy (HCM), and heart failure.

The SH2 domain-containing protein tyrosine phosphatase (PTP), Src homology protein 2 (SHP2), encoded by the *PTPN11* gene, is the first PTP to be directly implicated in cardiac disease (Nat Genet, 29(4):465–468, 2001; J Med Genet 39(8):571–574, 2002; J Med Genet 40(9):704–708, 2003) and is the first identified PTP found to have a critical role in adult cardiac function (Circulation 117(11):1423–1435, 2008; Mol Cell Biol 29(2):378–388, 2009). Indeed, differing mutations within SHP2 elicit distinct biochemical properties of the enzyme, each manifesting in a unique panoply of cardiac defects, including HCM.

Given the already identified key role for SHP2 in the heart, it is likely that SHP2 plays a significant role in cardiac metabolism as well. However, while the critical signaling pathways necessary for metabolic function in the heart overlap significantly with those known to be controlled by SHP2, a direct role for SHP2 in cardiac metabolism has not yet been elucidated. Here, we will discuss what is known about the functional role for SHP2 in the heart, how mutations in SHP2 can affect cardiac disease progression, and what direct or indirect mechanisms may exist for SHP2 regulation of cardiac metabolism.

M.I. Kontaridis, Ph.D. (✉)
Department of Medicine, Division of Cardiology, and Department of Biological Chemistry and Molecular Pharmacology, Beth Israel Deaconess Medical Center, Harvard Medical School, Boston, MA 02115, USA
e-mail: mkontari@bidmc.harvard.edu

E.V. Geladari, M.D. • C.V. Geladari, M.D.
Department of Medicine, Division of Cardiology, Beth Israel Deaconess Medical Center, Boston, MA 02115, USA

K.K. Bence (ed.), *Protein Tyrosine Phosphatase Control of Metabolism*,
DOI 10.1007/978-1-4614-7855-3_8, © Springer Science+Business Media New York 2013

Abbreviations

AMPK	Adenosine monophosphate-activated protein kinase
ATP	Adenosine triphosphate
CHDs	Congenital heart defects
CICR	Calcium-induced calcium release
CM	Cardiomyocyte
Db/Db	Diabetes/diabetes
DCM	Dilated cardiomyopathy
EGFR	Epidermal growth factor receptor
ERK	Extracellular regulated kinase
ET-1	Endothelin-1
ETC	Electron transport chain
$Ex3^{-/-}$	Exon 3 deleted
FA	Free fatty acids
FAK	Focal adhesion kinase
FOXO	Forkhead box transcription factors
FRS-2	Fibroblast growth factor receptor substrate-2
GAB-1	GRB2 associated binder-1
GLUT-1	Glucose transporter type-1
GLUT-4	Glucose transporter type-4
GOF	Gain-of-function
GPCR	G-protein coupled receptor
GSK3β	Glycogen synthase kinase 3β
HB-EGF	Heparin binding-epidermal growth factor
HCM	Hypertrophic cardiomyopathy
HF	Heart failure
IGF	Insulin-like growth factor
IGFR	Insulin-like growth factor receptor
IR	Insulin receptor
IRS-1	Insulin receptor substrate-1
JAK	Janus kinases
LS	Leopard syndrome
MAPK	Mitogen-activated protein kinase
MEFs	Mouse embryonic fibroblasts
MKK6	Mitogen-activated protein kinase kinase 6
MMPs	Metalloproteases
mTOR	Mammalian target of Rapamycin
NADH	Nicotinamide adenine dinucleotide
NADPH	Nicotinamide adenine dinucleotide phosphate
NF-kB	Nuclear factor B
NOS	Nitric oxide synthase
NRVMs	Neonatal rat ventricular myocytes
NS	Noonan syndrome

Ob/Ob	Obesity/obesity
OxPhos	Oxidative phosphorylation
PDGF	Platelet-derived growth factor
PI3K	Phosphatidylinositol 3-kinase
PKC	Protein kinase C
PTEN	Phosphatase and tensin homolog deleted on chromosome ten
PTP	Protein tyrosine phosphatase
PTPN11	Protein tyrosine phosphatase non-receptor type-11
pY	Phosphotyrosyl peptide
RhoA	Ras homolog gene family, member A
ROS	Reactive oxygen species
RTKs	Receptor tyrosine kinases
SFK	SRC family kinase
SH2	Src homology 2
SHP2	Src homology protein 2
STAT	Signal transducer activator of transcription
T-tubule	Transverse tubule
UT-II	Urotensin-II

SHP2: Structure, Regulation, and Signaling

The Src homology 2 (SH2) domain-containing PTP Src homology protein 2 (SHP2) is a ubiquitously expressed non-receptor protein tyrosine phosphatase (PTP) containing two SH2 domains, a PTP domain and a C-terminal tail with two tyrosine phosphorylation sites and a proline-rich motif [1]. The SH2 domains play a key role in both the conformational regulation and cellular targeting of the phosphatase. In the basal state, the backside loop of the N-SH2 of SHP2 is wedged into the PTP domain, rendering it both physically and chemically inactive [2] (Fig. 8.1). Binding of SHP2 to a phosphotyrosyl (pY) protein alters the conformation of its N-SH2 domain, preventing interaction of the backside loop with the PTP domain, thereby unfolding the enzyme and resulting in phosphatase activation [2–4]. Therefore, binding of the N-SH2 domain to a pY interacting protein is coupled to the activation of the phosphatase. The interaction of SHP2 with pY proteins, such as receptor tyrosine kinases (RTKs), cytokine receptors, scaffolding adapters [e.g., IRS-1 (Insulin Receptor Substrate-1), GAB-1 (GRB2 Associated Binder-1) and FRS-2 (Fibroblast growth factor Receptor Substrate-2) proteins], and immune inhibitory receptors mediates downstream signaling events that control important cellular processes such as proliferation, differentiation, and apoptosis [5].

SHP2 is intricately involved in regulatory cellular function. SHP2 is required for Ras/Extracellular Regulated Kinase (ERK) pathway activation by nearly all RTKs, cytokine receptors, and integrins [5–8]. SHP2 can bind either directly to RTKs, cytokine receptors, and/or integrins or indirectly through scaffolding or adapter proteins in order to become activated [5]. Cells expressing dominant negative SHP2 [9]

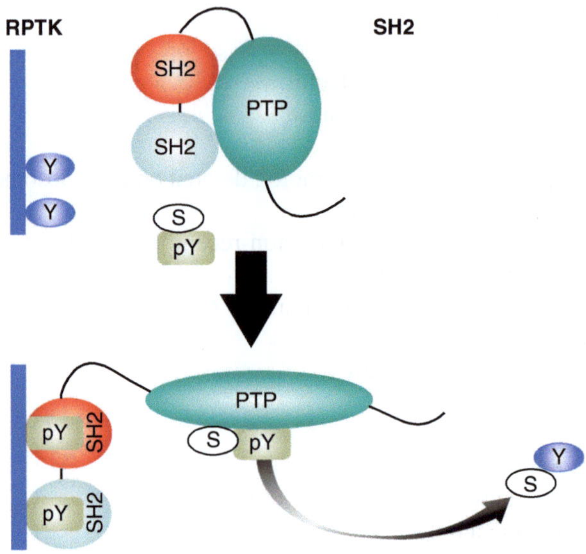

Fig. 8.1 SHP2 mechanism of action. In the basal state, the backside loop of the N-SH2 of SHP2 is wedged into the PTP domain, rendering the PTP domain physically and chemically inactive. Binding of SHP2 to an activated tyrosine kinase receptor and/or a tyrosyl phosphorylated scaffolding protein alters the conformation of SHP2, unfolding the enzyme, and resulting in phosphatase activation. This then allows access for binding of its substrates (S) to the catalytic pocket, where they can be dephosphorylated

or SHP2 Exon 3 deleted ($Ex3^{-/-}$) mouse embryonic fibroblasts (MEFs) [10] exhibited defective Ras activation, suggesting SHP2 acts upstream of Ras. However, other in vitro experiments showed that a catalytically inactive mutant of SHP2 could perturb (at least some) components of downstream signaling, even in the presence of a constitutively active Ras, suggesting that SHP2 may function in an upstream and/or parallel pathway to Ras [11, 12].

Despite extensive study, however, the mechanism(s) by which SHP2 mediates Ras/ERK activation remain controversial. SHP2 has been implicated in RTK-evoked Src family kinase (SFK) activation, with SFKs in turn being required for sustained Ras activation [13]. Other suggested targets include the RasGap binding sites on some RTKs [14, 15] and the regulatory pY sites on Sprouty proteins, a family of Ras inhibitors [16, 17]. Besides its positive signaling functions, SHP2 negatively regulates the phosphatidylinositol 3-kinase (PI3K)/AKT pathway in Epidermal Growth Factor Receptor (EGFR) signaling, but is required, at least in some cases, for Platelet-Derived Growth Factor (PDGF) and Insulin Growth Factor (IGF)-evoked AKT activity [12, 18]. Additional functional roles for SHP2 have been reported in Janus kinase (JAK)/Signal Transducer Activator of Transcription (STAT) [5], Nuclear Factor B (NF-kB) [7], and RhoA [19] signaling pathways.

Role for SHP2 in Cardiac Function

Because SHP2 can modulate signaling pathways downstream of multiple RTKs, integrin, and cytokine receptors, a defect in SHP2 can trigger complex cardiac disease phenotypes [20]. In an in vivo study, SHP2 was shown to be a critical regulator of normal adult cardiac function through its regulation of both the ERK and RhoA signaling pathways. Mice with muscle-specific deletion of *Ptpn11* rapidly developed severe dilated cardiomyopathy (DCM), without an intervening hypertrophic phase and with signs of cardiac dysfunction [21]. SHP2-deficient primary cardiomyocytes (CMs) isolated from these mice showed defective ERK activation in response to a variety of soluble agonists as well as to pressure overload, but also had hyperactivation of RhoA signaling, suggesting that both of these pathways contribute significantly to the dilated phenotype in SHP2-deficient hearts [21].

SHP2 has an already demonstrated role in controlling CM size in response to biomechanical stress by a mechanism that is dependent on focal adhesion kinase (FAK), SRC, and mTOR activities [22]. Low phosphorylation levels of FAK in nonstretched neonatal rat ventricular myocytes (NRVMs) coincided with a relatively high basal association of FAK with SHP2 and an increase in SHP2 phosphatase activity. In response to biomechanical stress, however, FAK/SHP2 association subsided and FAK phosphorylation increased, suggesting SHP2 negatively regulates hypertrophic signaling pathways in NRVMs. Indeed, depletion of SHP2 by specific small interfering RNA (siRNA) increased the phosphorylation of FAK and that of its downstream effectors, SRC, AKT, and mTOR, and induced a hypertrophic phenotype in nonstretched NRVMs. Inhibition of mTOR with rapamycin blunted both the hypertrophy observed in NRVMs depleted of SHP2 as well as the stretch-induced hypertrophy of NRVMs [22], suggesting that SHP2 is a critical regulator of the hypertrophic signaling process in CMs.

In addition, the role for SHP2 in human cardiac pathophysiology has recently begun to be elucidated. Missense *PTPN11* mutations were identified in ~50 % of cases of Noonan Syndrome (NS), a fairly common (1:1,000–1:2,500 live births) autosomal-dominant disorder characterized by multiple, variably penetrant defects, including proportionate short stature, facial dysmorphism, and skeletal anomalies [1, 23]. In addition, one or more congenital heart defects (CHDs) are usually present, including pulmonary valve stenosis with dysplastic leaflets, atrial/ventricular septal defects, and hypertrophic cardiomyopathy (HCM) [24]. However, genotype-phenotype correlation studies of NS patients showed that pulmonary valve stenosis and atrial septal defects were the two most common cardiac abnormalities attributed to SHP2 [25]; indeed, only 6 % of SHP2-associated NS patients present with HCM [26, 27], suggesting that HCM is not caused by most NS-specific SHP2 mutations.

Biochemically, nearly all NS-specific SHP2 mutations are activating mutants, i.e., they have increased PTP activity [28, 29]. Since most of the NS mutations in SHP2 reside within the N-SH2 domain interface that allows for the intermolecular interactions with the PTP domain, they disrupt the ability of SHP2 to retain the closed, inactive conformation without affecting the PTP domain. Therefore, NS mutants display increased basal activity (since they are unfolded) and behave as gain-of-function (GOF) alleles (since the PTP domain is open and exposed to substrate) [29]. Most, if not all, NS mutations are thought to cause disease by potentiating the Ras/ERK (and possibly other downstream) signaling pathway(s) [28, 30–34]. Indeed, in a heterozygous mouse model expressing the NS-specific SHP2 mutation D61G (D61G/+), embryos showed multiple cardiac defects, including enlarged endocardial cushions, all of which were attributed to the observed increase in ERK activation in these mouse hearts [31].

Dominant SHP2 missense mutations also cause a related, but less well-characterized disorder, LEOPARD syndrome (LS). LEOPARD is an acronym for multiple lentigines, electrocardiographic conduction abnormalities, ocular hypertelorism, pulmonary stenosis, abnormalities of genitalia, retardation of growth, and sensorineural deafness [35]. Because NS and LS share several features (pulmonic stenosis, ocular hypertelorism, growth retardation, genital abnormalities, and deafness), and each results from SHP2 mutations, they were originally thought to have similar disease pathogenesis. However, the biochemical properties of LS mutations were found to be distinct from *PTPN11* mutations that cause NS. Like the SHP2 mutations associated with NS, LS mutations perturb N-SH2/PTP domain interactions [36]. However, unlike NS, LS mutants typically affect conserved residues important for PTP catalysis (e.g., Y279, sets the depth of the catalytic cleft; T468, lies within/near the "signature motif.") [36]. Therefore, although LS alleles, like NS, are preferentially found in an open conformation, they are, unlike NS, catalytically impaired [12, 33, 36]. Indeed, LS mutations behave as dominant negatives and interfere with growth factor and downstream ERK-mediated signaling [36]. A recently generated heterozygous LS mouse model (Y279C/+) recapitulated nearly all aspects of the human disorder, including short stature, craniofacial dysmorphia, and morphologic, histologic, echocardiographic, and molecular evidence of HCM [37]. Heart and CM lysates isolated from Y279C/+ mice showed enhanced binding of SHP2 to IRS-1, decreased SHP2 catalytic activity, and abrogated agonist-evoked ERK signaling [37]. Interestingly, Y279C/+ mice also exhibited increased basal and agonist-induced AKT and mammalian target of rapamycin (mTOR) activities, and the cardiac defects in these mice were completely reversed by treatment with rapamycin, an inhibitor of mTOR [37]. Indeed, mice with cardiomyocyte-specific overexpression of another LS mutation, the Q510E, also developed HCM through a mechanism involving enhanced AKT and mTOR activities [38], together suggesting that the LS mutations in *PTPN11* enhance mTOR activity and that this pathway is critical for LS-associated HCM. Finally, and likely as a consequence of the increased AKT activity observed in LS mutant cells, glycogen synthase kinase-3β (GSK3β) hyper-phosphorylation was also observed in LS mutant fibroblasts [39, 40], suggesting a likely aberrant role for SHP2 loss-of-function mutations in glucose metabolism and energy storage in the heart, although this has yet to be determined directly.

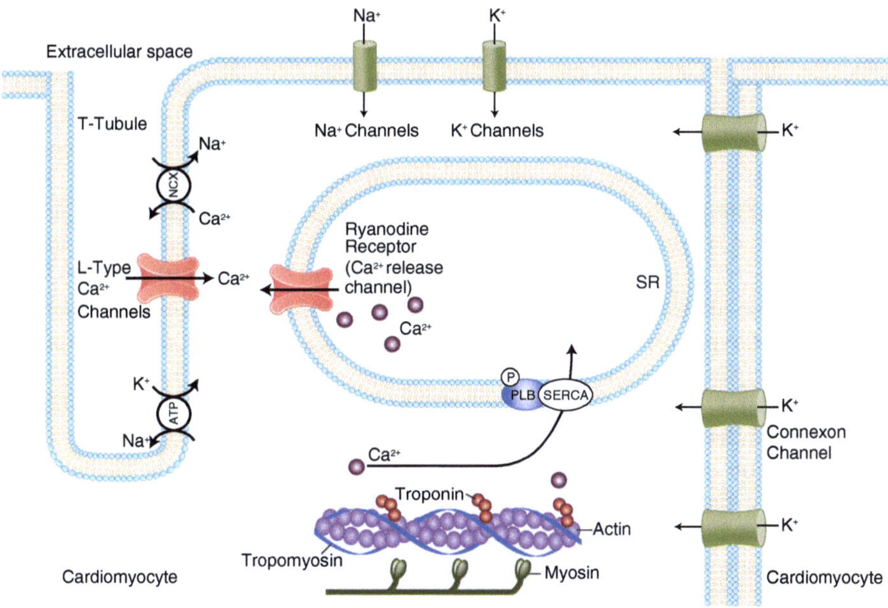

Fig. 8.2 Excitation-contraction coupling in cardiac muscle. The initial event in the cardiac muscle contraction is membrane depolarization, which occurs with ion entry through connexon channels from a neighboring cardiomyocyte followed by opening of voltage-gated Na^+ channels and Na^+ entry. The resultant rapid depolarization of the membrane inactivates Na^+ channels and opens both K^+ channels and Ca^{2+} channels. Entry of Ca^{2+} into the cell triggers the release of Ca^{2+} from the sarcoplasmic reticulum through the ryanodine channel. Ca^{2+} then binds to the troponin complex and activates the contractile apparatus. Cellular relaxation occurs on removal of Ca^{2+} from the cytosol by the Ca^{2+}-uptake pumps of the sarcoplasmic reticulum and by Na^+/Ca^{2+} exchange with the extracellular fluid

Importance of Cardiac Function and Regulation

The heart's primary function in the body is to maintain proper oxygenation through contraction. CMs are the cells that can regulate the contractile motion of the heart; they are critical for pumping blood efficiently to the lungs and to the rest of the body. To induce this action, the CMs undergo excitation-contraction coupling [41, 42] (Fig. 8.2). Initiation of the action potential in CMs stems from the entry of sodium ions across the sarcolemma. As the muscle impulse spreads from the sarcolemma to the transverse tubule (T-tubule), calcium ions are released into the sarcoplasm. In addition, an inward flux of extracellular calcium ions enters the cell through L-type calcium channels to sustain the depolarization and to maximize the duration of contraction [43]. Then, through a mechanism of calcium-induced calcium release (CICR), additional calcium is released into the sarcoplasm from the sarcoplasmic reticulum, increasing intracellular calcium levels. This increase in calcium flux inside the cell allows for the calcium ions to bind to the protein troponin on the actin filament, allowing it undergo a conformational change that exposes the binding sites for myosin on actin filaments [44]. Once myosin is bound to actin, it initiates "cross-bridge cycling," a process which regulates the rhythmic contraction of cardiac muscle cells [44].

In response to external stimuli and/or pathological stresses such as hypertension, heart muscle injury (myocardial infarction), or neurohumoral stimulation, additional signaling pathways (i.e., G-protein coupled receptor (GPCR), RTK, and/or cytokine[45–47]) become activated, enhancing excitation-contraction coupling, to allow the heart to compensate for the excess demand [43, 44, 48, 49]. The increased signaling also induces a pathological remodeling of the heart, making it abnormally thickened or hypertrophic (HCM) [50, 51]. Therefore, while onset of HCM appears initially compensatory, over time, when the heart can no longer keep up with the excess demand, it decompensates, transitions to dilation and, ultimately, to end-stage heart failure (HF) [50, 52], a serious life-threatening condition and one of the most common causes of death worldwide [53–55]. Importantly, the pathological remodeling events in the heart described above are tied to the aberrant up-regulation of pathways which are, in large part, controlled by SHP2, including GPCR, RTK, cytokine, and calcium-modulated pathways [45–47].

Mitochondria, Metabolism, and SHP2 in CMs

The heart optimally functions under conditions of aerobic metabolism, with only about 1 % of energy derived from anaerobic metabolism [56, 57]. Because of the heart's ability to delicately balance metabolic respiration, it is capable of compensating, at least to some degree, to a decreased oxygen supply. For example, under mild to moderate hypoxic conditions, the heart is able to increase anaerobic metabolism to ~10 %, in order to maintain proper contraction [58]. However, in cases of severe hypoxia, the heart loses its ability to produce enough energy via production of lactate and can no longer sustain ventricular contractions.

Importantly, under normal aerobic conditions, the majority of the energy (~70 %) needed for heart function comes from free fatty acid (FA) metabolism (fat and triglycerides), whereas only approximately 25 % is generated by glucose metabolism (carbohydrates) [58]. The remaining 5 % is mediated by amino acids and ketone bodies. Thus, nutritional and metabolic state is a critical determinant of proper cardiac function [58].

In the heart, it is the CMs that are the critical regulatory units of metabolism [59]. They utilize adenosine triphosphate (ATP) in order to contract, primarily through functional signaling effects mediated by mitochondria [56]. Indeed, CMs are largely made up of mitochondria, which occupy up to 40 % of the heart cell volume [56]. This enables the heart to undergo continuous aerobic respiration via oxidative phosphorylation (OxPhos), which accounts for more than 90 % of total ATP production in the heart [60]. Thus, the necessary demand for continuous energy of the heart can be achieved by aerobic respiration as mediated by CM mitochondrial function [61].

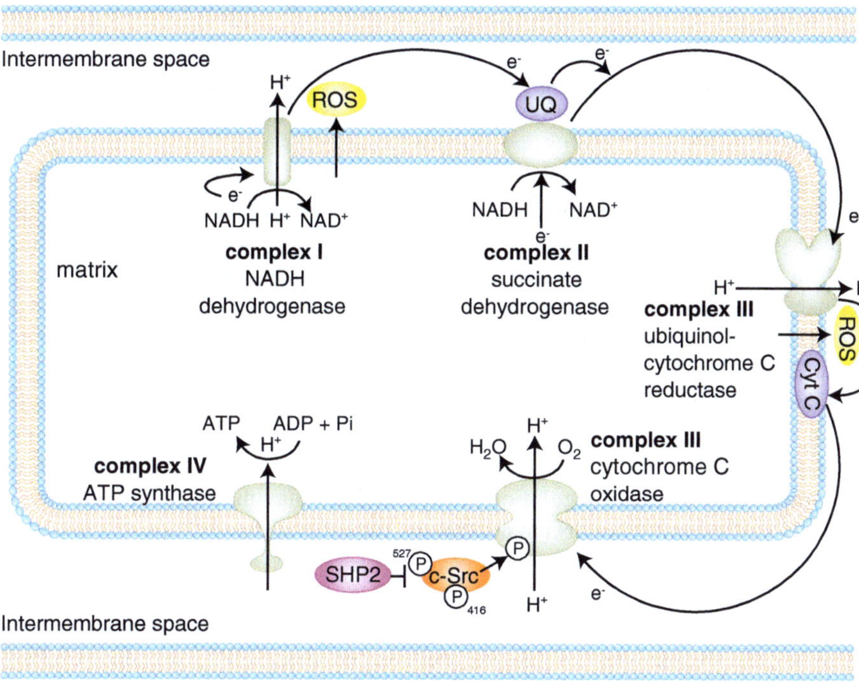

Fig. 8.3 SHP2 regulates OxPhos through regulation of c-SRC. Complex I accepts electrons from the Krebs cycle electron carrier nicotinamide adenine dinucleotide (NADH) and passes them to UQ (ubiquinone), which also receives electrons from complex II. UQ passes electrons to complex III which passes them to cytochrome C (Cyt C). Cyt C passes electrons to complex IV which uses the electrons and hydrogen ions to reduce molecular oxygen to water. SHP2 dephosphorylates Src at an inhibitory site (Tyr527) inducing its activation, and through this interaction, modulates the activity of cytochrome C oxidase and the function of OXPHOS machinery

SHP2 is abundantly expressed in the intercristae/intermembrane space of the mitochondria [62], and therefore, enzymes that comprise the OxPhos machinery and are found in the inner mitochondrial membrane are thought to be possible direct or indirect substrates of SHP2. Electron transport chain (ETC) and ATP synthase comprise the OxPhos machinery. The ETC is a series of electron transferring moieties that consist of nicotinamide adenine dinucleotide (NADH) dehydrogenase (complex I), succinate dehydrogenase (complex II), ubiquinol-cytochrome C reductase (complex III), and cytochrome C oxidase (complex IV) [63] (Fig. 8.3). Recent studies show that ROS produced by the mitochondrial OxPhos pathway may contribute to the pathology of NS, where reduced levels of cytochrome C oxidase and cytochrome C complexes are observed in MEFs containing the NS-specific SHP2 mutation D61G. Consequently, D61G MEFs showed increased mitochondrial damage, likely due to increased production of ROS and reduced ATP levels, that may, at least in part, contribute to the abnormalities observed in cardiac function in these patients. Finally, reduced cytochrome C oxidase (complex IV) was also observed in D61G MEFs, suggesting that SHP2 may modulate a compensatory role for increased cytochrome C oxidase activity. Likely, SHP2 may modulate the terminal step of the ETC, where transfer of the electron occurs from cytochrome C oxidase to cytochrome C and to molecular oxygen [64].

We cannot rule out the possibility that SHP2 may also indirectly affect mitochondrial function through targeting other identified phospho-proteins of the OxPhos machinery complex [65]. Previous studies during the last decade have shown that c-SRC, a non-RTK that plays a key role in cardiac mitochondria, is localized to the intermembrane space and activates cytochrome C oxidase by phosphorylation, in an effort to maintain cellular energy stores. CMs require activation of c-SRC for cytochrome C oxidase activity because of their requirement for high levels of ATP to maintain cardiac function [66]. Importantly, SHP2 has been shown to associate with SRC in multiple cell types. In vitro, SHP2 can mediate the dephosphorylation of SRC at its inhibitory site (Tyr527), thereby inducing its activation [67]. Therefore, SHP2 interaction with and regulation of SRC activity may modulate cytochrome C oxidase in cardiac mitochondria to control the function of the OxPhos machinery in the heart [64, 68] (Fig. 8.3).

SHP2 may also serve as a direct mediator of cardiac metabolism through its regulation of both mitogen-activated protein kinase (MAPK) and OxPhos regulation. While the mechanisms for this link are as yet unclear, SHP2 can reduce OxPhos complexes in hearts of mice that overexpress mitogen-activated protein kinase kinase 6 (MKK6), a p38 MAPK activator, thereby causing a subsequent reduction in respiration and a decrease in ROS production [69].

In addition, Urotensin-II (UT-II)-induced hypertrophy, ERK activity, and EGFR transactivation were also shown to be mediated by increased production of ROS and transient oxidation of SHP2 in CMs [70]. By its direct activation of the EGFR, UT-II, a vasoactive peptide that is secreted by endothelial cells, can mediate onset of cardiac fibrosis and cardiomyocyte hypertrophy in response to injury [71]. Briefly, high levels of UT-II stimulate matrix metalloproteases (MMPs) to induce the production of the pro-Heparin Binding-Epidermal Growth Factor (pro-HB-EGF) precursor, yielding the release of mature HB-EGF ligand, and subsequent transactivation of the EGFR [71]. Activation of the UT-II receptors also induce generation of ROS by nicotinamide adenine dinucleotide phosphate (NADPH) oxidases, leading to inhibition of SHP2 catalytic activity and subsequent induction of downstream mitogenic signaling pathways [71]. Together, these data suggest that UT-II may play a novel role in pulmonary hypertension by promoting cardiac remodeling [70]. Likely, production of ROS by UT-II induces transient oxidation of the catalytic cysteine residue of SHP2, suppressing its phosphatase activity, but not its capacity to serve as an adapter protein to bind the receptor, and thereby enhancing downstream signaling to ERK by EGFR [70] (Fig. 8.4). Similarly, other vasoactive peptides, such as angiotensin II [72] and endothelin-1 [73], show hypertrophic activity in CMs and are also associated with significant ERK phosphorylation via regulation of SHP2 activity.

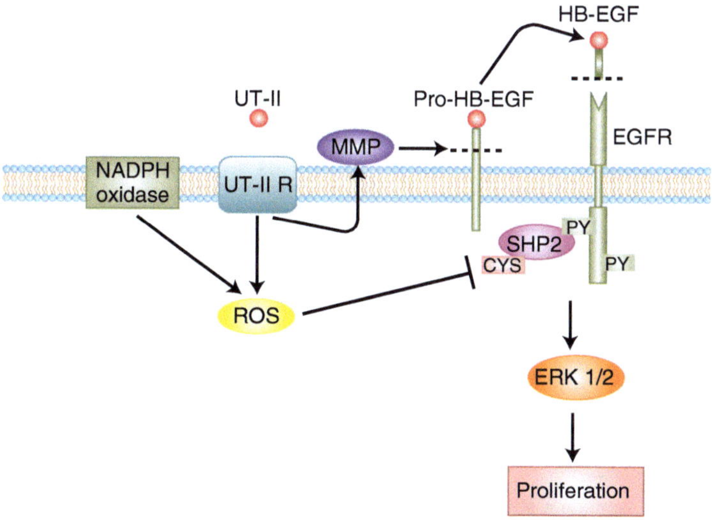

Fig. 8.4 Reactive oxygen species (ROS) inhibit the activity of SHP2. The phosphorylated EGFR is associated with SHP2 and is dephosphorylated. UT-II induces ROS generation concomitantly via UT-II receptors and NADPH oxidases and causes transient oxidation of the catalytic cysteine of SHP2 to inhibit the dephosphorylation activity of SHP2 on the receptor. This allows transactivated EGFR to transmit signals to the downstream hypertrophic signaling pathway

Cardiomyopathy and Insulin Resistance

The primary role of insulin in the heart is to induce glucose uptake and oxidation, modulate FA uptake, regulate protein synthesis, and modulate vascular function [74]. Dysregulation of the signaling pathways important for these metabolic processes leads to insulin resistance and aberrant glucose uptake [75–77]. For example, insulin resistance, primarily a result of obesity, can cause type 2 diabetes, hypertension, atherosclerosis, and/or metabolic syndrome [74]. More importantly, it also impairs the ability of the heart to adjust to changing energy demands by increasing delivery of FAs to the heart and by reducing the ability of the heart to use glucose, thereby shifting the heart toward a greater reliance on FAs for energy [74, 78, 79]. Consequently, secondary effects in the heart ensue, including hyperinsulinemia, hyperglycemia, and hyperlipidemia [75] and the heart undergoes cellular stress, with elevated reactive oxygen species (ROS) production, induced mitochondrial dysfunction, and increased apoptosis [74]. Ultimately, the combined metabolic, structural, and functional changes in the heart and vasculature lead to the onset of diabetic cardiomyopathy, coronary artery disease, myocardial ischemia, and HF [80, 81].

Normally, insulin is released from the pancreatic β-cells to induce glucose uptake in CMs [77, 82]. Binding of insulin to the insulin receptor (IR) or insulin-like growth factor receptor (IGFR) induces auto-phosphorylation and initiation of a downstream signaling cascade that allows for the recruitment of pY proteins to the membrane,

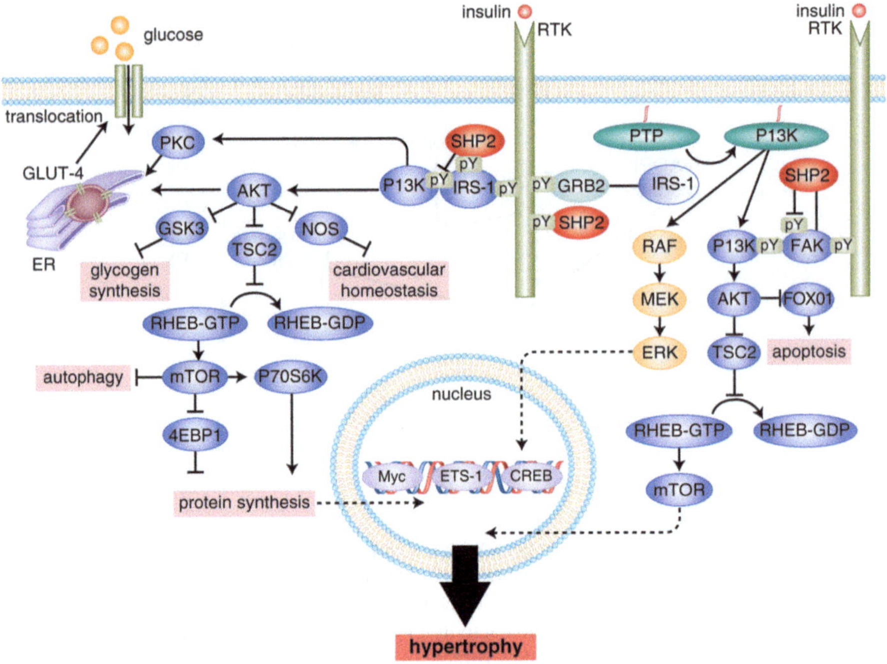

Fig. 8.5 SHP2-dependent signaling pathways leading to hypertrophy. Insulin binds to and activates the insulin receptor tyrosine kinase (RTK). Once transphosphorylated, the receptor recruits IRS-1, which can then mediate activation of the PI3K/AKT signaling pathway, leading to downstream activation of glycogen synthesis, autophagy, and protein synthesis, and culminating in mechanisms that can lead to cardiac hypertrophy. In addition, AKT activates nitric oxide synthase (NOS) to maintain cardiomyocyte homeostasis. AKT can also inhibit apoptosis through inhibition of FOXO-1. Moreover, stimulation of PKC and AKT by PI3K leads to translocation of GLUT-4 from the endoplasmic reticulum to the plasma membrane and mediates glucose metabolism in the cardiomyocyte; insulin resistance occurs as a consequence of the inability of GLUT4 to translocate to the membrane. SHP2 serves a critical role in this process; through its ability to inhibit PI3K signaling by dephosphorylating the PI3K binding site on IRS-1, SHP2 can, in essence, inhibit insulin signaling by this pathway. Upon activation of the insulin receptor, Focal Adhesion Kinase (FAK) can bind the RTK to stimulate PI3K/AKT downstream signaling. SHP2 can inhibit insulin signaling by binding to and inhibiting FAK activation, resulting in the apoptosis of the cardiomyocyte. Insulin can also mediate the activation of the Ras/ERK signaling cascade. Activation of the receptor recruits GRB2, SOS, and SHP2 to the membrane. SOS mediates the activation of Ras-GTP to mediate downstream PI3K/AKT and RAF/MEK/ERK activation. Consequently, ERK translocates to the nucleus, binds to transcription factors, and serves to ultimately induce hypertrophy

including the IRS-1 scaffolding/adapter protein [83] (Fig. 8.5). Briefly, in response to activation of the receptor, IRS-1, along with a panoply of other pY proteins, is recruited to the membrane to help foster activation of downstream signaling proteins, the most prominent of which is PI3K and its downstream effectors, AKT and protein kinase C (PKC) [83, 84] (Fig. 8.5). Activation of PKC, a serine/threonine kinase, in the heart serves a dual function. First, PKC acts as a negative feedback regulator of insulin signaling by increasing the serine phosphorylation of IRS-1 [85, 86]. Importantly, this regulation has been shown to be mediated by SHP2 activity [87]. Second, it leads to the translocation of both glucose transporter type 1 and type 4 (GLUT1 and GLUT4) to the membrane in order to facilitate glucose uptake in the cell [80, 88].

In addition to glucose uptake, insulin-mediated activation of PI3K and AKT regulates important cellular and functional processes including cardiac hypertrophy, protein translation, nitric oxide generation, apoptosis, and autophagy through activation of signaling intermediates, such as mTOR, S6K, and forkhead box transcription factors (FOXO), GSK3β, and nitric oxide synthase (NOS) (Fig. 8.4) [74, 89]. As expected, aberrant regulation of these signaling proteins also contributes to an increased risk of HCM and/or HF. For example, animal models on long-term high-fat diet or genetic models of obesity and severe insulin resistance, such as the *ob/ob* and *db/db* mice, showed impaired activation of AKT in response to insulin and went on to develop end-stage HF, suggesting inhibition of downstream insulin signaling is a predisposition for a more severe form of cardiac disease [90–92]. This was further supported by the phenotype of mice with cardiac-specific deletion of the IR, which displayed small hearts and reduced cardiomyocyte volumes [93]. Moreover, in response to pathological stimulation, these mice developed an exacerbated hypertrophy associated with mitochondrial dysfunction [94–96]. In contrast, however, rats fed a high-fat diet showed elevated basal levels of phosphorylated Akt and also developed cardiac contractile dysfunction [97].

Interestingly, cardiomyocyte-specific overexpression of the constitutively active catalytic subunit of PI3K, p110α, induced physiological, not pathological, hypertrophy [98]. Conversely, dominant negative p110α overexpression in the heart induced a non-pathological atrophy, even in the presence of IGFR overexpression and exercise training [98–100]. Cardiac-restricted loss of the lipid phosphatase, phosphatase, and tensin homolog deleted on chromosome ten (PTEN), which increases phosphorylation of AKT and GSK-β, also promoted heart growth and prevented the development of maladaptive ventricular remodeling, with preservation of angiogenesis and metabolic gene expression in response to pressure overload [101, 102].

Downstream of PI3K, deletion of AKT2 in gene-targeted mice caused insulin resistance, but had no other discernible phenotype [103]; in contrast, AKT1-null mice displayed growth retardation and were refractory to physiological cardiac hypertrophy when subjected to exercise training [104–106]. Similarly, overexpression of dominant negative AKT1 prevented hypertrophic growth of the heart [107]. Importantly, while acute cardiomyocyte-specific expression of a constitutively active AKT1 initially promoted a physiological type of hypertrophy, its prolonged activation ultimately proved pathological [107–110]. Taken together, these results indicate that acute AKT activation promotes an adaptive cellular growth program in the heart, but that sustained AKT signaling leads to pathological hypertrophy and heart failure [111].

Regulation of Insulin Signaling by SHP2

Genetic inactivation of insulin signaling in the heart contributes to HF by increasing mitochondrial dysfunction, decreasing angiogenesis, and increasing fibrosis in response to hemodynamic stress [74, 93–96, 112]. Because of its critical regulatory role in RTK signaling, SHP2 also plays a critical role in insulin signaling. Indeed, muscle-specific deletion of SHP2 leads to the development of insulin resistance and impaired glucose uptake [20]. Consequently, these mice also develop severe DCM and cardiac dysfunction [20, 21]. As described above, insulin resistance should lead to impaired PI3K/AKT signaling; however, muscle-specific deleted SHP2 mice show hyperactivation of the PI3K/AKT pathway [21]. One possible explanation for this apparent discrepancy is that, at least in response to some GFs, SHP2 can both regulate the strength and the duration of PI3K activation through its ability to bind to GAB-1 or IRS-1 and dephosphorylate the p85 binding sites for PI3K on these scaffolding proteins [18, 113]. In SHP2-deleted fibroblasts, an increased phosphorylation and association of the p85 with the scaffolding adapter Gab1 [18] and IRS-1 [114] was observed, thereby leading to increased downstream PI3K/AKT signaling. Indeed, increased pY-IRS-1 signals, increased IRS-1/p85 association, and increased AKT activity were all observed in SHP2-deficient muscle cells [20, 21]. Conversely, enhanced activity of SHP2 led to negative modulation of insulin signaling due to reduced pY of IRS-1 and decreased activation of PI3K, impairment of which can reduce the ability of insulin to stimulate glycogen synthesis [115]. Lastly, and as described above, the catalytically inactive Y279C/+ LS mice also displayed increased AKT activity in both CMs and whole hearts [37], with increased association and phosphorylation of IRS-1 and GAB-1 [37]. Similarly, GAB-1/PI3K complexes were found to be more abundant in fibroblasts from LS than control [39]. Consistently, purified recombinant LS mutants failed to dephosphorylate GAB1 PI3K-binding sites [39]. Taken together, these data suggest that in addition to the positive role for SHP2 in Ras/ERK signaling, SHP2 can also negatively regulate PI3K signaling in a cell-type and/or receptor-specific manner. Therefore, though SHP2 deficiency has a likely direct causative role in DCM, the additional development of insulin resistance likely further potentiates the progression to heart failure in those mice.

Downstream of PI3K: Cardiomyocyte Regulation of mTOR and Autophagy by SHP2

The downstream effector of PI3K/AKT activation is mTOR, a key regulator of cell growth and protein synthesis and negative regulator of a process called autophagy [116] (Fig. 8.5). CMs are intricately involved in autophagy, the catabolic process of regulating the synthesis, degradation, and recycling of a cell's own components through lysosomal machinery [117]. While autophagy is a normal process that occurs in cells, it can be further induced by cellular stress, such as nutrient deprivation [118]. The PI3K signaling pathway, through its activation of mTOR, is a key negative regulator of this process [119, 120].

Autolysosomal degradation of membrane lipids and proteins through autophagy generates FAs and amino acids, which are reused by the cell to maintain mitochondrial ATP production and protein synthesis, thereby promoting cell survival [118]. Interestingly, autophagy also promotes programmed cell death in some circumstances, although the mechanism for this dual-functionality in the heart remains unclear [121–123].

Autophagy plays a dual role in cardiac function as well. Increased autophagy is observed in acute and chronic ischemia, end-stage heart failure, and aging hearts [124]. However, in obesity and metabolic syndrome, increased levels of insulin, lipids, amino acids, and cytokines in high-fat diets induce activation of mTOR, thereby suppressing autophagy, but also leading to increased risk of cardiovascular disease, diabetes, cancer, and to a reduction in overall life expectancy [125]. It seems likely then that too much or too little autophagy proves to be equally deleterious. Acute activation of autophagy under a wide range of pathological conditions appears initially compensatory and critical for proper maintenance of cardiac metabolism and cellular homeostasis [126]. However, in response to chronic stress, when autophagy is excessive, heart failure ensues, likely due to the decline in both number and function of mitochondria, and a perturbation in cardiomyocyte metabolic flux [126]. Additional work is needed to fully define the metabolic profit and loss conferred by autophagy in cardiac hypertrophy and failure [126].

SHP2 plays an important role in mTOR regulation, and therefore (although not yet explicitly defined) autophagy, in the heart. It has been previously demonstrated that autophagy facilitates IFN-gamma-mediated Jak2-STAT1 activity through its ability to regulate SHP2 activity [127]. Without autophagy, ROS increases, causing SHP2-mediated STAT1 inactivation [127]. Inhibiting SHP2 reverses both the cellular inflammation and the IFN-gamma-induced activation of STAT1 in ATG5(–/–) MEFs, suggesting autophagy inhibits both the expression of ROS and SHP2 [127]. This is likely to play an important role in a pathologically stimulated heart.

Depletion of ATP also markedly inhibits mTOR through activation of adenosine monophosphate-activated protein kinase (AMPK), which functions independently of PI3K activation [128]. AMPK is an important metabolic sensor of energy stores in the heart [129]. When ATP production is decreased or when ATP consumption is increased, the relative AMP:ATP ratio increases, inducing the activation of AMPK to maintain ATP production and contractile function [74]. AMPK increases glucose and FA uptake and oxidation in the heart [130–133] and directly enhances insulin signaling in endothelial cells, protecting them from insulin resistance [133].

Notably, evidence exists for the regulation of AMPK by SHP2. Under conditions of growth factor deprivation, fibroblasts lacking SHP2 exhibited increased cell size, reduced cellular energy, and decreased AMPK activity, suggesting SHP2 functions to modulate cell growth during low-energy states [134]. In addition, pancreatic cells from mice with pancreas-specific deletion of SHP2 showed decreased levels of p-AMPK in response to both basal and high glucose, further suggesting that SHP2 may act as a negative regulator of ATP production [135]. Importantly, in mice with skeletal and cardiac muscle-specific deletion of SHP2, AMPK activation was remarkably decreased in response to metformin, as compared to controls, suggesting that the insulin resistance and impaired glucose uptake observed in these mice is, at least in part, due to an impaired SHP2-specific regulation of AMPK [20].

Role of SHP2 in Cardiac Metabolism: Future Directions

Metabolic processes are critical for normal heart, and specifically cardiomyocyte, function. Aberrant regulation of metabolic signaling pathways can lead to insulin resistance and cardiac dysfunction. Since SHP2 plays such a critical role in the normal homeostasis of the heart, it is likely that it is intricately involved in the normal functioning of many metabolic pathways. Here, we have discussed the known roles for SHP2 in the signaling pathways downstream of glucose uptake and insulin signaling and have suggested ways in which SHP2 may be involved in mediating proper metabolic control in the heart. Elucidation of the direct roles for SHP2 in cardiac metabolism will be critical going forward.

References

1. Tartaglia M et al (2001) Mutations in PTPN11, encoding the protein tyrosine phosphatase SHP-2, cause Noonan syndrome. Nat Genet 29(4):465–468
2. Hof P et al (1998) Crystal structure of the tyrosine phosphatase SHP-2. Cell 92(4):441–450
3. Barford D, Neel BG (1998) Revealing mechanisms for SH2 domain mediated regulation of the protein tyrosine phosphatase SHP-2. Structure 6(3):249–254
4. O'Reilly AM, Neel BG (1998) Structural determinants of SHP-2 function and specificity in Xenopus mesoderm induction. Mol Cell Biol 18(1):161–177
5. Neel BG, Gu H, Pao L (2003) The 'Shp'ing news: SH2 domain-containing tyrosine phosphatases in cell signaling. Trends Biochem Sci 28(6):284–293
6. Van Vactor D, O'Reilly AM, Neel BG (1998) Genetic analysis of protein tyrosine phosphatases. Curr Opin Genet Dev 8(1):112–126
7. Feng G (1999) SHP-2 tyrosine phosphatase: signaling one cell or many. Exp Cell Res 253:47–54
8. Tonks NK, Neel BG (2001) Combinatorial control of the specificity of protein tyrosine phosphatases. Curr Opin Cell Biol 13(2):182–195
9. Noguchi T et al (1994) Role of SH-PTP2, a protein-tyrosine phosphatase with src homology 2 domains, in insulin-stimulated ras activation. Mol Cell Biol 14:6674–6682
10. Shi ZQ et al (2000) Molecular mechanism for the Shp-2 tyrosine phosphatase function in promoting growth factor stimulation of Erk activity. Mol Cell Biol 20(5):1526–1536
11. Yamauchi K et al (1995) Protein-tyrosine-phosphatase SHPTP2 is a required positive effector for insulin downstream signaling. Proc Natl Acad Sci U S A 92:664–668
12. Lauriol J, Kontaridis MI (2011) PTPN11-associated mutations in the heart: has LEOPARD changed its RASpots? Trends Cardiovasc Med 21(4):97–104
13. Zhang SQ et al (2004) Shp2 regulates SRC family kinase activity and Ras/Erk activation by controlling Csk recruitment. Mol Cell 13(3):341–355
14. Klinghoffer RA, Kazlauskas A (1995) Identification of a putative Syp substrate, the PDGFß receptor. J Biol Chem 270(38):22208–22217
15. Agazie YM, Hayman MJ (2003) Molecular mechanism for a role of SHP2 in epidermal growth factor receptor signaling. Mol Cell Biol 23(21):7875–7886

16. Hanafusa H et al (2004) Shp2, an SH2-containing protein-tyrosine phosphatase, positively regulates receptor tyrosine kinase signaling by dephosphorylating and inactivating the inhibitor Sprouty. J Biol Chem 279(22):22992–22995
17. Jarvis LA et al (2006) Sprouty proteins are in vivo targets of Corkscrew/SHP-2 tyrosine phosphatases. Development 133(6):1133–1142
18. Zhang SQ et al (2002) Receptor-specific regulation of phosphatidylinositol 3′-kinase activation by the protein tyrosine phosphatase Shp2. Mol Cell Biol 22(12):4062–4072
19. Kontaridis MI et al (2004) SHP-2 positively regulates myogenesis by coupling to the Rho GTPase signaling pathway. Mol Cell Biol 24(12):5340–5352
20. Princen F et al (2009) Deletion of Shp2 tyrosine phosphatase in muscle leads to dilated cardiomyopathy, insulin resistance, and premature death. Mol Cell Biol 29(2):378–388
21. Kontaridis MI et al (2008) Deletion of Ptpn11 (Shp2) in cardiomyocytes causes dilated cardiomyopathy via effects on the extracellular signal-regulated kinase/mitogen-activated protein kinase and RhoA signaling pathways. Circulation 117(11):1423–1435
22. Marin TM et al (2008) Shp2 negatively regulates growth in cardiomyocytes by controlling focal adhesion kinase/Src and mTOR pathways. Circ Res 103(8):813–824
23. Tartaglia M et al (2002) PTPN11 mutations in Noonan syndrome: molecular spectrum, genotype-phenotype correlation, and phenotypic heterogeneity. Am J Hum Genet 70(6):1555–1563
24. Marino B et al (1999) Congenital heart diseases in children with Noonan syndrome: an expanded cardiac spectrum with high prevalence of atrioventricular canal. J Pediatr 135(6): 703–706
25. Yoshida R et al (2004) Protein-tyrosine phosphatase, nonreceptor type 11 mutation analysis and clinical assessment in 45 patients with Noonan syndrome. J Clin Endocrinol Metab 89(7):3359–3364
26. Nishikawa T et al (1996) Hypertrophic cardiomyopathy in Noonan syndrome. Acta Paediatr Jpn 38(1):91–98
27. Digilio MC et al (1998) Noonan syndrome and aortic coarctation. Am J Med Genet 80(2): 160–162
28. Keilhack H et al (2005) Diverse biochemical properties of Shp2 mutants. Implications for disease phenotypes. J Biol Chem 280(35):30984–30993
29. O'Reilly AM et al (2000) Activated mutants of SHP-2 preferentially induce elongation of Xenopus animal caps. Mol Cell Biol 20(1):299–311
30. Fragale A et al (2004) Noonan syndrome-associated SHP2/PTPN11 mutants cause EGF-dependent prolonged GAB1 binding and sustained ERK2/MAPK1 activation. Hum Mutat 23(3):267–277
31. Araki T et al (2004) Mouse model of Noonan syndrome reveals cell type- and gene dosage-dependent effects of Ptpn11 mutation. Nat Med 10(8):849–857
32. Niihori T et al (2005) Functional analysis of PTPN11/SHP-2 mutants identified in Noonan syndrome and childhood leukemia. J Hum Genet 50(4):192–202
33. Tartaglia M et al (2006) Diversity and functional consequences of germline and somatic PTPN11 mutations in human disease. Am J Hum Genet 78(2):279–290
34. Schubbert S et al (2005) Functional analysis of leukemia-associated PTPN11 mutations in primary hematopoietic cells. Blood 106(1):311–317
35. Gorlin RJ, Anderson RC, Moller JH (1971) The Leopard (multiple lentigines) syndrome revisited. Birth Defects Orig Artic Ser 07(4):110–115
36. Kontaridis MI et al (2006) PTPN11 (Shp2) mutations in LEOPARD syndrome have dominant negative, not activating, effects. J Biol Chem 281(10):6785–6792
37. Marin TM et al (2011) Rapamycin reverses hypertrophic cardiomyopathy in a mouse model of LEOPARD syndrome-associated PTPN11 mutation. J Clin Invest 121(3):1026–1043
38. Schramm C et al (2012) The PTPN11 loss-of-function mutation Q510E-Shp2 causes hypertrophic cardiomyopathy by dysregulating mTOR signaling. Am J Physiol Heart Circ Physiol 302(1):H231–H243

39. Edouard T et al (2010) Functional effects of PTPN11 (SHP2) mutations causing LEOPARD syndrome on epidermal growth factor-induced phosphoinositide 3-kinase/AKT/glycogen synthase kinase 3beta signaling. Mol Cell Biol 30(10):2498–2507
40. Ishida H et al (2011) LEOPARD-type SHP2 mutant Gln510Glu attenuates cardiomyocyte differentiation and promotes cardiac hypertrophy via dysregulation of Akt/GSK-3beta/beta-catenin signaling. Am J Physiol Heart Circ Physiol 301(4):H1531–H1539
41. Rapila R, Korhonen T, Tavi P (2008) Excitation-contraction coupling of the mouse embryonic cardiomyocyte. J Gen Physiol 132(4):397–405
42. Greenstein JL, Winslow RL (2011) Integrative systems models of cardiac excitation-contraction coupling. Circ Res 108(1):70–84
43. Bootman MD et al (2006) Calcium signalling during excitation-contraction coupling in mammalian atrial myocytes. J Cell Sci 119(Pt 19):3915–3925
44. McDowell SA et al (2004) Phosphoinositide 3-kinase regulates excitation-contraction coupling in neonatal cardiomyocytes. Am J Physiol Heart Circ Physiol 286(2):H796–H805
45. Seidman JG, Seidman C (2001) The genetic basis for cardiomyopathy: from mutation identification to mechanistic paradigms. Cell 104(4):557–567
46. Frey N, Olson EN (2003) Cardiac hypertrophy: the good, the bad, and the ugly. Annu Rev Physiol 65:45–79
47. Molkentin JD, Dorn IG II (2001) Cytoplasmic signaling pathways that regulate cardiac hypertrophy. Annu Rev Physiol 63:391–426
48. Fabiato A (1983) Calcium-induced release of calcium from the cardiac sarcoplasmic reticulum. Am J Physiol 245(1):C1–C14
49. Bu G et al (2009) Uniform action potential repolarization within the sarcolemma of in situ ventricular cardiomyocytes. Biophys J 96(6):2532–2546
50. Miner EC, Miller WL (2006) A look between the cardiomyocytes: the extracellular matrix in heart failure. Mayo Clin Proc 81(1):71–76
51. Maron BJ (2002) Hypertrophic cardiomyopathy: a systematic review. JAMA 287(10):1308–1320
52. Ly H et al (2007) Gene therapy in the treatment of heart failure. Physiology (Bethesda) 22:81–96
53. Banerjee P et al (2002) Diastolic heart failure: neglected or misdiagnosed? J Am Coll Cardiol 39(1):138–141
54. van Kraaij DJ et al (2002) Diagnosing diastolic heart failure. Eur J Heart Fail 4(4):419–430
55. Konstam MA (2003) "Systolic and diastolic dysfunction" in heart failure? Time for a new paradigm. J Card Fail 9(1):1–3
56. Merante F et al (1998) Myocardial aerobic metabolism is impaired in a cell culture model of cyanotic heart disease. Am J Physiol 275(5 Pt 2):H1673–H1681
57. Murgia M et al (2009) Controlling metabolism and cell death: at the heart of mitochondrial calcium signalling. J Mol Cell Cardiol 46(6):781–788
58. Kolwicz SC Jr, Tian R (2011) Glucose metabolism and cardiac hypertrophy. Cardiovasc Res 90(2):194–201
59. Malliri A et al (1998) The transcription factor AP-1 is required for EGF-induced activation of Rho-like GTPases, cytoskeletal rearrangements, motility, and in vitro invasion of A431 cells. J Cell Biol 143:1087–1099
60. Kayar SR, Banchero N (1987) Volume density and distribution of mitochondria in myocardial growth and hypertrophy. Respir Physiol 70(3):275–286
61. Vary TC, Reibel DK, Neely JR (1981) Control of energy metabolism of heart muscle. Annu Rev Physiol 43:419–430
62. Salvi M et al (2004) Tyrosine phosphatase activity in mitochondria: presence of Shp-2 phosphatase in mitochondria. Cell Mol Life Sci 61(18):2393–2404
63. Al Ghouleh I et al (2011) Oxidases and peroxidases in cardiovascular and lung disease: new concepts in reactive oxygen species signaling. Free Radic Biol Med 51(7):1271–1288
64. Lee I et al (2010) A suggested role for mitochondria in Noonan syndrome. Biochim Biophys Acta 1802(2):275–283

65. Helling S et al (2008) Phosphorylation and kinetics of mammalian cytochrome c oxidase. Mol Cell Proteomics 7(9):1714–1724
66. Miyazaki T et al (2003) Regulation of cytochrome c oxidase activity by c-Src in osteoclasts. J Cell Biol 160(5):709–718
67. Peng Z-Y, Cartwright CA (1995) Regulation of the Src tyrosine kinase and Syp tyrosine phosphatase by their cellular association. Oncogene 11:1955–1962
68. Salvi M et al (2002) Characterization and location of Src-dependent tyrosine phosphorylation in rat brain mitochondria. Biochim Biophys Acta 1589(2):181–195
69. Wall JA et al (2006) Alterations in oxidative phosphorylation complex proteins in the hearts of transgenic mice that overexpress the p38 MAP kinase activator, MAP kinase kinase 6. Am J Physiol Heart Circ Physiol 291(5):H2462–H2472
70. Liu JC et al (2009) Urotensin II induces rat cardiomyocyte hypertrophy via the transient oxidization of Src homology 2-containing tyrosine phosphatase and transactivation of epidermal growth factor receptor. Mol Pharmacol 76(6):1186–1195
71. Babinska M et al (2012) Is plasma urotensin II concentration an indicator of myocardial damage in patients with acute coronary syndrome? Arch Med Sci 8(3):449–454
72. Cheng TH et al (2004) Inhibitory effect of resveratrol on angiotensin II-induced cardiomyocyte hypertrophy. Naunyn Schmiedebergs Arch Pharmacol 369(2):239–244
73. Cheng TH et al (2005) Nitric oxide inhibits endothelin-1-induced cardiomyocyte hypertrophy through cGMP-mediated suppression of extracellular-signal regulated kinase phosphorylation. Mol Pharmacol 68(4):1183–1192
74. Abel ED, O'Shea KM, Ramasamy R (2012) Insulin resistance: metabolic mechanisms and consequences in the heart. Arterioscler Thromb Vasc Biol 32(9):2068–2076
75. Wright JJ et al (2009) Mechanisms for increased myocardial fatty acid utilization following short-term high-fat feeding. Cardiovasc Res 82(2):351–360
76. Cook SA et al (2010) Abnormal myocardial insulin signalling in type 2 diabetes and left-ventricular dysfunction. Eur Heart J 31(1):100–111
77. Lawrence SP, Holman GD, Koumanov F (2010) Translocation of the Na+/H+ exchanger 1 (NHE1) in cardiomyocyte responses to insulin and energy-status signalling. Biochem J 432(3):515–523
78. Carley AN, Severson DL (2005) Fatty acid metabolism is enhanced in type 2 diabetic hearts. Biochim Biophys Acta 1734(2):112–126
79. Dirkx E et al (2011) High fat diet induced diabetic cardiomyopathy. Prostaglandins Leukot Essent Fatty Acids 85(5):219–225
80. Gray S, Kim JK (2011) New insights into insulin resistance in the diabetic heart. Trends Endocrinol Metab 22(10):394–403
81. Bell DS (2003) Heart failure: the frequent, forgotten, and often fatal complication of diabetes. Diabetes Care 26(8):2433–2441
82. Bertrand L et al (2006) AMPK activation restores the stimulation of glucose uptake in an in vitro model of insulin-resistant cardiomyocytes via the activation of protein kinase B. Am J Physiol Heart Circ Physiol 291(1):H239–H250
83. Sun XJ et al (1991) Structure of the insulin receptor substrate IRS-1 defines a unique signal transduction protein. Nature 352(6330):73–77
84. Metz HE, Houghton AM (2011) Insulin receptor substrate regulation of phosphoinositide 3-kinase. Clin Cancer Res 17(2):206–211
85. Yu C et al (2002) Mechanism by which fatty acids inhibit insulin activation of insulin receptor substrate-1 (IRS-1)-associated phosphatidylinositol 3-kinase activity in muscle. J Biol Chem 277(52):50230–50236
86. Samuel VT, Shulman GI (2012) Mechanisms for insulin resistance: common threads and missing links. Cell 148(5):852–871
87. Mussig K et al (2005) Shp2 is required for protein kinase C-dependent phosphorylation of serine 307 in insulin receptor substrate-1. J Biol Chem 280(38):32693–32699
88. White MF, Kahn CR (1994) The insulin signaling system. J Biol Chem 269(1):1–4

89. Muniyappa R et al (2007) Cardiovascular actions of insulin. Endocr Rev 28(5):463–491
90. Mazumder PK et al (2004) Impaired cardiac efficiency and increased fatty acid oxidation in insulin-resistant ob/ob mouse hearts. Diabetes 53(9):2366–2374
91. Battiprolu PK et al (2012) Metabolic stress-induced activation of FoxO1 triggers diabetic cardiomyopathy in mice. J Clin Invest 122(3):1109–1118
92. Lee J et al (2010) Multiple abnormalities of myocardial insulin signaling in a porcine model of diet-induced obesity. Am J Physiol Heart Circ Physiol 298(2):H310–H319
93. Belke DD et al (2002) Insulin signaling coordinately regulates cardiac size, metabolism, and contractile protein isoform expression. J Clin Invest 109(5):629–639
94. Hu P et al (2003) Minimally invasive aortic banding in mice: effects of altered cardiomyocyte insulin signaling during pressure overload. Am J Physiol Heart Circ Physiol 285(3): H1261–H1269
95. Sena S et al (2009) Impaired insulin signaling accelerates cardiac mitochondrial dysfunction after myocardial infarction. J Mol Cell Cardiol 46(6):910–918
96. Boudina S et al (2009) Contribution of impaired myocardial insulin signaling to mitochondrial dysfunction and oxidative stress in the heart. Circulation 119(9):1272–1283
97. Ouwens DM et al (2007) Cardiac contractile dysfunction in insulin-resistant rats fed a high-fat diet is associated with elevated CD36-mediated fatty acid uptake and esterification. Diabetologia 50(9):1938–1948
98. Shioi T et al (2000) The conserved phosphoinositide 3-kinase pathway determines heart size in mice. EMBO J 19(11):2537–2548
99. McMullen JR et al (2004) The insulin-like growth factor 1 receptor induces physiological heart growth via the phosphoinositide 3-kinase(p110alpha) pathway. J Biol Chem 279(6): 4782–4793
100. McMullen JR et al (2003) Phosphoinositide 3-kinase(p110alpha) plays a critical role for the induction of physiological, but not pathological, cardiac hypertrophy. Proc Natl Acad Sci U S A 100(21):12355–12360
101. Crackower MA et al (2002) Regulation of myocardial contractility and cell size by distinct PI3K-PTEN signaling pathways. Cell 110(6):737–749
102. Oudit GY et al (2008) Loss of PTEN attenuates the development of pathological hypertrophy and heart failure in response to biomechanical stress. Cardiovasc Res 78(3):505–514
103. Cho H et al (2001) Insulin resistance and a diabetes mellitus-like syndrome in mice lacking the protein kinase Akt2 (PKB beta). Science 292(5522):1728–1731
104. Cho H et al (2001) Akt1/PKBalpha is required for normal growth but dispensable for maintenance of glucose homeostasis in mice. J Biol Chem 276(42):38349–38352
105. Chen WS et al (2001) Growth retardation and increased apoptosis in mice with homozygous disruption of the Akt1 gene. Genes Dev 15(17):2203–2208
106. DeBosch B et al (2006) Akt1 is required for physiological cardiac growth. Circulation 113(17):2097–2104
107. Shioi T et al (2002) Akt/protein kinase B promotes organ growth in transgenic mice. Mol Cell Biol 22(8):2799–2809
108. Matsui T et al (2002) Phenotypic spectrum caused by transgenic overexpression of activated Akt in the heart. J Biol Chem 277(25):22896–22901
109. Condorelli G et al (2002) Akt induces enhanced myocardial contractility and cell size in vivo in transgenic mice. Proc Natl Acad Sci U S A 99(19):12333–12338
110. Shiojima I et al (2005) Disruption of coordinated cardiac hypertrophy and angiogenesis contributes to the transition to heart failure. J Clin Invest 115(8):2108–2118
111. Maillet M, van Berlo JH, Molkentin JD (2013) Molecular basis of physiological heart growth: fundamental concepts and new players. Nat Rev Mol Cell Biol 14(1):38–48
112. McQueen AP et al (2005) Contractile dysfunction in hypertrophied hearts with deficient insulin receptor signaling: possible role of reduced capillary density. J Mol Cell Cardiol 39(6): 882–892
113. Kuhné MR et al (1993) The insulin receptor substrate 1 associates with the SH2-containing phosphotyrosine phosphatase Syp. J Biol Chem 268(16):11479–11481

114. Myers MGJ et al (1998) The COOH-terminal tyrosine phosphorylation sites on IRS-1 bind SHP-2 and negatively regulate insulin signaling. J Biol Chem 273:26908–26914
115. Ouwens DM, van der Zon GC, Maassen JA (2001) Modulation of insulin-stimulated glycogen synthesis by Src Homology Phosphatase 2. Mol Cell Endocrinol 175(1–2):131–140
116. Wullschleger S, Loewith R, Hall MN (2006) TOR signaling in growth and metabolism. Cell 124(3):471–484
117. Levine B, Klionsky DJ (2004) Development by self-digestion: molecular mechanisms and biological functions of autophagy. Dev Cell 6(4):463–477
118. Lum JJ, DeBerardinis RJ, Thompson CB (2005) Autophagy in metazoans: cell survival in the land of plenty. Nat Rev Mol Cell Biol 6(6):439–448
119. Mousavi SA et al (2003) Phosphoinositide 3-kinase regulates maturation of lysosomes in rat hepatocytes. Biochem J 372(Pt 3):861–869
120. Ravikumar B et al (2004) Inhibition of mTOR induces autophagy and reduces toxicity of polyglutamine expansions in fly and mouse models of Huntington disease. Nat Genet 36(6):585–595
121. Yu L et al (2004) Regulation of an ATG7-beclin 1 program of autophagic cell death by caspase-8. Science 304(5676):1500–1502
122. Shimizu S et al (2004) Role of Bcl-2 family proteins in a non-apoptotic programmed cell death dependent on autophagy genes. Nat Cell Biol 6(12):1221–1228
123. Matsui Y et al (2007) Distinct roles of autophagy in the heart during ischemia and reperfusion: roles of AMP-activated protein kinase and Beclin 1 in mediating autophagy. Circ Res 100(6):914–922
124. Hein S et al (2003) Progression from compensated hypertrophy to failure in the pressure-overloaded human heart: structural deterioration and compensatory mechanisms. Circulation 107(7):984–991
125. Sciarretta S, Volpe M, Sadoshima J (2012) Is reactivation of autophagy a possible therapeutic solution for obesity and metabolic syndrome? Autophagy 8(8):1252–1254
126. Wang ZV, Ferdous A, Hill JA (2012) Cardiomyocyte autophagy: metabolic profit and loss. Heart Fail Rev
127. Chang YP et al (2010) Autophagy facilitates IFN-gamma-induced Jak2-STAT1 activation and cellular inflammation. J Biol Chem 285(37):28715–28722
128. Dennis PB et al (2001) Mammalian TOR: a homeostatic ATP sensor. Science 294(5544):1102–1105
129. Wong AK et al (2009) AMP-activated protein kinase pathway: a potential therapeutic target in cardiometabolic disease. Clin Sci (Lond) 116(8):607–620
130. Russell RR III et al (1999) Translocation of myocardial GLUT-4 and increased glucose uptake through activation of AMPK by AICAR. Am J Physiol 277(2 Pt 2):H643–H649
131. Luiken JJ et al (2003) Contraction-induced fatty acid translocase/CD36 translocation in rat cardiac myocytes is mediated through AMP-activated protein kinase signaling. Diabetes 52(7):1627–1634
132. Marsin AS et al (2000) Phosphorylation and activation of heart PFK-2 by AMPK has a role in the stimulation of glycolysis during ischaemia. Curr Biol 10(20):1247–1255
133. Ruderman NB et al (2003) Malonyl-CoA and AMP-activated protein kinase (AMPK): possible links between insulin resistance in muscle and early endothelial cell damage in diabetes. Biochem Soc Trans 31(Pt 1):202–206
134. Zito CI et al (2007) SHP-2 regulates cell growth by controlling the mTOR/S6 kinase 1 pathway. J Biol Chem 282(10):6946–6953
135. Zhang SS et al (2009) Coordinated regulation by Shp2 tyrosine phosphatase of signaling events controlling insulin biosynthesis in pancreatic beta-cells. Proc Natl Acad Sci U S A 106(18):7531–7536

Chapter 9
Metabolic Effects of Neural and Pancreatic Shp2

Zhao He, Sharon S. Zhang, Jianxiu Yu, and Gen-Sheng Feng

Abstract Shp2, encoded by *Ptpn11*, is a non-receptor tyrosine phosphatase that contains two Src-homology 2 domains within its structure. Previous biochemical and biological analyses indicate that this enzyme participates in signaling events elicited by various growth factors, cytokines, antigens, and hormones. Recent experiments using the cell type-specific gene knockout approaches in mice have revealed its critical roles in control of metabolism in mammals. In the central nervous system, Shp2 positively transmits leptin signals and may do so by association with the estrogen receptor α, influencing control of food intake, energy expenditure, and body weight. In pancreatic β-cells, Shp2 coordinates multiple pathways to regulate biosynthesis and secretion of insulin in control of glucose homeostasis. Herein we summarize the metabolic functions of Shp2 in the hypothalamus and pancreas.

Z. He, Ph.D. (✉)
Department of Pathology and Division of Biological Sciences, University of California, San Diego, La Jolla, CA 92093-0864, USA

School of Food Science and Technology, Jiangnan University, Wuxi, Jiangsu 214122, China
e-mail: zhaohe@jiangnan.edu.cn

S.S. Zhang, M.D., Ph.D.
Department of Pathology, University of Chicago Medical Center, Chicago, IL 60637, USA

J. Yu, Ph.D.
Department of Biochemistry and Molecular Cell Biology, Shanghai Jiao Tong University School of Medicine, Shanghai, China, 200025

G.-S. Feng, Ph.D.
Department of Pathology, University of California, San Diego, San Diego, CA 92121, USA

K.K. Bence (ed.), *Protein Tyrosine Phosphatase Control of Metabolism*,
DOI 10.1007/978-1-4614-7855-3_9,

Abbreviations

AAV	Adeno-associated virus
α-MSH	Alpha-Melanocyte-stimulating hormone
Akt	Ak thymoma viral proto-oncogene
BDNF	Brain-derived neurotrophic factor
CamKIIα	$Ca2^+$/calmodulin-dependent protein kinase II α
Cbp	Csk-binding protein
ChIP	Chromatin immunoprecipitation
Cre	Cre recombinase
CRE3	Pan-neuronal cre-transgenic line
Csk	C-Src tyrosine kinase
Cys	Cysteine
db/db	Leptin receptor-deficient mice
E2	Estradiol-17β
ELISA	Enzyme-linked immunosorbent assay
ERα	Estrogen receptor α
Erk	Extracellular signal-regulated kinase
FoxO1	Forkhead box protein O1
Glut2	Glucose transporter 2
GM-CSF	Granulocyte-macrophage colony-stimulating factor
GTP	Guanosine-5′-triphosphate
HFD	High-fat diet
Hnf1α	Hepatocyte nuclear factor 1 alpha
Hnf4α	Hepatocyte nuclear factor 4 alpha
IGF1	Insulin-like growth factor 1
Ins1	*Insulin 1* gene
Ins2	*Insulin 2* gene
IRS1	Insulin receptor substrate 1
IRS2	Insulin receptor substrate 2
Jak2	Janus kinase 2
LepRb	Leptin receptor long form
Leu	Leucine
MAPK	Mitogen-activated protein kinase
Mek	Mitogen-activated protein kinase kinase
MODY	Maturity onset diabetes of the young
NFATc1	Nuclear factor of activated T cells c1
NIRKO	Neural-specific insulin receptor knockout
NPY	Neuropeptide Y
PAG	Phosphoprotein associated with glycosphingolipid-enriched microdomains
Panc	Pancreas
PDGF	Platelet-derived growth factor
Pdx1	Pancreatic and duodenal homeobox 1
PI3K	Phosphatidylinositol 3-kinase
PKB	Protein kinase B

POMC	Proopiomelanocortin
PTP1B	Protein tyrosine phosphatase 1B
PTP-1D	Protein tyrosine phosphatase 1D
PTP-2C	Protein tyrosine phosphatase 2C
ptpn11	Protein tyrosine phosphatase non-receptor type 11
PTPs	Protein tyrosine phosphatases
Raf	Rapidly accelerated fibrosarcoma
RasGAP	Ras GTPase activating protein
RT-PCR	Reverse transcription polymerase chain reaction
RTK	Receptor tyrosine kinase
SH2	Src-homology 2
Shp2	SH2 domain-containing protein tyrosine phosphatase-2
Ser	Serine
Socs3	Suppressor of cytokine signaling 3
STAT3	Signal transducer and activator of transcription 3
Thr	Threonine
Tyr	Tyrosine

Introduction

The Src-homology 2 (SH2)-containing protein tyrosine phosphatases (PTPs) constitute a small subfamily of non-receptor PTPs, with two members Shp1 and Shp2 [18, 50]. The molecular architecture of these two enzymes is very similar; both contain two SH2 domains (SH2-N and SH2-C) at the N-terminus, a central PTPase domain and a C-terminal tail. Shp1 is predominantly expressed in hematopoietic and lymphocytic cells, whereas Shp2 is ubiquitously expressed in various cell types. Shp2, previously named as SH-PTP2, PTP1D, PTP2C, or Syp, is encoded by protein-tyrosine phosphatase non-receptor type 11 (*Ptpn11*) and is the mammalian homologue of the *corkscrew* gene in *Drosophila*.

The crystal structure of Shp2 protein (lacking its C-terminal tail) reveals that this enzyme normally maintains an auto-inhibitory state, in which the SH2-N domain interacts with the catalytic domain via multiple non-covalent contacts, to prevent the access of substrates. Inactive Shp2 can be transformed into an active state by SH2-mediated binding to pTyr residues on growth factor/cytokine receptors or intracellular phosphoproteins, to release the PTP domain. For example, Tyr1009 in the C-terminal tail region of platelet-derived growth factor (PDGF) receptor was identified as a specific docking site for Shp2 (Fig. 9.1) [31]. Tyr1172 and Tyr1222 on insulin receptor substrate 1 (IRS1) were shown to bind the two SH2 domains of Shp2 [12], and a phospho-peptide that contains both of these two sites was much more potent than peptides with either one site in activating the phosphatase *in vitro* [61]. Similarly, two tandemly arranged binding sites for Shp2 have been localized to the C-terminal tail in IRS2-4 and Gab1-3 (Grb2-associated binder 1-3) proteins [38], suggesting that binding and activating Shp2 is a common mechanism for these adaptor/scaffolding proteins in mediating cell signaling events (Fig. 9.1). Like other

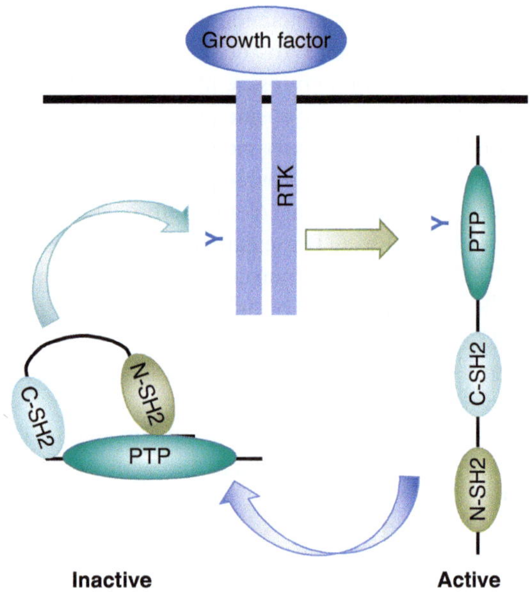

Fig. 9.1 The activation mechanism of non-receptor tyrosine phosphatase Shp2. Without stimulation, Shp2 maintains an inactive state in an auto-inhibitory manner. Upon activation, Shp2 releases its catalytic domain PTP to dephosphorylate its substrates. It is a reversible process from inactive to active state

PTPs, Cys459 in the catalytic center of Shp2 is essential for the phosphatase activity, and a C459S mutant is catalytically inactive and displays a dominant negative effect when expressed in cells.

Deletion or point mutations of *Ptpn11*/Shp2 have been associated with several genetic disorders. Autosomal dominant mutations in *Ptpn11* have been detected in nearly half of Noonan syndrome patients [63]. Functional analyses suggest that these mutants exhibit excessive phosphatase activity due to disruption of the auto-inhibitory mechanism, which leads to enhanced signaling through the Ras/Raf/Mek/Erk pathway [6, 24]. On the other hand, inherited mutations resulting in decreased catalytic activity of Shp2 were detected in patients with LEOPARD syndrome and metachondromatosis [6, 24]. Targeted deletion of different exons at the *ptpn11*/Shp2 locus in mice leads to embryonic lethality phenotype in homozygotes, with variable severity of phenotypes [60, 66]. In more recent experiments, several groups have employed the conditional gene targeting approach in mice, which has revealed important physiological functions of Shp2 in various tissues/organs. In this chapter, we discuss and review Shp2 functions in regulation of metabolic signaling in the brain and pancreas.

Shp2 Participates in Metabolic Signaling

Immediately after cloning of the cDNA for Shp2, several groups demonstrated that insulin treatment of cells induced physical association of IRS1 with Shp2, in a tyrosine phosphorylation-dependent manner [35, 61]. In these experiments, the SH2 domains of Shp2 were shown to mediate its association with pTyr-containing molecules. A phospho-peptide containing the pTyr site and the surrounding amino acid sequence recognized by the SH2 domains of Shp2 is a potent activator of the phosphatase activity *in vitro*, as demonstrated by using a purified recombinant protein [61]. By expressing a catalytically inactive mutant of Shp2 in 3T3 cells, Saltiel's group demonstrated a positive effect of Shp2 in promoting extracellular signal-regulated kinase (Erk) activation by insulin [44]. Consistently, Kasuga's group also showed that expression of a dominant negative mutant of Shp2 suppressed insulin-stimulated Ras activation in CHO-IR cells [51]. Indeed, genetic and biochemical analyses in a variety of model systems indicate that Shp2 acts to promote Ras-Erk signaling elicited by hormones, cytokines, and growth factors [38, 50]. Ironically, the molecular mechanism for Shp2 promotion of the Erk pathway is not yet fully understood. One notion is that Shp2 may act to prevent a premature shut-down of the activated Erk pathway, so that the active Erk signal is sustained long enough to elicit a cellular response of physiological significance. Multiple substrates may be involved in this process, and the dephosphorylation event mediated by Shp2 apparently leads to inactivation of a negative regulator for the Ras-Erk pathway directly or indirectly. One proposal is that Shp2 can dephosphorylate a pTyr site on cell surface receptors for binding of Ras GTPase activating protein (RasGAP), thereby preventing RasGAP recruitment to the plasma membrane and maintaining Ras in its guanosine-5′-triphosphate (GTP)-bound active state [1, 16]. Another possible mechanism is that Shp2 dephosphorylates a transmembrane protein PAG/Cbp, which recruits Csk, a negative regulator of Src-family kinases, to the plasma membrane [69]. If this model is correct, Shp2 is positioned upstream of Src kinases, in mediating activation of the Ras-Erk cascade [56, 69]. Experimental data also suggest that Sprouty proteins, negative feedback regulators of the Ras-Erk pathway, may be Shp2 target molecules [25, 53].

More recent experiments using conditional gene knockout approaches in mice suggest that Shp2 negatively regulates insulin-elicited signaling events in the liver, as hepatocyte-specific deletion of Shp2 resulted in up-regulation of insulin response in the liver [43]. Using either Albumin-Cre-mediated chronic Shp2 deletion in hepatocytes or acute Shp2 deletion by tail vein injection of Ad5-Cre, Haj and colleagues showed that hepatic Shp2 deficiency led to improved insulin sensitivity and enhanced glucose tolerance. This phenotype is likely due to removal of a direct dephosphorylation effect by Shp2, resulting in increased association of PI-3-kinase with IRS1 and IRS2 and enhanced Ak thymoma viral proto-oncogene (Akt) activation by insulin [43]. Further phenotypic analysis of this mutant mouse line by the same group indicated that Shp2 deficiency in the liver caused decreased hepatosteatosis, enhanced insulin-induced suppression of hepatic glucose production, and impaired development of insulin resistance following high-fat diet challenge [48].

Shp2 and Leptin Signaling Pathway

Leptin, an adipokine secreted by adipose tissue, binds to the leptin receptor long form (LepRb) in the hypothalamus to control food intake, activity, and metabolism. Following leptin stimulation, LepRb activates Janus kinase 2 (Jak2) that is physically associated with the receptor. Activated Jak2 in turn phosphorylates the receptor on tyrosine residues in the cytoplasmic domain including Tyr985 and Tyr1138 [29, 55]. The first evidence for Shp2 involvement in leptin signaling was the demonstration that p-Tyr985 is recognized by the SH2 domains of Shp2 [10, 41]. To directly fish out intracellular signaling molecules that act immediately downstream of LepRb, Friedman's group prepared affinity chromatography beads with a phospho-peptide spanning Tyr985 of LepRb. This approach led to identification of Shp2 in the lysates of bovine and mouse hypothalamus that bind directly to activated LepRb [41]. In co-transfection experiments with 293T cells, the same group showed that leptin treatment induced association of LepRb with Shp2, which is dependent on phosphorylation of Tyr985. Upon binding to LepRb, Shp2 also gets tyrosine-phosphorylated apparently by receptor-associated Jak2 kinase [10, 41]. Mutation of Tyr985 on LepRb disrupts the interaction between LepRb and Shp2, which reduces the phosphorylation of Shp2 induced by leptin. Moreover, expression of wild-type Shp2 suppresses the phosphorylation of Jak2 in response to leptin stimulation, whereas expression of inactive Shp2 mutant enhances the level of Jak2 phosphorylation [10, 41].

Signal transducer and activator of transcription 3 (Stat3) docks on Tyr1138 of LepRb when it is phosphorylated, and Stat3 activation is thought to be a crucial signaling event of leptin action, which regulates energy balance and metabolism [19]. Deletion of neural Stat3 leads to hyperphagia, morbid obesity, diabetes, and infertility, indicating that STAT3 activation in the brain positively transduces leptin signal to regulate energy balance and metabolism [22]. Two groups have generated knockin mice, mutating the Stat3-binding site on LepRb, Y1138S [5] and Y1138F [29], respectively. Both of these mutants display leptin resistance and obesity phenotypes. However, caution needs to be taken for data interpretation of the Y1138S mutant, as it is unclear whether the introduced serine residue becomes an unintended target for protein Ser/Thr kinases. Knockin mouse lines with mutated Tyr985 of LepRb were also generated as Y985F or Y985L mutant [7, 67]. The phenotype analysis of the Tyr985 mutants could be complicated by the fact that this pTyr residue is also a docking site for Socs3 (suppress of cytokine signaling 3), a negative regulator of Stat3 signaling [45]. Also, it remains to be determined if the introduced Tyr985-to-Leu mutation can lead to distortion of the LepRb structure [7].

Shp2 was first predicted to be a negative regulator of leptin signal, based on the observations that Shp2 acts to downregulate leptin-induced Stat3 activation and a mutant of LepRb lacking the Shp2 binding site pTyr985 exhibits increased leptin response *in vitro*. Unexpectedly, disruption of Shp2 in forebrain neuronal cells leads to early onset obesity and leptin resistance, and the mutant mice display increased serum levels of leptin, insulin, glucose, and triglycerides [68]. The mutant mouse line (CaSKO) was generated by crossing *Shp2* floxed allele with *CamKIIα-Cre* transgenic mice, in which the Cre recombinase is predominantly expressed in neuronal cells in the forebrain region after postnatal day 5. Homozygous deletion of Shp2 in this brain region did not affect the brain development and there was no obvious phenotypic difference between wt controls and mutants at the time of weaning. However, the CaSKO animals exhibit accelerated body weight increase immediately after weaning, which overlaps with the time window of onset for leptin action around postnatal day 28 in mice. The abnormal body weight gain is evidently due to development of obesity, based on detection of significantly increased white adipose tissue mass. Furthermore, dramatically increased serum leptin levels in CaSKO mice suggest that leptin resistance contributes to the obese phenotype. Consistent with the previous biochemical data *in vitro*, deletion of Shp2 in neuronal cells results in up-regulation of leptin-induced Stat3 activation in the hypothalamus *in vivo*, supporting the notion that Shp2 acts as a negative regulator of Stat3. However, Shp2 removal abolishes leptin-stimulated Erk activation, which may contribute, at least in part, to the leptin resistance phenotype. Quantitative reverse transcription polymerase chain reaction (RT-PCR) analysis showed no significant difference in the expression levels of *proopiomelanocortin* (*POMC*) mRNA between control and mutant mice under fed or fasting status. However, Shp2 ablation abolished the increase of *neuropeptide Y* (*NPY*) mRNA expression under fasting condition. Collectively, these genetic and biochemical data provide a physiological evidence for Shp2 action in the leptin signaling pathway. Of course, one cannot exclude the possibility that defects in other hormone or cytokine signaling pathways due to Shp2 removal also contribute to the obese phenotype in CaSKO animals. Although hyperinsulinemia was detected, it is unlikely that the obese phenotype of mutant animals is induced primarily due to direct interference of insulin signaling in the brain. Previous experiments in the Kahn lab showed that mice with neural-specific insulin receptor knockout (NIRKO) exhibited normal body weight within 6 months if fed regular chow diet [8]. Shp2 has been shown to act downstream of the receptor for brain-derived neurotrophic factor (BDNF), and deletion of BDNF in the postnatal brain causes an obese phenotype [33, 57, 65]. There is no significant change in the BDNF expression levels in the brain of CaSKO mice as compared to controls, suggesting that obesity in this mouse line is not associated with BDNF resistance or defect.

The positive role of Shp2 in leptin signaling has also been demonstrated by phenotypic analysis of another mutant mouse line, in which Shp2 is removed broadly in neural cells in the brain [34]. This line was generated by breeding *Shp2* floxed mice with a transgenic mouse line (CRE3) in which Cre is widely expressed in the central nervous system [34]. Interestingly, despite enormous biochemical data suggesting Shp2 involvement in multiple signaling pathways elicited by growth factors and cytokines, the most prominent phenotype was significant gain of body weight, and development of obesity and diabetes in homozygous mutants under different genetic backgrounds. Unlike the CaSKO mice, homozygous mutants in this line showed hyperphagia in addition to accelerated body weight gain. Secondary to obesity, the mutant mice developed progressively deteriorating diabetes and associated pathological complications, including vasculitis, diabetic nephropathy, urinary bladder infections, gastric paresis, prostatitis, and impaired spermatogenesis. Consistent with the obese and hyperleptinemia phenotype, acute leptin responses were altered in the hypothalamic neurons in the mutant animals. Together, these observations argue that Shp2 is a critical regulator of intracellular signaling in neuronal control of metabolism, food intake, and energy expenditure in adult mammals. This physiological function of Shp2 in mediating metabolic activity is primarily due to its regulation of leptin signaling in the hypothalamus.

The role of Shp2 in regulation of leptin signaling in the central nervous system has been further analyzed by deleting Shp2 directly in POMC-expressing neurons (POMC-*Shp2*$^{-/-}$ mice) [4]. POMC deficiency leads to development of obesity in human subjects and mouse models [13]. Selective ablation of LepRb in POMC neurons leads to excessive gain of body weight and adiposity, hyperleptinemia, and abnormal expression of hypothalamic neuropeptides [3]. Furthermore, selective reexpression of LepRb in LepRb-deficient *db/db* mice largely restores leptin response in the hypothalamus, as evidenced by marked decrease in food intake, modest reduction in body weight, and normalization of blood glucose levels [28]. Together, these results support a critical role of POMC neurons in mediating leptin signaling in the hypothalamus for control of energy balance and metabolism. In contrast to the resistance to diet-induced obesity in mice with POMC-specific deletion of protein tyrosine phosphatase 1B (PTP1B), a negative regulator of leptin signaling, Shp2 loss in POMC neurons leads to development of obesity and leptin resistance. Also detected in POMC-*Shp2*$^{-/-}$ mice were significantly reduced expression of *POMC* mRNA and alpha-Melanocyte-stimulating hormone (α-MSH) peptide and increased levels of diet-induced liver triglyceride. On either low-fat or high-fat diet, POMC-*Shp2*$^{-/-}$ mice showed elevated body weight (with increased epididymal fat pad weight and fat mass) than age- and sex-matched controls, although there was no difference in the lean mass between the two groups of animals. Consistent with the phenotype in CaSKO mice, Shp2 loss in POMC neurons does not alter food intake but decreases energy expenditure. Decreased sensitivity to leptin is evident in POMC-*Shp2*$^{-/-}$ mice, as demonstrated by their impaired response to a high dose of leptin injection, with regard to suppression of body weight and food intake, in comparison to control mice.

Shp2 Couples Leptin and Estrogen Signals

As discussed above, targeted gene deletion of Shp2 in POMC neurons, forebrain region, or broadly in the brain results in attenuation of leptin signals and development of leptin resistance and obesity in rodents, suggesting a positive role of Shp2 in mediating leptin response in the brain. Most recent data from transgenic expression of a dominant-active Shp2^{D61A} mutant in forebrain neurons offers further support to a notion that Shp2 acts to amplify leptin signal [26]. Interestingly, this experiment has also unexpectedly revealed a molecular link by Shp2 between the leptin and estrogen signaling pathways.

It has been long recognized that estradiol-17β (E2), a reproductive hormone, also plays a critical role in control of body weight and energy balance [14]. Estrogen deficiency is well correlated with increased risk for development of obesity and type 2 diabetes in ovariectomized rodents or postmenopausal women [11, 58]. Hormone treatment in postmenopausal women is efficient for prevention of obesity progression, and improvement of insulin sensitivity and glucose tolerance [59]. Estrogen replacement in ovariectomized animals leads to a decrease in food intake and an increase in energy expenditure, resulting in suppression of obesity development [21]. Targeted deletion of the aromatase that catalyzes the synthesis of estrogen, or disruption of estrogen receptor α (ERα), leads to more severe age-dependent obesity in females than in males, indicating that estrogen deficiency is responsible for progression of sexually dimorphic obesity [27, 30]. Targeted deletion of ERα or knockdown of ERα expression by microinjection of siRNA into the ventromedial nucleus of hypothalamus leads to obesity and metabolic defects in mice [27, 46]. Indeed, estrogen has a leptin-like effect in activation of intracellular signaling pathways, including Erk, Akt, and Stat3, in hypothalamic melanocortin cells [20], although the underlying molecular mechanism for cross-talk between estrogen and leptin pathways remains to be elucidated.

Although the expression level of $Shp2^{D61A}$ mutant was similar between male and female animals, a gender effect was observed in response to high-fat diet (HFD)-induced body weight gain. Female, but not male, transgenic mice were resistant to the effects of high calorie diet challenge, exhibiting only a modest increase of body weight following feeding for 20 weeks. Of note, this effect was alleviated in ovariectomized animals, suggesting a role for estrogen in this sexually dimorphic obese phenotype induced by HFD. Expression of $Shp2^{D61A}$ in the brain resulted in decreased gonadal fat pad mass and reduced adipocyte size in female mice fed HFD. The female transgenic mice displayed lower levels of hepatic steatosis fed either regular or HFD. Consistent with the lower body temperature of mice with Shp2 deletion in forebrain neurons [68], expression of $Shp2^{D61A}$ mutant leads to increased heat production, elevated O_2 consumption, and CO_2 release, with decreased food intake in female transgenic mice. After feeding with HFD, transgenic female mice also showed nearly normal blood glucose and insulin levels and maintained insulin sensitivity and glucose tolerance. Consistent with the sexually dimorphic phenotype observed in transgenic mice, microinjection of adeno-associated virus (AAV) expressing $Shp2^{D61A}$ mutant into the mediobasal hypothalamus of C57BL/6 mice also induced an anti-obese effect in female but not in male mice when placed on HFD.

The anti-obesity phenotype is clearly associated with the increased leptin sensitivity in female transgenic mice, as demonstrated by more prolonged body weight loss following intraperitoneal injection of leptin. The gender effect also suggests that the $Shp2^{D61A}$ mutant molecule acts to enhance leptin sensitivity, by working in concert with a sex hormone pathway in the hypothalamus. Indeed, immunofluorescent staining and co-immunoprecipitation experiments revealed co-localization of Shp2 with ERα and a leptin signaling component (by using leptin-stimulated pY-Stat3 signal as a marker). Estrogen treatment also leads to enhanced association of Shp2 with ERα in cell culture *in vitro*. Combined treatment with leptin and estrogen induced higher levels of p-Erk, p-Akt, and pY-Stat3 in the hypothalamus, and the synergistic effect is more prominent in female than in male mice. In biochemical analysis, knockdown of either Shp2 or ERα had similar inhibitory effects on leptin induction of p-Erk, and expression of $Shp2^{D61A}$ leads to more profound and sustained Erk activation by leptin and estrogen in MCF-7 cells. Thus, Shp2 mediates the cross-talk between leptin and estrogen signaling pathways via binding to ERα, and the interaction of Shp2 with ERα is enhanced by the ligand estrogen. Enhanced

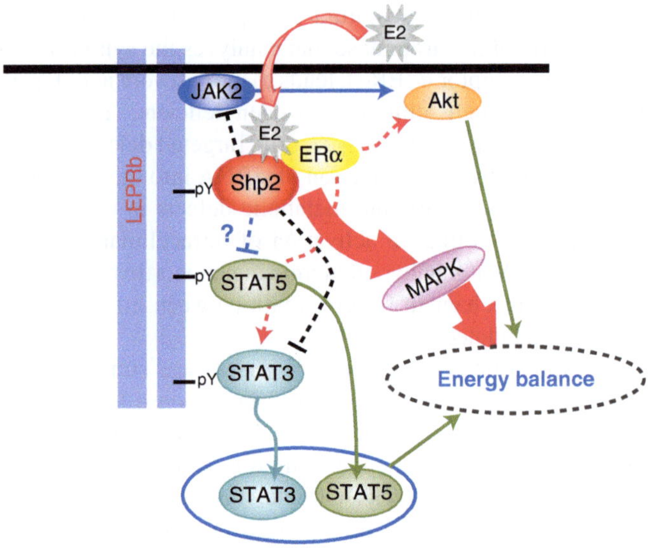

Fig. 9.2 A model of leptin/Shp2/ERα signaling pathway for energy balance. Shp2 mediates leptin action through inhibiting the activation of JAK2, STAT3, and STAT5 and enhancing the MAPK signaling pathway. Estrogen activates Akt, STAT3 and MAPK via ERα. Leptin and estrogen synergistically induce the activation of MAPK, but not Akt or STAT3. Shp2 and ERα are required for MAPK activity by estrogen or leptin stimulation

activation of Erk by leptin and estrogen mediated by a Shp2/ERα complex may be an important mechanism in control of energy homeostasis in the hypothalamus (Fig. 9.2) [26]. It remains to be elucidated how the interaction of Shp2 with ERα mediates the cellular response to leptin.

Leptin binding to LepRb also stimulates the phosphorylation and nuclear translocation of Stat5 [23, 42, 47], and deletion of Stat5 in neural cells leads to leptin resistance, hyperphagia, and obesity via modulating granulocyte-macrophage colony-stimulating factor (GM-CSF) signaling [40]. It will be interesting to determine if Shp2 regulates leptin-stimulated Stat5 activation in the hypothalamus, as this phosphatase has been shown to promote prolactin-induced Stat5 signaling in the mammary gland [32].

A Role for Shp2 in Control of Insulin Secretion from Pancreatic β-Cells

The molecular mechanisms underlying β-cell failure are not fully understood, although it is widely recognized that β-cell dysfunction plays a critical role in the development of both forms of diabetes. Whereas autoimmune destruction of β-cells is a major factor in the etiology of type 1 diabetes, type 2 diabetes is often secondary to peripheral insulin resistance and eventually develops due to progressive exhaustion and failure of β-cells under pressure to secret more insulin in compensation for insulin resistance. Recent experimental data suggest critical roles of receptors for insulin and insulin-like growth factor 1 (IGF1) in regulation of β-cell mass and functions [36, 37]. Interestingly, β-cells lacking insulin receptor fail to secret insulin properly in response to glucose stimulation, suggesting an unanticipated role of the insulin pathway in β-cell sensing of blood glucose levels.

To determine whether Shp2 plays a role in β-cell development and function, a mouse line ($Shp2^{panc-/-}$) was generated by crossing *Shp2* floxed mice with *Pdx1-Cre* transgenic mice, in which Shp2 is selectively ablated in the pancreas [70]. Immunofluorescent staining confirmed deletion of Shp2 in pancreatic β-cells in this mutant mouse line. Homozygous mutants are born with the Mendelian frequency and display normal postnatal survival and development. Further characterization of this mutant mouse line reveals a critical role of Shp2 as a regulator and coordinator of multiple signaling pathways in β cells.

$Shp2^{panc-/-}$ mice exhibited progressive glucose intolerance and increased body weight gain as compared to controls, accompanied by a mild decrease in serum insulin levels. Consistently, following a glucose injection, acute phase insulin secretion was ablated in $Shp2^{panc-/-}$ mice, suggesting a defect in β-cell response to rising blood glucose levels. In human subjects with defective glucose tolerance or in the early stages of type 2 diabetes, loss of acute phase insulin release is frequently detected, despite enhancement of the second-phase secretion [64]. A failure in the rapid first-phase insulin secretion may be due to abnormalities in glucose entry, glucose metabolism, availability of readily releasable insulin granules, exocytosis of insulin-containing granules, and/or alterations in intracellular glucose/lipid balance within β cells. In addition, dramatically decreased β-cell mass may also lead to ablation of the acute phase insulin secretion. Anatomical examination and immunostaining results suggest that deletion of Shp2 in the pancreas did not cause aberrant change in islet morphology or β-cell mass in mice fed regular chow diet. However, immunostaining and enzyme-linked immunosorbent assay (ELISA) detected markedly decreased insulin content within islets of $Shp2^{panc-/-}$ mice, as compared to controls. Therefore, the loss of first-phase insulin secretion in mutant mice is likely due to a defect in insulin biosynthesis in β-cells of $Shp2^{panc-/-}$ mice. siRNA-mediated knockdown of Shp2 expression in INS-1 832/13 cells also resulted in decreased insulin secretion in both basal and high glucose conditions *in vitro*, without affecting cell proliferation. Together, these results suggest a unique role of Shp2 in β-cell synthesis and secretion of insulin for glucose homeostasis.

Removal of Shp2 from pancreatic β-cells results in an age-dependent inability to handle a glucose challenge, indicating that appropriate Shp2 expression in β cells is crucial for control of glucose metabolism. It has been widely recognized that intact first-phase insulin secretion is of importance in maintaining normal glucose tolerance. Analysis of human subjects with impaired glucose tolerance showed multiple defects in both qualitative and quantitative measures of insulin secretion [9]. Reduced early-phase insulin secretion in response to oral glucose challenge was detected in subjects with normal glucose tolerance who are relatives of type 2 diabetic patients, and therefore measurement of insulin secretion has been used as a predictor of future development of type 2 diabetes [15]. Impaired glucose tolerance in *Shp2*$^{panc-/-}$ mice points to a functional defect in β-cells devoid of Shp2.

Glucose entry is a rate-limiting step governing β cell response during glucose-stimulated insulin secretion. Genetic analysis has revealed a correlation between type 2 diabetes and polymorphisms in genes coding for proteins involved in the glucose-sensing machinery [2]. A variety of monogenic forms of diabetes are associated with disrupted glucose-sensing and glucose-signaling pathways and consequently the loss of first-phase insulin secretion from β-cells [54]. In the classical view, the glucose-sensing system in β cells consists of a high capacity, low affinity glucose transporter-2 (Glut2), which balances the extra- and intracellular glucose levels, and a high KM glucokinase, which phosphorylates glucose to glucose-6-phosphate at a rate that is very sensitive to the change of circulating glucose levels. Progressively decreasing expression level of Glut2 is a critical factor for both type 1 and type 2 diabetes. On the other hand, glucokinase plays a prominent role in glucose sensing by exerting tight control over the rate of glycolysis [62]. In *Shp2*$^{panc-/-}$ mice, the expression levels of both Glut2 and glucokinase were significantly downregulated in pancreatic islets, and a similar phenotype was observed in an *in vitro* study which utilized an Shp2 gene silencing method. The impaired expression of glucose sensors in Shp2-deficient β cells is directly connected to loss of first-phase insulin secretion and consequently manifestation of glucose intolerance.

Shp2 Promotes Insulin Gene Transcription and Biosynthesis in β Cells

Insulin gene (*Ins1* and *Ins2*) transcription levels and insulin protein amounts are markedly downregulated in pancreatic β cells in *Shp2*$^{panc-/-}$ mice *in vivo* or following Shp2 gene silencing in cell lines *in vitro*. Molecular analysis suggests that the impaired insulin gene transcription and biosynthesis are mostly due to decreased expression level and lower activity of pancreatic and duodenal homeobox 1 (Pdx1) in Shp2-deficient β cells. Pdx1 is known to play a critical role in modulation of *Ins1* and *Ins2* gene expression and acts in concert with other transcription factors. Thus, reduced Pdx1 levels and activity may have a profound impact on insulin gene transcription. Consistent with this notion, forced expression of exogenous Pdx1 in Shp2-knockdown INS-1 832/13 cells improved insulin biosynthesis, which argues that Pdx1 is a major downstream component in a pathway regulated by Shp2 for up-regulation of insulin biosynthesis.

Previous experiments by several laboratories have suggested that optimal expression of key transcription factors is required for normal pancreatic development and proper β-cell function. Inherited mutations in genes encoding these transcription factors are directly linked to pathogenesis of type 2 diabetes. For example, mutations in *Pdx1*, *glucokinase*, hepatocyte nuclear factor 4 alpha (*Hnf4α*), or hepatocyte nuclear factor 1 alpha (*Hnf1α*) can lead to monogenic forms of type 2 diabetes known as maturity onset diabetes of the young (MODY) [52]. Quantitative RT-PCR analysis revealed that the expression profiles of several β-cell specific genes, including *Ins1*, *Ins2*, *Glut2*, *Pdx1*, *glucokinase*, nuclear factor of activated T cells c1 (*NFATc1*), *Hnf1*α and *Hnf4*α, were disrupted in both Shp2-deficient islets and Shp2-knockdown INS-1 832/13 cells. Reduced expression of these genes likely contributes to the impaired insulin production and secretion in Shp2-deficient and -knockdown β cells.

The signaling molecules acting upstream of Pdx1 in linking Shp2 action to Pdx1 expression and activation remain to be determined. Experimental results suggest that glucose-stimulated Akt/FoxO1 and Erk pathways are altered in Shp2-deficient cells. In previous studies [17], several groups detected physical association of Shp2 with IRS1 and other IRS family members. These adaptor/scaffolding proteins share two tyrosine residues at the C-terminal tail that can bind Shp2 via two SH2 domains. Consistently, co-immunoprecipitation assay detected a complex consisting of Shp2 and IRS2 in INS-1 832/13 cells, and the interaction of these two proteins was enhanced by cell exposure to high glucose levels in the culture medium. Shp2 is also associated with the p85 subunit of phosphatidylinositol 3-kinase (PI3K), suggesting a putative role for Shp2 in the IRS2-PI3K pathway in β-cells. Decreased levels of p-Y-IRS2 and reduced amounts of IRS2 and p85 complex were detected in Shp2-knockdown cells. Consistently, impaired levels of p-Akt (Ser-473) and p-FoxO1 (Ser-256), components downstream of the IRS2 and PI3K pathway, were observed in Shp2-knockdown cells exposed to either 3 or 15 mM glucose, in comparison with controls. Therefore, downregulation of Shp2 expression results in inhibition of glucose-stimulated activation of Akt in INS-1 832/13 cells.

Increased content of forkhead box protein O1 (FoxO1) in the nucleus of Shp2-knockdown INS-1 832/13 cells was noted by immunostaining assay, as compared to controls where a large portion of FoxO1 was found in the cytoplasm. Consistently, chromatin immunoprecipitation (ChIP) assay detected significantly increased FoxO1 binding to *cis*-acting elements within the Pdx1 promoter in Shp2-knockdown cells compared to controls. Previous studies showed that expression of a FoxO1 mutant that was retained in the nucleus repressed Pdx1 expression [49]. Therefore, nuclear localization of FoxO1 in Shp2-knockdown cells likely contributes to Pdx1 repression and impaired *Ins1* and *Ins2* gene transcription. Indeed, markedly reduced Pdx1 expression was detected in Shp2-deficient islets by immunostaining analysis and also in Shp2 knockdown cells by immunoblotting. In parallel with the Akt/FoxO1 pathway, activated Erk1/2 kinases have been shown to activate Pdx1, resulting in up-regulation of insulin gene expression in β-cells [39]. Shp2 knockdown leads to remarkably lower p-Erk1/2 signals in INS-1 832/13 cells following stimulation with a high concentration of glucose. In co-immunoprecipitation assays, Shp2 was found in association with Sprouty 1 in both low and high glucose conditions, and Shp2 knockdown resulted in elevated tyrosine phosphorylation of Sprouty 1. These results suggest that Sprouty 1 is a substrate for Shp2 and dephosphorylation of Sprouty, a feedback inhibitor of the Erk pathway, likely leads to enhanced Erk1/2 signaling. Consistently, expression of a dominant negative mutant of Shp2 ($Shp2^{C459S}$) suppressed impaired p-Erk1/2 and p-Akt signals, and consequently insulin biosynthesis, suggesting that the catalytic activity of Shp2 is required for its biological function in β-cells.

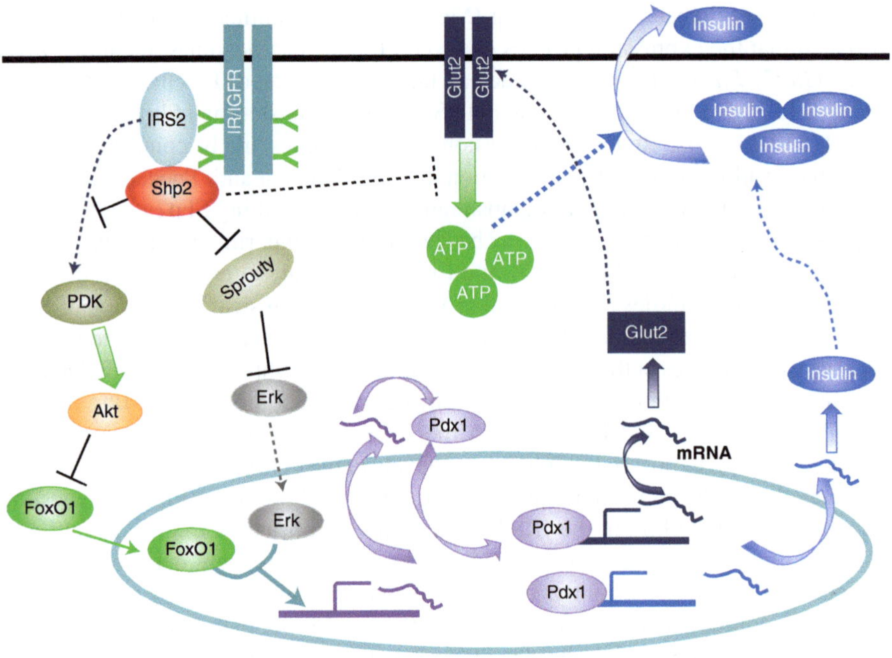

Fig. 9.3 A model for Shp2 orchestration of signaling events controlling insulin biosynthesis in pancreatic β-cells. Shp2 acts downstream of IRS2 in coordinated regulation of glucose and insulin signals in pancreatic β-cells. Shp2 promotes signaling through both the PI3K-Akt-FoxO1 and Erk pathways in control of Pdx1 expression

Taken together, the impaired insulin expression in Shp2-deficient β-cells is a direct effect of reduced expression and activation of Pdx1, a key transcription factor in directing *Ins1* and *Ins2* gene transcription. Shp2 acts to promote Pdx1 expression and activity through modulation of PI3K/Akt/FoxO1 and Erk1/2 activation, as summarized in Fig. 9.3. However, this is likely a simplified model, and it is important to point out that deficiency of other transcription factors may also contribute to the defect in insulin production in Shp2-deficient β-cells.

Deletion of Shp2 in the pancreas, while not causing diabetes directly, may increase susceptibility to diabetes-inducing factors. Previous studies showed that mice devoid of insulin or IGF1 receptor in β-cells display normal β-cell proliferation and development, indicating that insulin or IGF1 signal is not crucial for early development of islet β-cells [36, 37]. Characterization of *Shp2*$^{panc-/-}$ mice also suggests that Shp2 has a limited role in pancreatic development but is a critical regulator of β-cell function in adults. However, it is also possible that a failure to observe developmental defects in *Shp2*$^{panc-/-}$ mice could be due to progressive deletion of a floxed target gene by Pdx1-Cre in conditional gene knockout mouse models. Furthermore, to define a direct role of Shp2 in pancreatic β-cells, it is necessary to generate another mouse line in which Shp2 is selectively deleted in β-cells only.

Conclusion

Shp2, a widely expressed non-receptor tyrosine phosphatase, has been implicated in regulation of physiological activities in a variety of cell types. Discussion in this chapter focuses on its function in metabolic signaling events, in particular the amplification of leptin signaling in the hypothalamus for control of body weight and energy balance and the regulation of insulin synthesis in pancreatic β-cells for glucose homeostasis. This has largely been accomplished by the generation and phenotypic analyses of cell type-specific gene knockout mouse models. Further studies are clearly needed to elucidate the signaling cascades in these important physiological processes and their defects in obesity and diabetes, in order to develop new and more effective pharmaceutic intervention strategies.

Acknowledgment The authors thank colleagues for helpful discussion and critical reading of the manuscript and apologize for omitting citation of many other important reports in this short chapter. Work in the Feng laboratory was supported by NIH grants DK073945, DK075916, and HL096125.

References

1. Agazie YM, Hayman MJ (2003) Molecular mechanism for a role of SHP2 in epidermal growth factor receptor signaling. Mol Cell Biol 23:7875–7886
2. Ashcroft FM, Rorsman P (2004) Molecular defects in insulin secretion in type-2 diabetes. Rev Endocr Metab Disord 5:135–142
3. Balthasar N, Coppari R, McMinn J, Liu SM, Lee CE, Tang V, Kenny CD, McGovern RA, Chua SC Jr, Elmquist JK et al (2004) Leptin receptor signaling in POMC neurons is required for normal body weight homeostasis. Neuron 42:983–991
4. Banno R, Zimmer D, De Jonghe BC, Atienza M, Rak K, Yang W, Bence KK (2010) PTP1B and SHP2 in POMC neurons reciprocally regulate energy balance in mice. J Clin Invest 120:720–734
5. Bates SH, Stearns WH, Dundon TA, Schubert M, Tso AW, Wang Y, Banks AS, Lavery HJ, Haq AK, Maratos-Flier E et al (2003) STAT3 signalling is required for leptin regulation of energy balance but not reproduction. Nature 421:856–859
6. Binder G (2009) Noonan syndrome, the Ras-MAPK signalling pathway and short stature. Horm Res 71(suppl 2):64–70
7. Bjornholm M, Munzberg H, Leshan RL, Villanueva EC, Bates SH, Louis GW, Jones JC, Ishida-Takahashi R, Bjorbaek C, Myers MG Jr (2007) Mice lacking inhibitory leptin receptor signals are lean with normal endocrine function. J Clin Invest 117:1354–1360
8. Bruning JC, Gautam D, Burks DJ, Gillette J, Schubert M, Orban PC, Klein R, Krone W, Muller-Wieland D, Kahn CR (2000) Role of brain insulin receptor in control of body weight and reproduction. Science 289:2122–2125
9. Byrne MM, Sturis J, Sobel RJ, Polonsky KS (1996) Elevated plasma glucose 2 h postchallenge predicts defects in beta-cell function. Am J Physiol 270:E572–E579
10. Carpenter LR, Farruggella TJ, Symes A, Karow ML, Yancopoulos GD, Stahl N (1998) Enhancing leptin response by preventing SH2-containing phosphatase 2 interaction with Ob receptor. Proc Natl Acad Sci U S A 95:6061–6066
11. Carr MC (2003) The emergence of the metabolic syndrome with menopause. J Clin Endocrinol Metab 88:2404–2411

12. Case RD, Piccione E, Wolf G, Benett AM, Lechleider RJ, Neel BG, Shoelson SE (1994) SH-PTP2/Syp SH2 domain binding specificity is defined by direct interactions with platelet-derived growth factor beta-receptor, epidermal growth factor receptor, and insulin receptor substrate-1-derived phosphopeptides. J Biol Chem 269:10467–10474
13. Coll AP, Farooqi IS, Challis BG, Yeo GS, O'Rahilly S (2004) Proopiomelanocortin and energy balance: insights from human and murine genetics. J Clin Endocrinol Metab 89:2557–2562
14. Dubuc PU (1985) Effects of estrogen on food intake, body weight, and temperature of male and female obese mice. Proc Soc Exp Biol Med 180:468–473
15. Elbein SC, Hasstedt SJ, Wegner K, Kahn SE (1999) Heritability of pancreatic beta-cell function among nondiabetic members of Caucasian familial type 2 diabetic kindreds. J Clin Endocrinol Metab 84:1398–1403
16. Fambrough D, McClure K, Kazlauskas A, Lander ES (1999) Diverse signaling pathways activated by growth factor receptors induce broadly overlapping, rather than independent, sets of genes. Cell 97:727–741
17. Feng GS (1999) Shp-2 tyrosine phosphatase: signaling one cell or many. Exp Cell Res 253: 47–54
18. Feng GS, Pawson T (1994) Phosphotyrosine phosphatases with SH2 domains: regulators of signal transduction. Trends Genet 10:54–58
19. Friedman JM (2004) Modern science versus the stigma of obesity. Nat Med 10:563–569
20. Gao Q, Horvath TL (2008) Cross-talk between estrogen and leptin signaling in the hypothalamus. Am J Physiol Endocrinol Metab 294:E817–E826
21. Gao Q, Mezei G, Nie Y, Rao Y, Choi CS, Bechmann I, Leranth C, Toran-Allerand D, Priest CA, Roberts JL et al (2007) Anorectic estrogen mimics leptin's effect on the rewiring of melanocortin cells and Stat3 signaling in obese animals. Nat Med 13:89–94
22. Gao Q, Wolfgang MJ, Neschen S, Morino K, Horvath TL, Shulman GI, Fu XY (2004) Disruption of neural signal transducer and activator of transcription 3 causes obesity, diabetes, infertility, and thermal dysregulation. Proc Natl Acad Sci U S A 101:4661–4666
23. Gong Y, Ishida-Takahashi R, Villanueva EC, Fingar DC, Munzberg H, Myers MG Jr (2007) The long form of the leptin receptor regulates STAT5 and ribosomal protein S6 via alternate mechanisms. J Biol Chem 282:31019–31027
24. Grossmann KS, Rosario M, Birchmeier C, Birchmeier W (2010) The tyrosine phosphatase Shp2 in development and cancer. Adv Cancer Res 106:53–89
25. Hanafusa H, Torii S, Yasunaga T, Matsumoto K, Nishida E (2004) Shp2, an SH2-containing protein-tyrosine phosphatase, positively regulates receptor tyrosine kinase signaling by dephosphorylating and inactivating the inhibitor Sprouty. J Biol Chem 279:22992–22995
26. He Z, Zhang SS, Meng Q, Li S, Zhu HH, Raquil MA, Alderson N, Zhang H, Wu J, Rui L et al (2012) Shp2 controls female body weight and energy balance by integrating leptin and estrogen signals. Mol Cell Biol 32:1867–1878
27. Heine PA, Taylor JA, Iwamoto GA, Lubahn DB, Cooke PS (2000) Increased adipose tissue in male and female estrogen receptor-alpha knockout mice. Proc Natl Acad Sci U S A 97: 12729–12734
28. Huo L, Gamber K, Greeley S, Silva J, Huntoon N, Leng XH, Bjorbaek C (2009) Leptin-dependent control of glucose balance and locomotor activity by POMC neurons. Cell Metab 9:537–547
29. Jiang L, You J, Yu X, Gonzalez L, Yu Y, Wang Q, Yang G, Li W, Li C, Liu Y (2008) Tyrosine-dependent and -independent actions of leptin receptor in control of energy balance and glucose homeostasis. Proc Natl Acad Sci U S A 105:18619–18624
30. Jones ME, Thorburn AW, Britt KL, Hewitt KN, Wreford NG, Proietto J, Oz OK, Leury BJ, Robertson KM, Yao S et al (2000) Aromatase-deficient (ArKO) mice have a phenotype of increased adiposity. Proc Natl Acad Sci U S A 97:12735–12740
31. Kazlauskas A, Feng GS, Pawson T, Valius M (1993) The 64-kDa protein that associates with the platelet-derived growth factor receptor beta subunit via Tyr-1009 is the SH2-containing phosphotyrosine phosphatase Syp. Proc Natl Acad Sci U S A 90:6939–6943

32. Ke Y, Lesperance J, Zhang EE, Bard-Chapeau EA, Oshima RG, Muller WJ, Feng GS (2006) Conditional deletion of Shp2 in the mammary gland leads to impaired lobulo-alveolar outgrowth and attenuated Stat5 activation. J Biol Chem 281:34374–34380
33. Kernie SG, Liebl DJ, Parada LF (2000) BDNF regulates eating behavior and locomotor activity in mice. EMBO J 19:1290–1300
34. Krajewska M, Banares S, Zhang EE, Huang X, Scadeng M, Jhala US, Feng GS, Krajewski S (2008) Development of diabesity in mice with neuronal deletion of Shp2 tyrosine phosphatase. Am J Pathol 172:1312–1324
35. Kuhne MR, Pawson T, Lienhard GE, Feng GS (1993) The insulin receptor substrate 1 associates with the SH2-containing phosphotyrosine phosphatase Syp. J Biol Chem 268: 11479–11481
36. Kulkarni RN, Bruning JC, Winnay JN, Postic C, Magnuson MA, Kahn CR (1999) Tissue-specific knockout of the insulin receptor in pancreatic beta cells creates an insulin secretory defect similar to that in type 2 diabetes. Cell 96:329–339
37. Kulkarni RN, Holzenberger M, Shih DQ, Ozcan U, Stoffel M, Magnuson MA, Kahn CR (2002) beta-cell-specific deletion of the Igf1 receptor leads to hyperinsulinemia and glucose intolerance but does not alter beta-cell mass. Nat Genet 31:111–115
38. Lai LA, Zhao C, Zhang EE, Feng GS (2004) The Shp-2 tyrosine phosphatase. In: Arino J, Alexander D (eds) Protein phosphatases. Springer, Berlin, Heidelberg, pp 275–299
39. Lawrence M, Shao C, Duan L, McGlynn K, Cobb MH (2008) The protein kinases ERK1/2 and their roles in pancreatic beta cells. Acta Physiol (Oxf) 192:11–17
40. Lee JY, Muenzberg H, Gavrilova O, Reed JA, Berryman D, Villanueva EC, Louis GW, Leinninger GM, Bertuzzi S, Seeley RJ et al (2008) Loss of cytokine-STAT5 signaling in the CNS and pituitary gland alters energy balance and leads to obesity. PLoS One 3:e1639
41. Li C, Friedman JM (1999) Leptin receptor activation of SH2 domain containing protein tyrosine phosphatase 2 modulates Ob receptor signal transduction. Proc Natl Acad Sci U S A 96:9677–9682
42. Martens N, Uzan G, Wery M, Hooghe R, Hooghe-Peters EL, Gertler A (2005) Suppressor of cytokine signaling 7 inhibits prolactin, growth hormone, and leptin signaling by interacting with STAT5 or STAT3 and attenuating their nuclear translocation. J Biol Chem 280:13817–13823
43. Matsuo K, Delibegovic M, Matsuo I, Nagata N, Liu S, Bettaieb A, Xi Y, Araki K, Yang W, Kahn BB et al (2010) Altered glucose homeostasis in mice with liver-specific deletion of Src homology phosphatase 2. J Biol Chem 285:39750–39758
44. Milarski KL, Saltiel AR (1994) Expression of catalytically inactive Syp phosphatase in 3T3 cells blocks stimulation of mitogen-activated protein kinase by insulin. J Biol Chem 269:21239–21243
45. Mori H, Hanada R, Hanada T, Aki D, Mashima R, Nishinakamura H, Torisu T, Chien KR, Yasukawa H, Yoshimura A (2004) Socs3 deficiency in the brain elevates leptin sensitivity and confers resistance to diet-induced obesity. Nat Med 10:739–743
46. Musatov S, Chen W, Pfaff DW, Mobbs CV, Yang XJ, Clegg DJ, Kaplitt MG, Ogawa S (2007) Silencing of estrogen receptor alpha in the ventromedial nucleus of hypothalamus leads to metabolic syndrome. Proc Natl Acad Sci U S A 104:2501–2506
47. Mutze J, Roth J, Gerstberger R, Hubschle T (2007) Nuclear translocation of the transcription factor STAT5 in the rat brain after systemic leptin administration. Neurosci Lett 417:286–291
48. Nagata N, Matsuo K, Bettaieb A, Bakke J, Matsuo I, Graham J, Xi Y, Liu S, Tomilov A, Tomilova N et al (2012) Hepatic Src homology phosphatase 2 regulates energy balance in mice. Endocrinology 153(7):3158–3169
49. Nakae J, Biggs WH III, Kitamura T, Cavenee WK, Wright CV, Arden KC, Accili D (2002) Regulation of insulin action and pancreatic beta-cell function by mutated alleles of the gene encoding forkhead transcription factor Foxo1. Nat Genet 32:245–253
50. Neel BG, Gu H, Pao L (2003) The 'Shp'ing news: SH2 domain-containing tyrosine phosphatases in cell signaling. Trends Biochem Sci 28:284–293

51. Noguchi T, Matozaki T, Horita K, Fujioka Y, Kasuga M (1994) Role of SH-PTP2, a protein-tyrosine phosphatase with Src homology 2 domains, in insulin-stimulated Ras activation. Mol Cell Biol 14:6674–6682
52. O'Rahilly S, Barroso I, Wareham NJ (2005) Genetic factors in type 2 diabetes: the end of the beginning? Science 307:370–373
53. Pan Y, Carbe C, Powers A, Feng GS, Zhang X (2010) Sprouty2-modulated Kras signaling rescues Shp2 deficiency during lens and lacrimal gland development. Development 137: 1085–1093
54. Porter JR, Barrett TG (2005) Monogenic syndromes of abnormal glucose homeostasis: clinical review and relevance to the understanding of the pathology of insulin resistance and beta cell failure. J Med Genet 42:893–902
55. Ren D, Li M, Duan C, Rui L (2005) Identification of SH2-B as a key regulator of leptin sensitivity, energy balance, and body weight in mice. Cell Metab 2:95–104
56. Ren Y, Meng S, Mei L, Zhao ZJ, Jove R, Wu J (2004) Roles of Gab1 and SHP2 in paxillin tyrosine dephosphorylation and Src activation in response to epidermal growth factor. J Biol Chem 279:8497–8505
57. Rios M, Fan G, Fekete C, Kelly J, Bates B, Kuehn R, Lechan RM, Jaenisch R (2001) Conditional deletion of brain-derived neurotrophic factor in the postnatal brain leads to obesity and hyperactivity. Mol Endocrinol 15:1748–1757
58. Rogers NH, Perfield JW II, Strissel KJ, Obin MS, Greenberg AS (2009) Reduced energy expenditure and increased inflammation are early events in the development of ovariectomy-induced obesity. Endocrinology 150:2161–2168
59. Salpeter SR, Walsh JM, Ormiston TM, Greyber E, Buckley NS, Salpeter EE (2006) Meta-analysis: effect of hormone-replacement therapy on components of the metabolic syndrome in postmenopausal women. Diabetes Obes Metab 8:538–554
60. Saxton TM, Henkemeyer M, Gasca S, Shen R, Rossi DJ, Shalaby F, Feng GS, Pawson T (1997) Abnormal mesoderm patterning in mouse embryos mutant for the SH2 tyrosine phosphatase Shp-2. EMBO J 16:2352–2364
61. Sugimoto S, Wandless TJ, Shoelson SE, Neel BG, Walsh CT (1994) Activation of the SH2-containing protein tyrosine phosphatase, SH-PTP2, by phosphotyrosine-containing peptides derived from insulin receptor substrate-1. J Biol Chem 269:13614–13622
62. Sweet IR, Matschinsky FM (1997) Are there kinetic advantages of GLUT2 in pancreatic glucose sensing? Diabetologia 40:112–119
63. Tartaglia M, Gelb BD (2005) Noonan syndrome and related disorders: genetics and pathogenesis. Annu Rev Genomics Hum Genet 6:45–68
64. Unger RH (1991) Diabetic hyperglycemia: link to impaired glucose transport in pancreatic beta cells. Science 251:1200–1205
65. Xu B, Goulding EH, Zang K, Cepoi D, Cone RD, Jones KR, Tecott LH, Reichardt LF (2003) Brain-derived neurotrophic factor regulates energy balance downstream of melanocortin-4 receptor. Nat Neurosci 6:736–742
66. Yang W, Klaman LD, Chen B, Araki T, Harada H, Thomas SM, George EL, Neel BG (2006) An Shp2/SFK/Ras/Erk signaling pathway controls trophoblast stem cell survival. Dev Cell 10:317–327
67. You J, Yu Y, Jiang L, Li W, Yu X, Gonzalez L, Yang G, Ke Z, Li C, Liu Y (2010) Signaling through Tyr985 of leptin receptor as an age/diet-dependent switch in the regulation of energy balance. Mol Cell Biol 30:1650–1659
68. Zhang EE, Chapeau E, Hagihara K, Feng GS (2004) Neuronal Shp2 tyrosine phosphatase controls energy balance and metabolism. Proc Natl Acad Sci U S A 101:16064–16069
69. Zhang SQ, Yang W, Kontaridis MI, Bivona TG, Wen G, Araki T, Luo J, Thompson JA, Schraven BL, Philips MR et al (2004) Shp2 regulates SRC family kinase activity and Ras/Erk activation by controlling Csk recruitment. Mol Cell 13:341–355
70. Zhang SS, Hao E, Yu J, Liu W, Wang J, Levine F, Feng GS (2009) Coordinated regulation by Shp2 tyrosine phosphatase of signaling events controlling insulin biosynthesis in pancreatic beta-cells. Proc Natl Acad Sci U S A 106:7531–7536

Chapter 10
Protein Tyrosine Phosphatase Epsilon as a Regulator of Body Weight and Glucose Metabolism

Ari Elson

Abstract The obesity pandemic has focused attention in recent years to the physiological and molecular mechanisms that regulate body weight and glucose metabolism. Key signaling pathways that regulate both parameters depend heavily on reversible phosphorylation of proteins on tyrosine residues, a process regulated by the opposing activities of tyrosine kinases and tyrosine phosphatases. Here we review the roles of protein tyrosine phosphatase epsilon (PTPe) in regulating the leptin and insulin signaling pathways and through them—body weight and glucose metabolism. Mice lacking PTPe are leptin-hypersensitive and are protected from weight gain that follows a high-fat diet. PTPe helps downregulate leptin receptor signaling in the hypothalamus by dephosphorylating JAK2 following activation of the leptin receptor, thus inhibiting the receptor post-activation. PTPe is induced to perform this function after undergoing leptin receptor-induced phosphorylation at its C-terminal Y695. Mice lacking PTPe are also insulin-hypersensitive, indicating that PTPe downregulates signaling by this receptor as well. Studies in muscle cells confirm that PTPe inhibits insulin receptor signaling, possibly by targeting the receptor itself. These studies identify PTPe as a physiological inhibitor of both signaling pathways and as a factor in supporting the resistance to leptin and insulin that is established in obesity and in type-II diabetes, respectively.

A. Elson, Ph.D. (✉)
Department of Molecular Genetics, The Weizmann Institute of Science,
Rehovot 76100, Israel
e-mail: ari.elson@weizmann.ac.il

K.K. Bence (ed.), *Protein Tyrosine Phosphatase Control of Metabolism*,
DOI 10.1007/978-1-4614-7855-3_10,

Abbreviations

AgRP	Agouti-related protein
BHK	Baby hamster kidney
CD	Chow diet
Cyt-PTPe	Cytosolic isoform of PTPe
EKO	PTPe-knockout
ERK	Extracellular signal-regulated kinase
GSK	Glycogen synthase kinase
HFD	High-fat diet
PEPCK	Phosphoenolpyruvate carboxykinase
PKB	Protein kinase B
POMC	Pro-opiomelanocortin
PTK	Protein tyrosine kinase
PTP	Protein tyrosine phosphatase
PTPe	Protein tyrosine phosphatase epsilon
RPTPe	Receptor-type isoform of PTPe
RPTPs	Receptor-type PTPs
WT	Wild-type

Protein Tyrosine Phosphatases

Reversible phosphorylation of proteins is one of the better-studied molecular mechanisms for regulation of protein structure and function in vivo. Although the majority of protein phosphorylation events occur in eukaryotic cells on serine or threonine residues, tyrosine phosphorylation plays a major and well-established role in regulating cellular processes [1, 2]. Tyrosine phosphorylation is a reversible process that is regulated by the generically opposing activities of protein tyrosine kinases (PTKs) and protein tyrosine phosphatases (PTPs). The human genome contains 90 PTK genes and 107 genes that encode PTPs; of these, 85 and 81 genes, respectively, produce products that target protein substrates [3]. The numbers of PTKs and PTPs are then similar and are dwarfed by the much larger numbers of known and hypothesized tyrosine phosphorylation sites in cellular proteins. As a result, each PTK or PTP targets on average several substrates and fulfills multiple distinct physiological roles in different cell types.

The 38 members of the "classical" PTP family, which target exclusively phosphotyrosine residues in proteins, form the core of the PTP superfamily. Classical PTPs contain one or two PTP domains of approximately 240 amino acids each in their cytosolic regions and dephosphorylate their substrates by a two-step mechanism. In the first step, the PTP forms a covalent bond with the phosphate group of the substrate phosphotyrosine, displacing the phosphate group from this tyrosine in the process and releasing the dephosphorylated protein. The PTP-phosphate bond is then hydrolyzed, regenerating the active PTP and releasing the phosphate group [4]. Classical PTPs can be classified into two subgroups: the receptor-type PTPs (RPTPs), which number 21 genes, and the remaining 17 genes that encode non-RPTPs. Each subgroup can be subdivided further according to structural and sequence similarities. RPTPs contain an extracellular domain of varying length and structure, a membrane-spanning domain, and one or two cytosolic PTP domains. Although the structure of the extracellular domains of RPTPs resemble those of PTKs, few ligands of RPTPs are known and most are orphan receptors. Non-RPTPs are typically comprised of a single PTP domain that is flanked by other protein domains, which control the subcellular localization of the enzyme or regulate its activity [3, 5].

Recent studies have elucidated the molecular roles of specific PTPs in regulating discrete physiological processes such as regulation of body weight, glucose homeostasis, neural development, immune function, bone structure, and others. The reader is referred to several recent reviews that discuss the physiological roles of PTPs and their modes of regulation in detail [5–10]. Advances in understanding the molecular roles of discrete PTPs in physiology have led to attempts to inhibit PTPs for pharmaceutical use, either by design of specific inhibitors (e.g., [11, 12]) or by other means (e.g., [13]). Major challenges in the field include identifying PTPs that are suitable candidates for pharmaceutical intervention in a given context and designing inhibitors that can enter cells and function in vivo in a specific manner, without targeting related PTPs whose activities may be beneficial.

PTP Epsilon

Protein tyrosine phosphatase epsilon (PTPe) exists as a small family of proteins, all produced from the single *Ptpre* gene (*PTPRE* in humans; Fig. 10.1a). PTPe was described originally as a receptor-type PTP (RPTPe, tm-PTPe, PTPepsilonM) that includes a short and heavily glycosylated extracellular domain and two cytosolic PTP domains of which only the first, membrane-proximal one, is active [14–16]. A second major form of PTPe was subsequently discovered. This cytosolic isoform of PTPe (cyt-PTPe) (=PTPepsilonC) contains a short hydrophilic sequence of 12 amino acids that replaces the membrane-spanning and extracellular domains of RPTPe [17, 18]. The two major PTPe proteins are products of distinct transcripts of the *Ptpre* gene; each transcript is expressed from a separate promoter and exhibits a unique pattern of expression among cells and tissues [17–19]. Both transcripts share most of their sequence and differ only at their 5′ end; as a result, their protein products RPTPe and cyt-PTPe differ only at their N termini (Fig. 10.1a, b). The *Ptpre* gene gives rise to two additional but less-abundant proteins, p67 and p65. p67 is transcribed from the RPTPe or cyt-PTPe mRNAs starting at an initiation codon that is located in the region common to both mRNAs. In contrast, p65 is produced by calpain-mediated post-translational processing of RPTPe, cyt-PTPe, or p67 proteins [20, 21] (Fig. 10.1a). The four PTPe proteins differ only at their amino termini (Fig. 10.1b), hence each has its own pattern of subcellular localization. RPTPe is exclusively membranal, cyt-PTPe is predominantly cytosolic, but is found to some extent associated with the membrane or within the cell nucleus, while p67 and p65 are exclusively cytosolic [18, 20–22].

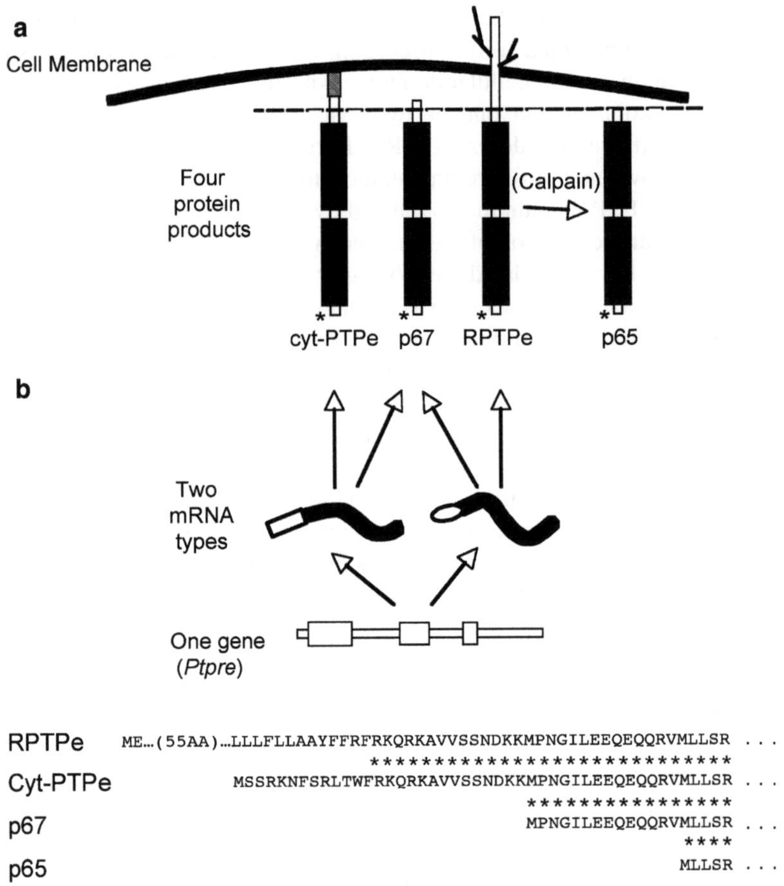

Fig. 10.1 The PTPe proteins. (**a**) Schematic outline of the four known forms of PTPe protein. The single *Ptpre* gene gives rise to two distinct mRNA species via separate promoters. The 5′ sequences of these mRNAs (*white oval* and *rectangle*) are unique, but downstream sequences are identical. Each mRNA produces a major protein—RPTPe or cyt-PTPe. Both mRNAs also produce p67 via initiation of translation at an ATG codon located in the sequences common to both mRNAs. Finally, all three PTPe proteins can be cleaved by calpain to generate p65. All PTPe proteins contain the same PTP domains (*black rectangles*); the N terminus of cyt-PTPe is unique to this form, (*striped rectangle*) *Asterisk* denote C-Terminal phosphorylation site. (**b**) N terminal sequences of all 4 forms of PTPe protein. *Asterisks* denote identical protein sequences

Physiological Roles of PTPe

In a manner similar to other PTPs that have several substrates, PTPe has been shown to regulate several divergent physiological systems; in some of these the role of PTPe is inhibitory, while in others it activates signaling events:

(a) *Breast cancer*: Analysis of mammary tumors that were induced in mice due to expression of oncogenes in their mammary epithelium revealed that RPTPe is expressed specifically in tumors initiated by activated (V664E) Neu or activated

(G12R) Ras, but not by c-Myc or several other onco-proteins [14]. Further studies indicated that RPTPe is not an oncogene in its own right [23], but that it supports the transformed phenotype induced by Neu in mammary epithelial tumor cells. RPTPe performs this role by linking Neu with its downstream effector Src; Neu phosphorylates RPTPe at its C-terminal Y695, which drives RPTPe to dephosphorylate and activate Src [24]. Accordingly, mammary tumor cells induced by Neu that lack RPTPe exhibit reduced Src activity and appear less transformed than similar cells that express RPTPe [25].

(b) *Peripheral nerve myelination*: cyt-PTPe is expressed in Schwann cells, which carry out myelination of axons in the peripheral nervous system. The delayed-rectifier, voltage-gated potassium channels Kv2.1 and Kv1.2 are physiological substrates of cyt-PTPe in these cells [26]. Membrane depolarization triggers Kv2.1 and induces it to facilitate exit of potassium cations from the cell; this process is potentiated by phosphorylation of Kv2.1 at Y124 by Src or Fyn. Cyt-PTPe antagonizes the activity of Src or Fyn by dephosphorylating Y124 of Kv2.1, thus downregulating channel activity both in vitro and in vivo [27]. In agreement, mice that lack PTPe exhibit a severe delay in sciatic nerve myelination and Kv2.1 is hyper-phosphorylated in Schwann cells derived from these mice [26]. A similar mechanism may act in the central nervous system, since expression of an inactive "substrate trapping" mutant of RPTPe in transgenic mice delays optic nerve myelination [28].

(c) *Osteoclast activity*: Female mice that lack PTPe exhibit mild osteopetrosis (gain of bone mass), which is due to reduced adhesion and activity of bone-resorbing osteoclasts. Osteoclasts express cyt-PTPe; lack of this PTP significantly disrupts the structure, cellular distribution, and stability of podosomes, the adhesion structures of these cells, leading to reduced bone-resorbing activity [29]. In agreement, mobilization of hematopoietic precursor cells from the bone marrow to the general circulation, in which osteoclasts participate, is defective in PTPe-deficient female mice [30]. At the molecular level, cyt-PTPe links signaling by integrins, the mechano-sensory receptors present on osteoclasts, with downstream activation of Src. Integrin activation induces phosphorylation of cyt-PTPe at its C-terminal Y638 (=Y695 in RPTPe), an event that is central in enabling cyt-PTPe to dephosphorylate and activate Src [31]. Accordingly, Src activity is reduced in osteoclasts that lack cyt-PTPe. As proper activity of Src is absolutely required for osteoclast-mediated bone resorption [32, 33], this strongly suggests that lack of cyt-PTPe disrupts osteoclast activity via reducing Src activity. Indeed, increasing Src activity in PTPe-deficient osteoclasts can correct the podosomal phenotype of PTPe-deficient osteoclasts [31].

(d) *Other roles*: PTPe can downregulate mitogenic signaling induced by MAP kinase [34, 35] or by JAK-STAT signaling in M1 leukemia cells [36–38]. Macrophages from PTPe-deficient mice are impaired in their respiratory burst response and produce reduced amounts of cytokines in response to bacterial lipopolysaccharide [39], and PTPe negatively regulates proliferation of endothelial cells [40]. RPTPe also affects erythrocyte morphology and downregulates activity of Ca^{2+}-activated potassium channels in these cells [41]. In what follows we describe a novel role for PTPe in regulating body mass and whole-body glucose homeostasis.

Regulation of Body Weight by Protein Tyrosine Phosphatases

The obesity pandemic has focused attention on the physiological and molecular mechanisms that regulate body weight and on possible approaches to manipulate them for therapeutic gain. Leptin, a hormone that is produced in white adipose tissue, is a well-established regulator of energy balance and body weight [42]. Leptin affects body weight by activating its receptor in the hypothalamus; this leads to activation of the receptor-associated PTK JAK2, to subsequent phosphorylation of downstream effector molecules, and to activation of the STAT3, PI3-kinase, and extracellular signal-regulated kinase (ERK) signaling pathways (Fig. 10.2). Activation of the leptin receptor ultimately increases production of the anorexigenic (appetite-depressing) neuropeptide pro-opiomelanocortin (POMC) and inhibits production of the orexigenic (appetite-stimulating) neuropeptides agouti-related protein (AgRP) and neuropeptide Y. The combined effect of leptin is therefore to decrease appetite and stimulate energy expenditure, which is consistent with presence within the organism of sufficient energy deposits in the form of body fat [43–46]. Interestingly, obesity is often associated in humans with resistance to the effects of leptin that is caused by poor activation of hypothalamic leptin signaling by the hormone. As a result, obese individuals often display elevated concentrations of leptin in circulation and respond poorly in terms of weight loss to exogenously administered leptin.

The central role of the JAK2 PTK in hypothalamic leptin signaling strongly suggests that tyrosine phosphorylation events are critical in regulation of body weight. In agreement, several PTPs have been shown to participate in regulation of leptin receptor signaling in this context (reviewed elsewhere within this volume).

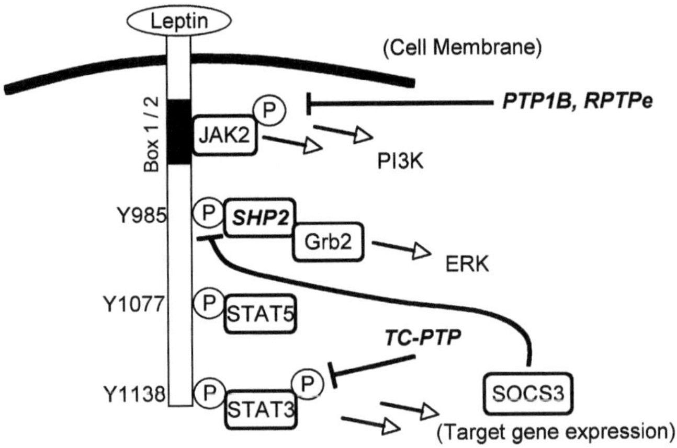

Fig. 10.2 Schematic outline of hypothalamic leptin receptor signaling. The hypothalamic leptin receptor is associated with the JAK2 PTK. Upon activation by leptin, the receptor activates JAK2, which autophosphorylates and trans-phosphorylates the receptor at tyrosines 985, 1077, and 1138. Phosphorylated tyrosines bind SHP2, STAT5, and STAT3, respectively, and activate the signaling pathways shown. Inhibition of leptin signaling is achieved by SOCS3, a target gene of STAT3, and by tyrosine phosphatases such as PTP1B, TC-PTP, and PTPe. PTPs are shown in *boldface italic* type

The non-receptor PTP PTP1B dephosphorylates and inactivates hypothalamic JAK2; accordingly, mice lacking PTP1B are hypersensitive towards the effects of leptin and exhibit reduced adiposity [47–49]. The closely related PTP TCPTP also inhibits leptin receptor signaling. Mice lacking TCPTP in neurons exhibit leptin hypersensitivity and are resistant to weight gain induced by high-fat food. In a manner distinct from PTP1B, TCPTP targets STAT3; accordingly, combined deletion of both PTPs induces stronger protection from diet-induced obesity than loss of either PTP alone [50]. The SH2 domain-containing PTP SHP2 affects leptin signaling in several ways. SHP2 promotes leptin signaling by binding the leptin receptor and activating downstream ERK signaling; however, it also inhibits JAK2 and STAT3 signaling, thereby downregulating receptor activity. The overall effect of SHP2 is to upregulate hypothalamic leptin receptor signaling, as shown by the increased adiposity, decreased leptin sensitivity, and decreased energy expenditure in mice that lack hypothalamic SHP2 [51, 52]. In agreement, expression of a dominant-active mutant of SHP2 in forebrain neurons of mice induced resistance to weight gain caused by high-fat food and improved whole-body glucose homeostasis [53]. Interestingly, this latter effect was detected in females, suggesting that SHP2 links estrogen and leptin signaling [53]. Inhibition of PTPs that downregulate leptin receptor signaling should improve the response of the receptor to its ligand and promote weight loss. However, such inhibition of PTPs should be specific and avoid inhibiting other PTPs, such as SHP2, that perform opposite roles in hypothalamic leptin receptor signaling [54] and whose inhibition would most likely increase body weight. In this context, we describe the roles of PTPe in regulation of hypothalamic leptin signaling.

Lack of PTPe Protects Mice from Weight Gain and Increases Basal Metabolic Rate

Mice that lack PTPe do not express any form of PTPe protein (PTPe-knockout [EKO] mice; [26]). EKO mice of both genders are born in Mendelian ratios and gain weight normally throughout development. However, when placed on a high-fat diet (HFD), female EKO mice gain less weight than their wild-type (WT) counterparts; after 11 weeks on fat-rich food, EKO female mice gained 55 % less weight than controls [55]. Significant protection from weight gain was observed also in mice that had been ovariectomized, a procedure that also results in massive weight gain in WT mice. Moreover, elderly female EKO mice (age >1 year) weighed significantly less than matched WT controls [55]. Reduced weight of EKO mice was associated with reduced mass of white adipose tissue deposits. Interestingly, protection from weight gain was detected in female EKO mice; little protection was observed in male EKO mice. Further studies by indirect calorimetry revealed that when fed regular lab chow (in which 18 % of calories are derived from fat), female EKO mice tended to produce more body heat, consume more oxygen, and produce more carbon dioxide, but this did not reach the level of statistical significance. In contrast, when mice were shifted to high-fat food (59 % of calories from fat), body heat production, oxygen consumption, and carbon dioxide production were significantly increased in EKO mice and were significantly higher than in control WT mice. In all of these studies EKO mice and matched controls consumed similar amounts of food and their physical activities were similar. It then appears that EKO female mice gain less weight when fed fat-rich food due to increased basal metabolic rate, which reduces the amount of excess food-derived energy stored as fat [55].

Loss of PTPe Induces Leptin Hypersensitivity in the Hypothalamus

Female mice lacking PTPe exhibit a strong trend for reduced concentrations of leptin in circulation [55]. When mice were subjected to a HFD, ovariectomy, or were analyzed at an advanced age—all paradigms in which EKO mice protected from weight gain—EKO mice exhibited significantly reduced levels of leptin in circulation. As was the case with weight gain protection, circulating levels of leptin were normal in male EKO mice; fat-rich food increased circulating levels of leptin significantly but similarly in both EKO and WT male mice. Reduced leptin in circulation could arise from a primary defect in adipose tissue that reduces leptin synthesis or secretion. Alternatively, the primary defect could reside in the hypothalamus: feedback from increased hypothalamic leptin sensitivity could reduce synthesis or secretion of leptin by adipose tissue in order to maintain normal overall hypothalamic leptin signaling. Strong data in favor of the latter explanation was provided when female EKO mice were shown to be leptin-hypersensitive when challenged with leptin in acute and in chronic leptin stimulation protocols. Acute injection of leptin to female EKO mice resulted in increased phosphorylation of hypothalamic STAT3, while chronic administration of leptin over several days resulted in a greater drop in food intake in these mice than in matched controls [55]. Coupled with high levels of RPTPe expression in the hypothalamus and very low PTPe expression in adipose tissue, these results indicate that lack of PTPe upregulates leptin receptor signaling in the hypothalamus. Indeed, further studies revealed co-localization of PTPe and leptin receptor mRNAs in neurons in the arcuate nucleus of the hypothalamus [55].

PTPe Dephosphorylates JAK2 and Downregulates Leptin Receptor Signaling

Molecular studies in heterologous systems indicated that RPTPe can reduce leptin-induced phosphorylation of JAK2 (at Y1007/1008) and STAT3 (at Y705). Moreover, RPTPe and the leptin receptor co-immunoprecipitate together with JAK2, and purified RPTPe can dephosphorylate purified JAK2. In all, the data indicate that RPTPe physically associates with the leptin receptor and JAK2, and that JAK2 is a substrate of RPTPe in the context of leptin receptor signaling [55]. It is then reasonable to conclude that lack of RPTPe prevents sufficient dephosphorylation of hypothalamic JAK2 following leptin stimulation, does not inhibit leptin signaling sufficiently, and results in leptin hypersensitivity.

Regulation of RPTPe by Leptin-Induced Phosphorylation

As indicated above, the abilities of both major forms of PTPe, RPTPe and cyt-PTPe, to carry out their physiological function can be affected by phosphorylation of their C-terminal tyrosine residue (Y695 in RPTPe=Y638 in cyt-PTPe, Fig. 10.1). Accordingly, phosphorylation of Y695 is critical for RPTPe to activate Src in Neu-induced mammary tumor cells [24], while similar phosphorylation of cyt-PTPe supports its activation of Src in osteoclasts. Phosphorylation of cyt-PTPe is also required for its EGF-induced association with tubulin, which inhibits the PTP [31, 56]. These findings suggest the existence of a signaling module, in which activation of a cellular receptor induces C-terminal phosphorylation of RPTPe or cyt-PTPe that induces the PTP to dephosphorylate a downstream molecule as part of its physiological function. Receptor-induced phosphorylation may be direct (most likely as is the case of Neu in mammary tumors, [24]) or indirect via a surrogate kinase (such as when activated integrins in osteoclasts partially activate Src to phosphorylate cyt-PTPe [31]). The molecular mechanism by which phosphorylation affects PTPe is unknown, but it most likely does not involve altering the kinetic parameters of its catalytic activity [24]. Rather, it might affect the ability of PTPe to interact with other molecules.

The above results suggested that C-terminal phosphorylation of RPTPe may play a role in regulating its activity in leptin signaling. In agreement with the above model, intra-peritoneal injection of leptin into female WT mice induced transient phosphorylation of RPTPe at Y695 (Fig. 10.3a) [55]. Feeding female EKO mice high-fat food induces phenotypes in these mice (higher basal metabolic rates, protection from diet-induced weight gain), suggesting that the importance of RPTPe function is increased when mice are challenged by fat-rich food. Leptin-induced phosphorylation of RPTPe was higher in mice fed fat-rich food than in mice fed regular lab chow (Fig. 10.3b), suggesting that RPTPe phosphorylation is important for its function. Further studies indicated that JAK2 can phosphorylate RPTPe at Y695 upon leptin stimulation in cells, and that purified JAK2 can phosphorylate purified RPTPe in vitro [55]. Importantly, the non-phosphorylatable mutant Y695F RPTPe is significantly impaired in its ability to dephosphorylate JAK2 following leptin stimulation, confirming the functional importance of this phosphorylation event [55]. In all, the data suggest a model (Fig. 10.4) by which RPTPe is associated with the leptin receptor and JAK2. Upon leptin binding to its receptor, JAK2 phosphorylates several downstream substrates, among them RPTPe at Y695. RPTPe then dephosphorylates and inactivates JAK2, thus participating in a negative-feedback regulatory loop that helps return leptin signaling to its pre-stimulation levels. PTPe auto-dephosphorylates at Y695, thus possibly limiting its own activity in this pathway. In this respect PTPe functions similarly to PTP1B and to the related

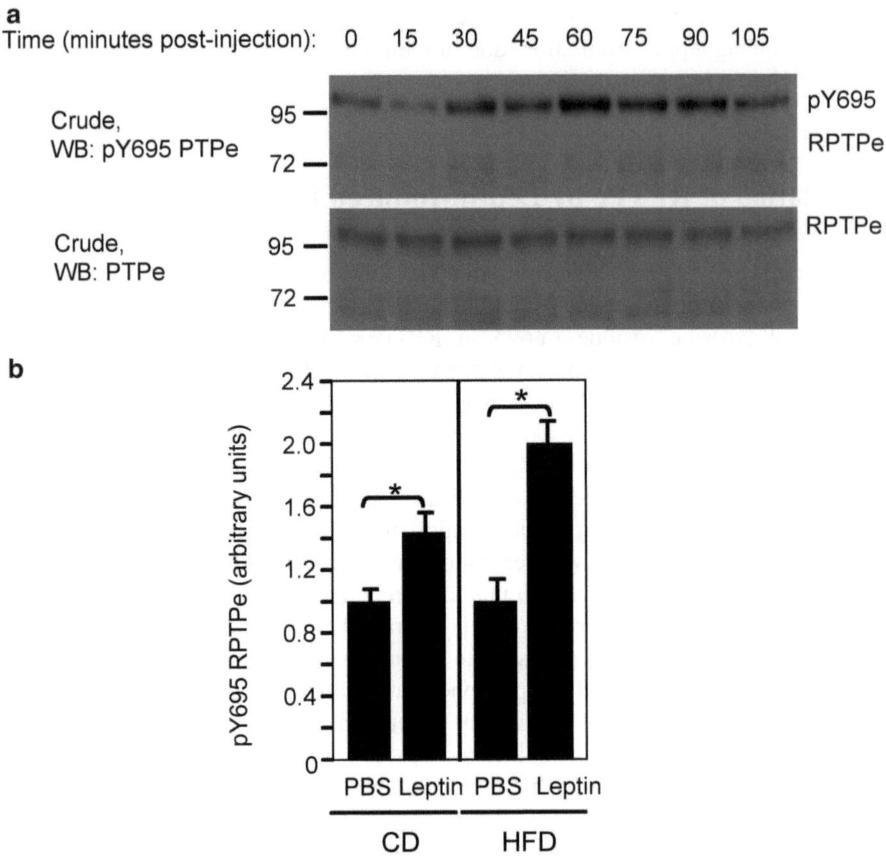

Fig. 10.3 Activation of the leptin receptor induces phosphorylation of hypothalamic RPTPe at Y695. (**a**) Four-month-old female WT mice were injected intraperitoneally with leptin (5 μg/g body weight). Hypothalami were isolated and analyzed with a pY695-PTPe-specific antibody. (**b**) Bar diagram showing increased phosphorylation of RPTPe at Y695 at $t=60$ min following leptin injection in mice fed chow diet (CD) or fatty food (HFD). $N=7$–10 mice per bar; *Asterisks*, $p \leq 0.0065$ by Student's *t*-test. Reprinted from Cell Metabolism, 13(5), Rousso-Noori et al., Protein tyrosine phosphatase epsilon affects body weight by downregulating leptin signaling in a phosphorylation-dependent manner, pp. 562–572, Copyright 2011, with permission from Elsevier

TCPTP, which inhibit hypothalamic leptin receptor signaling; absence of any of these PTPs induces leptin hypersensitivity in mice [47–50]. Although both PTP1B and RPTPe target JAK2, only RPTPe appears to be regulated by C-terminal phosphorylation [48, 49, 55]. This indicates that PTP1B and RPTPe may inhibit leptin signaling in distinct physiological contexts. Phosphorylation-dependent inhibition of RPTPe also implies that RPTPe might play a role in shutting down the leptin receptor post-activation rather than in preventing its inappropriate activation beforehand. Finally, loss of RPTPe affects the body weight of females in particular. Estrogen signaling in the hypothalamus affects body weight, and it is possible that the physiological role of RPTPe is affected by both estrogen and leptin signaling events. Further studies are required to clarify this issue.

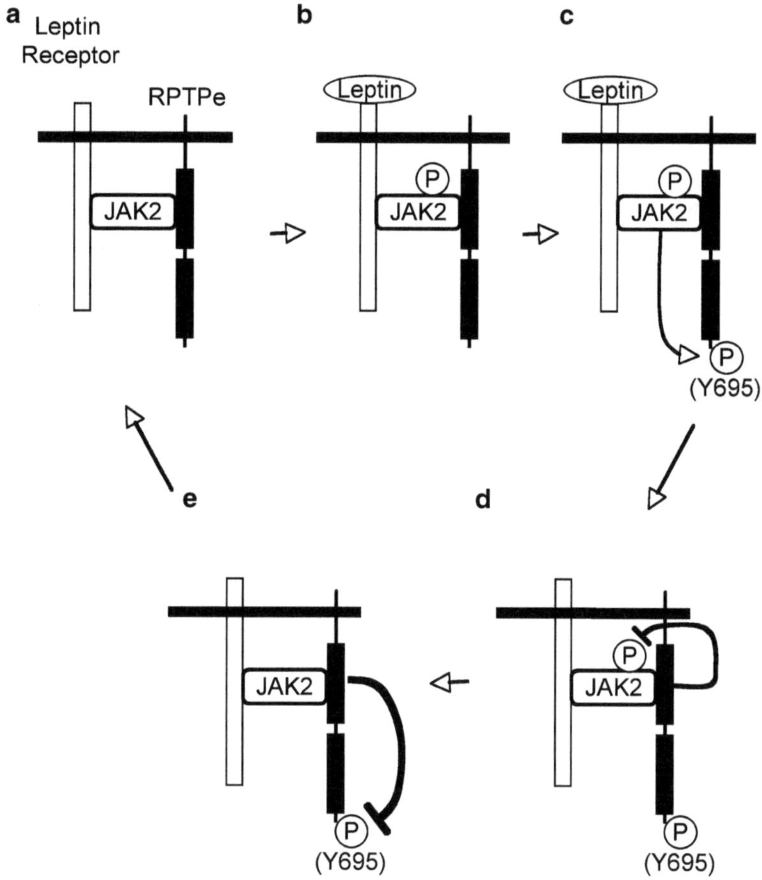

Fig. 10.4 Inhibition of hypothalamic leptin receptor signaling by RPTPe. In this schematic model, prior to initiation of leptin signaling the leptin receptor, JAK2 and RPTPe form part of a molecular complex (**a**). Following binding of leptin to its receptor, JAK2 undergoes autophosphorylation (**b**) and trans-phosphorylates RPTPe at Y695 (**c**). RPTPe then dephosphorylates JAK2 and inactivates it (**d**), after which RPTPe auto-dephosphorylates (**e**) and returns the system to its original state (**a**). Model scheme adapted from Cell Metabolism, 13(5), Rousso-Noori et al., Protein tyrosine phosphatase epsilon affects body weight by downregulating leptin signaling in a phosphorylation-dependent manner, pp. 562–572, Copyright 2011, with permission from Elsevier

PTPe as an Inhibitor of Insulin Receptor Signaling

Some of the physiological functions of the receptors for insulin and for leptin are closely related. Obesity and diabetes are both part of the metabolic syndrome and, if left to follow its clinical course, obesity is often followed by disruptions to glucose homeostasis and diabetes. Moreover, both insulin and leptin play similar roles in the hypothalamus with respect to regulation of body weight [57]. Leptin is believed to function predominantly in regulating body weight of females, while insulin is more dominant in this respect in males [58, 59]. Several studies have suggested that PTPe inhibits insulin receptor signaling. A screen of Baby Hamster Kidney (BHK) cells that overexpressed the insulin receptor identified RPTPe as a phosphatase that can dephosphorylate the receptor and downregulate its activity [60]. In this system the non-receptor form of PTPe, cyt-PTPe, did not affect insulin receptor signaling, although subsequent studies in other systems indicate that cyt-PTPe can also target the receptor (see below). Later studies in primary hepatocytes confirmed that RPTPe targets the insulin receptor and can inhibit downstream signaling events, such as activation of protein kinase B (PKB), ERK, and glycogen synthase kinase 3 (GSK3), as well as insulin-induced glycogen synthesis and insulin-induced suppression of phosphoenolpyruvate carboxykinase expression (PEPCK) [61]. Adenoviral-induced expression of RPTPe in mice also reduced insulin-induced inhibition of PEPCK expression in the liver [61].

More recent studies have shown that PTPe, in particular cyt-PTPe, can inhibit insulin signaling in skeletal muscle cells. The L6 rat skeletal muscle cell line expresses predominantly cyt-PTPe. Expression of GFP-tagged cyt-PTPe in L6 cells revealed that after stimulating the cells with insulin, cyt-PTPe co-localizes with the internalized receptor [62]. Over-expression of cyt-PTPe in these cells reduced phosphorylation of the insulin receptor at tyrosines 972 and 1162/1163; cyt-PTPe also inhibited phosphorylation of IRS1, PKB, and GSK and reduced insulin-induced glucose uptake into the cells. As expected, downregulation of cyt-PTPe expression by RNAi induced the opposite results [62]. Increased phosphorylation of the insulin receptor and IRS1 were also observed when primary skeletal muscle cells from EKO mice were stimulated with insulin [62]. In all, it appears that both RPTPe and cyt-PTPe can inhibit insulin receptor phosphorylation and downregulate downstream signaling events in cultured cells.

These conclusions were borne out in vivo when whole-body glucose homeostasis and insulin signaling were evaluated in EKO mice. Fasting glucose and fasting insulin levels were normal in both male and female EKO mice that had been fed regular chow. However, male EKO mice that had been fed a HFD exhibited reductions of 21 % and 45 % in their circulating levels of glucose and insulin, respectively, after an overnight fast [55]. Improved performance in regulating blood glucose levels in the presence of reduced circulating insulin indicates that EKO mice are insulin-hypersensitive. In agreement, male EKO mice showed better control of their blood glucose levels than WT controls when injected with a bolus of glucose; similar results were obtained in both lean and obese mice [55]. Interestingly, male EKO mice presented a stronger phenotype in this respect than female EKO mice. Injection of insulin to mice fed regular lab chow induced increased hyperphosphorylation of PKB in muscles and in livers of EKO mice compared to control mice, providing additional evidence in favor of PTPe inhibiting insulin receptor signaling in vivo [55]. In all, the results presented here indicate that both major forms of PTPe, RPTPe and cyt-PTPe, inhibit insulin receptor signaling. Although the precise mechanism by which PTPe exerts this effect is not proven, the fact that phosphorylation of the insulin receptor itself is affected by changes in PTPe expression (e.g., [62]) strongly suggests that the receptor is a substrate of PTPe.

Gender Effects

An intriguing aspect of the effects of PTPe on leptin and on insulin signaling is their gender specificity. Loss of PTPe affects hypothalamic leptin signaling and protects from weight gain induced by a variety of physiological stimuli primarily in females, while it is mainly male EKO mice that enjoy improved glucose homeostasis. The basis for these effects is not clear at present. It has been suggested that, while both leptin and insulin signal in the hypothalamus to affect body weight, leptin is dominant in this respect in females while insulin is dominant in males [59]. This may occur due to gender-specific effects on signaling processes (e.g., by sex hormones) or to gender-specific variations in co-expression of PTPe and receptors for leptin or insulin in subsets of neurons. These interpretations are consistent with the fact that loss of hypothalamic PTPe increases leptin sensitivity and results in a female-specific body weight phenotype. The fact that a similar phenotype is not observed in EKO male mice may indicate that insulin receptor signaling is not affected in hypothalamic neurons. Similarly, the finding that EKO male mice display more significant improvement in glucose homeostasis than EKO females suggests that insulin receptor signaling in the periphery is affected by loss of PTPe predominantly in males. Further studies are required to challenge these interpretations and to uncover their molecular bases.

Future Directions

Studies outlined in this chapter indicate that PTPe plays a role in regulating signaling by the leptin and insulin receptors, thus affecting body weight and glucose homeostasis. While the basic role of PTPe in these signaling pathways is clear, further work is required to understand how PTPe functions in each tissue and cell type, and how these distinct effects combine in a whole organism to generate the overall picture observed in whole-body EKO mice described here.

From a broader perspective studies of PTPe and of other PTPs demonstrate that PTPs in general play pivotal roles in regulating leptin and insulin signaling and affect body weight and whole-body glucose homeostasis. The increasing spread of the metabolic syndrome pandemic makes it extremely tempting to design specific inhibitors of select PTPs to help regain control of leptin and insulin signaling in vivo for therapeutic gain. Demonstrated successes in designing specific inhibitors of PTKs for treatment of diseases in the past decade make this approach even more appealing. For this to occur, more complete basic understanding of the roles of various PTPs in regulating the relevant physiological pathways in vivo needs to be obtained. The difficulties encountered in designing PTP inhibitors that are specific and effective in vivo are, unfortunately, not trivial. However, studies such as those summarized here help lay the scientific foundation for this, by identifying specific PTPs as well as target cells and tissues where targeting should occur for maximal gain.

References

1. Hunter T (2000) Signaling—2000 and beyond. Cell 100(1):113–127
2. Lemmon MA, Schlessinger J (2010) Cell signaling by receptor tyrosine kinases. Cell 141(7):1117–1134
3. Alonso A, Sasin J, Bottini N, Friedberg I, Osterman A, Godzik A et al (2004) Protein tyrosine phosphatases in the human genome. Cell 117(6):699–711
4. Barford D, Das AK, Egloff MP (1998) The structure and mechanism of protein phosphatases: insights into catalysis and regulation. Annu Rev Biophys Biomol Struct 27:133–164
5. Tonks NK (2006) Protein tyrosine phosphatases: from genes, to function, to disease. Nat Rev Mol Cell Biol 7(11):833–846
6. Navis AC, van den Eijnden M, Schepens JT, Hooft van Huijsduijnen R, Wesseling P, Hendriks WJ (2010) Protein tyrosine phosphatases in glioma biology. Acta Neuropathol 119(2): 157–175
7. Vang T, Miletic AV, Arimura Y, Tautz L, Rickert RC, Mustelin T (2008) Protein tyrosine phosphatases in autoimmunity. Annu Rev Immunol 26:29–55
8. Hendriks WJ, Elson A, Harroch S, Stoker AW (2008) Protein tyrosine phosphatases: functional inferences from mouse models and human diseases. FEBS J 275(5):816–830
9. Pulido R, Hooft van Huijsduijnen R (2008) Protein tyrosine phosphatases: dual-specificity phosphatases in health and disease. FEBS J 275(5):848–866
10. den Hertog J, Ostman A, Bohmer FD (2008) Protein tyrosine phosphatases: regulatory mechanisms. FEBS J 275(5):831–847
11. Barr AJ (2010) Protein tyrosine phosphatases as drug targets: strategies and challenges of inhibitor development. Future Med Chem 2(10):1563–1576
12. Scott LM, Lawrence HR, Sebti SM, Lawrence NJ, Wu J (2010) Targeting protein tyrosine phosphatases for anticancer drug discovery. Curr Pharm Des 16(16):1843–1862
13. Haque A, Andersen JN, Salmeen A, Barford D, Tonks NK (2011) Conformation-sensing antibodies stabilize the oxidized form of PTP1B and inhibit its phosphatase activity. Cell 147(1): 185–198
14. Elson A, Leder P (1995) Protein-tyrosine phosphatase epsilon. An isoform specifically expressed in mouse mammary tumors initiated by v-Ha-ras OR neu. J Biol Chem 270(44):26116–26122
15. Krueger NX, Streuli M, Saito H (1990) Structural diversity and evolution of human receptor-like protein tyrosine phosphatases. EMBO J 9:3241–3252
16. Lim KL, Lai DS, Kalousek MB, Wang Y, Pallen CJ (1997) Kinetic analysis of two closely related receptor-like protein-tyrosine-phosphatases, PTP alpha and PTP epsilon. Eur J Biochem 245(3):693–700
17. Nakamura K, Mizuno Y, Kikuchi K (1996) Molecular cloning of a novel cytoplasmic protein tyrosine phosphatase PTP epsilon. Biochem Biophys Res Commun 218:726–732
18. Elson A, Leder P (1995) Identification of a cytoplasmic, phorbol ester-inducible isoform of protein tyrosine phosphatase epsilon. Proc Natl Acad Sci U S A 92(26):12235–12239
19. Tanuma N, Nakamura K, Kikuchi K (1999) Distinct promoters control transmembrane and cytosolic protein tyrosine phosphatase epsilon expression during macrophage differentiation. Eur J Biochem 259:46–54
20. Gil-Henn H, Volohonsky G, Elson A (2001) Regulation of protein-tyrosine phosphatases alpha and epsilon by calpain-mediated proteolytic cleavage. J Biol Chem 276(34):31772–31779
21. Gil-Henn H, Volohonsky G, Toledano-Katchalski H, Gandre S, Elson A (2000) Generation of novel cytoplasmic forms of protein tyrosine phosphatase epsilon by proteolytic processing and translational control. Oncogene 19(38):4375–4384
22. Kraut J, Volohonsky G, Toledano-Katchalski H, Elson A (2002) Nuclear localization of non-receptor protein tyrosine phosphatase epsilon is regulated by its unique N-terminal domain. Exp Cell Res 281(2):182–189

23. Elson A (1999) Protein tyrosine phosphatase epsilon increases the risk of mammary hyperplasia and mammary tumors in transgenic mice. Oncogene 18(52):7535–7542
24. Berman-Golan D, Elson A (2007) Neu-mediated phosphorylation of protein tyrosine phosphatase epsilon is critical for activation of Src in mammary tumor cells. Oncogene 26(49): 7028–7037
25. Gil-Henn H, Elson A (2003) Tyrosine phosphatase-epsilon activates Src and supports the transformed phenotype of Neu-induced mammary tumor cells. J Biol Chem 278(18): 15579–15586
26. Peretz A, Gil-Henn H, Sobko A, Shinder V, Attali B, Elson A (2000) Hypomyelination and increased activity of voltage-gated K(+) channels in mice lacking protein tyrosine phosphatase epsilon. EMBO J 19:4036–4045
27. Tiran Z, Peretz A, Attali B, Elson A (2003) Phosphorylation-dependent regulation of Kv2.1 Channel activity at tyrosine 124 by Src and by protein-tyrosine phosphatase epsilon. J Biol Chem 278(19):17509–17514
28. Muja N, Lovas G, Romm E, Machleder D, Ranjan M, Gallo V et al (2004) Expression of a catalytically inactive transmembrane protein tyrosine phosphatase epsilon (tm-PTP epsilon) delays optic nerve myelination. Glia 48(4):278–297
29. Chiusaroli R, Knobler H, Luxenburg C, Sanjay A, Granot-Attas S, Tiran Z et al (2004) Tyrosine phosphatase epsilon is a positive regulator of osteoclast function in vitro and in vivo. Mol Biol Cell 15(1):234–244
30. Kollet O, Dar A, Shivtiel S, Kalinkovich A, Lapid K, Sztainberg Y et al (2006) Osteoclasts degrade endosteal components and promote mobilization of hematopoietic progenitor cells. Nat Med 12(6):657–664
31. Granot-Attas S, Luxenburg C, Finkelshtein E, Elson A (2009) PTP epsilon regulates integrin-mediated podosome stability in osteoclasts by activating Src. Mol Biol Cell 20(20):4324–4334
32. Boyce BF, Yoneda T, Lowe C, Soriano P, Mundy GR (1992) Requirement of pp 60c-src expression for osteoclasts to form ruffled borders and resorb bone in mice. J Clin Invest 90(4):1622–1627
33. Destaing O, Sanjay A, Itzstein C, Horne WC, Toomre D, De Camilli P et al (2008) The tyrosine kinase activity of c-Src regulates actin dynamics and organization of podosomes in osteoclasts. Mol Biol Cell 19(1):394–404
34. Toledano-Katchalski H, Kraut J, Sines T, Granot-Attas S, Shohat G, Gil-Henn H et al (2003) Protein tyrosine phosphatase epsilon inhibits signaling by mitogen-activated protein kinases. Mol Cancer Res 1(7):541–550
35. Wabakken T, Hauge H, Finne EF, Wiedlocha A, Aasheim H (2002) Expression of human protein tyrosine phosphatase epsilon in leucocytes: a potential ERK pathway-regulating phosphatase. Scand J Immunol 56(2):195–203
36. Tanuma N, Nakamura K, Shima H, Kikuchi K (2000) Protein-tyrosine phosphatase PTPepsilon C inhibits Jak-STAT signaling and differentiation induced by interleukin-6 and leukemia inhibitory factor in M1 leukemia cells. J Biol Chem 275:28216–28221
37. Tanuma N, Shima H, Nakamura K, Kikuchi K (2001) Protein tyrosine phosphatase epsilonC selectively inhibits interleukin-6- and interleukin- 10-induced JAK-STAT signaling. Blood 98:3030–3034
38. Tanuma N, Shima H, Shimada S, Kikuchi K (2003) Reduced tumorigenicity of murine leukemia cells expressing protein-tyrosine phosphatase, PTPepsilon C. Oncogene 22(12): 1758–1762
39. Sully V, Pownall S, Vincan E, Bassal S, Borowski AH, Hart PH et al (2001) Functional abnormalities in protein tyrosine phosphatase epsilon-deficient macrophages. Biochem Biophys Res Commun 286:184–188
40. Thompson LJ, Jiang J, Madamanchi N, Runge MS, Patterson C (2001) PTP-epsilon, a tyrosine phosphatase expressed in endothelium, negatively regulates endothelial cell proliferation. Am J Physiol Heart Circ Physiol 281:H396–H403

41. De Franceschi L, Biondani A, Carta F, Turrini F, Laudanna C, Deana R et al (2008) PTPepsilon has a critical role in signaling transduction pathways and phosphoprotein network topology in red cells. Proteomics 8(22):4695–4708
42. Spiegelman BM, Flier JS (2001) Obesity and the regulation of energy balance. Cell 104(4):531–543
43. Wauman J, Tavernier J (2011) Leptin receptor signaling: pathways to leptin resistance. Front Biosci 17:2771–2793
44. Myers MG, Cowley MA, Munzberg H (2008) Mechanisms of leptin action and leptin resistance. Annu Rev Physiol 70:537–556
45. Myers MG Jr, Munzberg H, Leinninger GM, Leshan RL (2009) The geometry of leptin action in the brain: more complicated than a simple ARC. Cell Metab 9(2):117–123
46. Morris DL, Rui L (2009) Recent advances in understanding leptin signaling and leptin resistance. Am J Physiol Endocrinol Metab 297(6):E1247–E1259
47. Bence KK, Delibegovic M, Xue B, Gorgun CZ, Hotamisligil GS, Neel BG et al (2006) Neuronal PTP1B regulates body weight, adiposity and leptin action. Nat Med 12(8):917–924
48. Cheng A, Uetani N, Simoncic PD, Chaubey VP, Lee-Loy A, McGlade CJ et al (2002) Attenuation of leptin action and regulation of obesity by protein tyrosine phosphatase 1B. Dev Cell 2(4):497–503
49. Zabolotny JM, Bence-Hanulec KK, Stricker-Krongrad A, Haj F, Wang Y, Minokoshi Y et al (2002) PTP1B regulates leptin signal transduction in vivo. Dev Cell 2(4):489–495
50. Loh K, Fukushima A, Zhang X, Galic S, Briggs D, Enriori PJ et al (2011) Elevated hypothalamic TCPTP in obesity contributes to cellular leptin resistance. Cell Metab 14(5):684–699
51. Krajewska M, Banares S, Zhang EE, Huang X, Scadeng M, Jhala US et al (2008) Development of diabesity in mice with neuronal deletion of Shp2 tyrosine phosphatase. Am J Pathol 172(5):1312–1324
52. Zhang EE, Chapeau E, Hagihara K, Feng GS (2004) Neuronal Shp2 tyrosine phosphatase controls energy balance and metabolism. Proc Natl Acad Sci U S A 101(45):16064–16069
53. He Z, Zhang SS, Meng Q, Li S, Zhu HH, Raquil MA et al (2012) Shp2 controls female body weight and energy balance by integrating leptin and estrogen signals. Mol Cell Biol 32(10):1867–1878
54. Banno R, Zimmer D, De Jonghe BC, Atienza M, Rak K, Yang W et al (2010) PTP1B and SHP2 in POMC neurons reciprocally regulate energy balance in mice. J Clin Invest 120(3):720–734
55. Rousso-Noori L, Knobler H, Levy-Apter E, Kuperman Y, Neufeld-Cohen A, Keshet Y et al (2011) Protein tyrosine phosphatase epsilon affects body weight by downregulating leptin signaling in a phosphorylation-dependent manner. Cell Metab 13(5):562–572
56. Sines T, Granot-Attas S, Weisman-Welcher S, Elson A (2007) Association of tyrosine phosphatase epsilon with microtubules inhibits phosphatase activity and is regulated by the epidermal growth factor receptor. Mol Cell Biol 27(20):7102–7112
57. Belgardt BF, Bruning JC (2010) CNS leptin and insulin action in the control of energy homeostasis. Ann N Y Acad Sci 1212:97–113
58. Clegg DJ, Riedy CA, Smith KA, Benoit SC, Woods SC (2003) Differential sensitivity to central leptin and insulin in male and female rats. Diabetes 52(3):682–687
59. Shi H, Seeley RJ, Clegg DJ (2009) Sexual differences in the control of energy homeostasis. Front Neuroendocrinol 30(3):396–404
60. Andersen JN, Elson A, Lammers R, Romer J, Clausen JT, Moller KB et al (2001) Comparative study of protein tyrosine phosphatase-epsilon isoforms: membrane localization confers specificity in cellular signalling. Biochem J 354(Pt 3):581–590
61. Nakagawa Y, Aoki N, Aoyama K, Shimizu H, Shimano H, Yamada N et al (2005) Receptor-type protein tyrosine phosphatase epsilon (PTPepsilonM) is a negative regulator of insulin signaling in primary hepatocytes and liver. Zoolog Sci 22(2):169–175
62. Aga-Mizrachi S, Brutman-Barazani T, Jacob AI, Bak A, Elson A, Sampson SR (2008) Cytosolic protein tyrosine phosphatase-epsilon is a negative regulator of insulin signaling in skeletal muscle. Endocrinology 149(2):605–614

Chapter 11
The Role of LMPTP in the Metabolic Syndrome

Stephanie M. Stanford, Massimo Bottini, and Nunzio Bottini

Abstract The low molecular weight class of protein tyrosine phosphatases (PTPs) has been implicated as key modulators of pathways controlling human susceptibility to a host of disorders. As a result, these enzymes are emerging as novel targets for the treatment of a variety of ailments, ranging from cardiovascular disorders to neoplastic syndromes to infectious diseases. For decades evidence has been accumulating that the human low molecular weight PTP (LMPTP) is a key regulator of the metabolic conditions that accompany obesity and can put humans at risk for type 2 diabetes, coronary artery disease, and other, sometimes lethal, complications. While the LMPTP is ubiquitously expressed and involved in numerous signaling pathways controlling cell growth and differentiation, this review will focus on the critical role of LMPTP in regulating insulin resistance and its implications for metabolic homeostasis.

Abbreviations

ACP1	Acid phosphatase locus 1
ADA	Adenosine deaminase
ASO	Antisense oligonucleotide
BMI	Body mass index
BPTP	Bovine heart protein tyrosine phosphatase
CAD	Coronary artery disease
DIO	Diet-induced obese

S.M. Stanford, Ph.D. • N. Bottini, M.D., Ph.D. (✉)
Division of Cellular Biology, La Jolla Institute for Allergy and Immunology,
La Jolla, CA 92037, USA
e-mail: nunzio@liai.org

M. Bottini, Ph.D.
Infectious and Inflammatory Diseases Center, Sanford Burnham Medical Research Institute,
La Jolla, CA 92037, USA

K.K. Bence (ed.), *Protein Tyrosine Phosphatase Control of Metabolism*,
DOI 10.1007/978-1-4614-7855-3_11,

f	Fast isoform
hAAP	Human adipocyte acid phosphatase
HCPTP	Human cytosolic low-molecular-weight protein tyrosine phosphatase
IR	Insulin receptor
LDL	Low-density lipoprotein
LMPTP	Low molecular weight protein tyrosine phosphatase
PTP	Protein tyrosine phosphatase
s	Slow isoform
SNP	Single nucleotide polymorphism
SV3	Splicing variant 3
T1D	Type 1 diabetes
T2D	Type 2 diabetes
ZAP-70	ζ-chain-associated protein tyrosine phosphatase of 70 kDa

Introduction

Human obesity is frequently accompanied by metabolic abnormalities that increase risk for type 2 diabetes (T2D) and coronary artery disease (CAD), called the metabolic syndrome [56]. This so-called metabolic syndrome includes conditions such as insulin resistance, hypertriglyceridemia, and hypercholesterolemia, which frequently occur together and greatly expose individuals to the risk of obesity-related complications and lethality. As the prevalence of obesity is rapidly climbing in developed countries, targets for attenuation of the metabolic syndrome are currently of great therapeutic relevance. Studies by multiple groups strongly suggest that human LMPTP is a key predisposing risk factor and potential target for treatment of the metabolic syndrome [6].

The Class II cysteine-based subfamily of PTPs is represented by the low molecular weight PTPs (LMPTPs, also abbreviated as LMW-PTPs), unique enzymes of roughly 18 kDa expressed in all eukaryotes, from yeasts to humans [1]. These PTPs contain a high degree of conservation of the primary sequence among organisms, and similar proteins with tyrosine phosphatase activity have been isolated from prokaryotic organisms, including gram-negative [13, 34, 75] and gram-positive [57] bacteria and Archaea [42]. Although this subfamily contains the CX_5R signature motif found in the P-loop of all PTPs and efficiently hydrolyzes phospho-Tyr-based substrates with the same catalytic mechanism, these enzymes contain little primary sequence homology to other PTP subfamilies [62].

The Discovery of LMPTP

LMPTP was originally identified as an acid phosphatase from red blood cells [23, 38] and found to be expressed in many other mammalian tissues at high concentrations [37, 77]. Historically, LMPTP has been referred to by a number of names, including

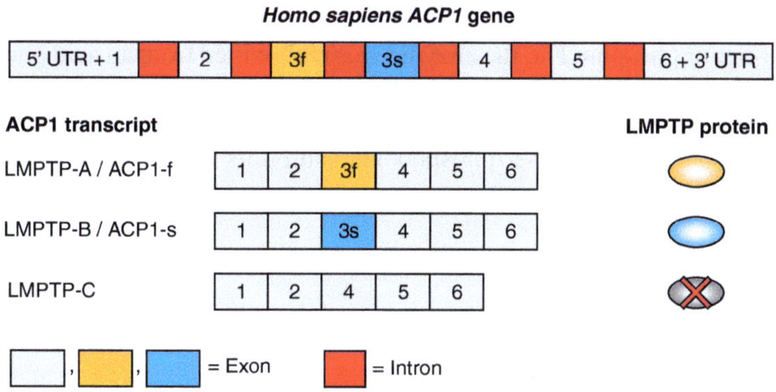

Fig. 11.1 Isoforms of LMPTP. Alternative splicing of the ACP1 pre-mRNA leads to three productive transcripts, encoding for the LMPTP-A (ACP1-f; *yellow*), LMPTP-B (ACP1-s; *blue*), and LMPTP-C (*gray*) isoforms. LMPTP-A and LMPTP-B result from the mutually exclusive retention of the third exon (called exon 3f; present in LMPTP-A), or the fourth exon (called exon 3s; present in LMPTP-B). LMPTP-C results from removal of both exons 3f and 3s and encodes for a catalytically inactive protein

the cytosolic low molecular weight protein (cLMWPTPase [8] or LMW-PTP [62], human adipocyte acid phosphatase (hAAP) [65], human cytosolic low-molecular-weight protein tyrosine phosphatase (HCPTP) [39], and bovine heart protein tyrosine phosphatase (BPTP) [83]. The human *ACP1* gene, located on chromosome 2p25, contains 7 exons and 6 introns [11]. Four isoforms of LMPTP have been reported to arise as a result of an alternative splicing of a primary transcript. Three productive mRNAs encode for LMPTP-A, -B, and -C (Fig. 11.1). The two major isoforms, LMPTP-A and LMPTP-B, result from mutually exclusive retention of exon 3 (present in LMPTP-A) or exon 4 (present in LMPTP-B) [44, 64]. The isoenzymes consist of single polypeptide chains of 157 amino acids, which both undergo removal of the Met1 and acetylation of the N-terminal Ala2. They show an identical sequence except for the region containing amino acids 40–73, called the variable loop, which contains only 41 % identity [24, 25, 77]. This loop flanks the catalytic site and confers specificity for substrates and modulating ligands [19]. The two isoforms also contain translationally silent base-pair substitutions at nucleotides encoding amino acids 74 and 75. LMPTP-A and LMPTP-B have historically been known as the fast (f) and slow isoforms (s) of the acid phosphatase ACP1, respectively, due to their resolution in non-denaturing starch gel electrophoresis [38]. Further characterization revealed that the mass of the f isoform is slightly higher than the s isoform [27]. Analysis of the exon structure of the homologous gene among various species revealed that this mutually exclusive alternative splicing of the transcript is retained among all vertebrates tested, including mammals and fish, while yeasts express only a single isoform, suggesting the origin of this gene predates early vertebrates [63]. LMPTP-C is a catalytically inactive isoform, resulting from excision of both exons 3 and 4 from the transcript, and may act as a dominant negative form of the phosphatase [74]. The fourth mRNA isoform, called SV3 (splicing variant 3), is a nonproductive transcript [54].

Once it was discovered that LMPTP dephosphorylates phospho-Tyr, but not phospho-Ser or phospho-Thr, the enzyme was reclassified from an acid phosphatase to a PTP. Indeed, LMPTP is inhibited by the PTP inhibitors vanadate, Zn^{2+}, and phenylarsine oxide, but is not inhibited by the classical acid phosphatase inhibitor tartrate, or by the phospho-Ser/phospho-Thr inhibitors EDTA, sodium fluoride, or okadaic acid [14, 65].

The Regulation of LMPTP

The crystal structures of both human LMPTP-A [52, 81] and LMPTP-B have been solved [80] and reveal that the LMPTP active site is lined by the variable loop (residues 40–73) that distinguishes the two isoforms and confers their specificity for different ligands (reviewed in [50, 72]). For example, the two isoforms are differentially modulated by purine derivatives. LMPTP-A is inhibited by adenine and activated by hypoxanthine, while LMPTP-B is activated by adenine and unaffected by hypoxanthine [26, 76]. Cyclic GMP was also shown to strongly induce the activity of LMPTP-B from rat liver, with little effect on the LMPTP-A isoform [18, 19].

LMPTP activity is regulated by modifications of specific amino acid residues. Similar to several other PTPs [4], LMPTP can be regulated through a redox mechanism. A disulfide bond is reversibly formed upon oxidation of the thiol groups of Cys17 and the catalytic Cys12 by either H_2O_2 or nitric oxide, a modification that can be reversed by removal of oxidative conditions or by presence of reducing agents such as dithioreitol or reduced glutathione [72]. As is also the case with many PTPs, LMPTP is reversibly regulated by phosphorylation on tyrosine residues. Phosphorylation of LMPTP on Tyr131 and Tyr132 by the Src family kinases has been demonstrated to modulate the activity of the phosphatase and the binding to other interactors. Phosphorylation on Tyr131 strongly enhances the enzymatic activity of LMPTP, whereas phosphorylation on Tyr132 does not affect the activity, but instead provides a docking site for SH2-containing proteins such as Grb2, which regulates substrate access to the LMPTP active site [12, 31, 73].

LMPTP as a Regulator of Insulin Signaling

A variety of substrates and physiological functions have been proposed for LMPTP, and for comprehensive reviews of the function of LMPTP, the reader is referred to the following authoritative publications [6, 50, 67]. Decades of studies have demonstrated the involvement of LMPTP in growth factor signaling through dephosphorylation of growth factor receptors, including the platelet-derived growth factor receptor [16, 17], fibroblast growth factor receptor [61], and ephrin receptor [43].

In T cells, LMPTP promotes T cell receptor-mediated signaling through dephosphorylation of the ζ-chain-associated protein tyrosine phosphatase of 70 kDa (ZAP-70) on the negative regulatory Tyr292 [10].

The insulin receptor (IR) was first suggested as a substrate when the LMPTP-A and LMPTP-B homologs from rat liver (called AcP1 and AcP2) were isolated and tested for in vitro activity against a series of phospho-Tyr containing peptides [70]. In this study both isoforms dephosphorylated a phospho-peptide comprised of the amino acid sequence of 1,146–1,158 of the human IRβ chain phosphorylated at Tyr1151. A role for LMPTP as a negative regulator of insulin signaling was first proposed in a report of results from over-expression of a dominant-negative mutant (Cys12Ser) of LMPTP in NIH-3T3 cells stably over-expressing the insulin receptor (IR) [15]. In this study, cells expressing the mutant LMPTP displayed increased tyrosine phosphorylation of the IRβ upon insulin stimulation, and mutant LMPTP from these cells co-immunoprecipitated with the IR in an insulin-dependent manner. This interaction was also shown in an in vitro binding assay and could be competed out by the competitive PTP inhibitor orthovanadate. Cells over-expressing the mutant LMPTP showed increased insulin-stimulated Src kinase activity and underwent increased insulin-stimulated mitosis and glucose incorporation into glycogen, providing the first suggestion that LMPTP affects both the mitogenic and metabolic effects of insulin.

Confirmation of LMPTP as a key regulator of insulin signaling occurred in 2007, when it was reported that treatment of obese mice with an anti-LMPTP targeted antisense oligonucleotide (ASO) led to improved insulin sensitivity and hyperglycemia [60]. In this report, the authors showed that ASO-induced knock-down of LMPTP in primary mouse hepatocytes increased insulin-induced phosphorylation of the IRβ chain and of the downstream serine-threonine kinase Akt on Ser473. They proceeded to demonstrate that knock-down of LMPTP expression in the liver and fat of both diet-induced obese (DIO) and genetically obese leptin-deficient *ob*/*ob* mice increased insulin signaling, improved insulin resistance, lowered blood insulin and glucose levels, and improved glucose tolerance compared to control ASO-treated mice [60]. ASO treatment in the DIO mice significantly reduced plasma cholesterol levels and liver triglyceride levels. LMPTP knock-down also improved hepatic steatosis and lowered plasma triglycerides, but not cholesterol, in the *ob*/*ob* mice. The ASO treatment, interestingly, did not affect body weight or increase metabolic rate of either mouse model. Western blot analysis showed that the ASO treatment led to increased insulin-induced tyrosine phosphorylation of the IRβ in the liver of DIO and *ob*/*ob* mice and the fat of *ob*/*ob* mice. The LMPTP-ASO-treated mice also exhibited decreased basal pTyr-IRβ levels in the fat of both models, attributed to their decreased blood insulin levels. Increased insulin-induced phosphophorylation of Akt on Ser473 was also elevated in liver and fat of LMPTP-ASO-treated mice compared to the control ASO-treated mice. Not only did this study suggest that LMPTP may be a key regulator of insulin-mediated signaling, but also provided evidence that LMPTP may be a relevant therapeutic target to relieve the metabolic complications that often accompany obesity.

Polymorphisms of the ACP1 Gene

In humans the *ACP1* gene is highly polymorphic, with three common codominant alleles, termed *ACP1**A, *ACP1**B, and *ACP1**C [38]. The *ACP1**A and *ACP1**B alleles are present at high frequencies in all populations. The *ACP1**C allele is common in Caucasians [35]. At the population level, other more rare alleles include the *ACP1**R allele, found in Southern African populations [20], the *ACP1**TIC-1 allele, found in the Ticuna population of the Central Amazons [79], and the *ACP1**GUA-1 allele found in the Guaymi population of Panama and Costa Rica [55]. The *ACP1**A, *B, and *C polymorphisms were originally identified from their patterns in starch gel electrophoresis of red blood cell lysates [38]. Since then, these alleles have been sequenced and well characterized, and found to affect not only total LMPTP enzymatic activity, but also the ratio between the A/f and B/s isoforms. Figure 11.2 describes the three single nucleotide polymorphisms (SNPs) defining these alleles. The A/G SNP in exon 5 encodes an Arg at amino acid 105 (CGA) *A allele, which is a Gln (CAA) in *B and *C. Two C/T substitutions in exons 3f and 3s, in the codons for amino acids 43 and 44, are synonymous SNPs that do not change the encoded amino acid but affect splicing of the mRNA transcript and the resulting ratio between the A/f and B/s isoforms [22]. The ratio of A/f to B/s conferred by a particular allele does not appear to vary from tissue to tissue [3, 71]. The *A allele confers the lowest LMPTP enzymatic activity, while the *C allele leads to the highest activity. A summary of the biochemical differences between these genotypes is shown in Fig. 11.3. The total enzymatic activity of these six genotypes determined by these alleles is *A/*A < *A/*B < (*B/*B, *A/*C) < *B/*C < *C/*C [68].

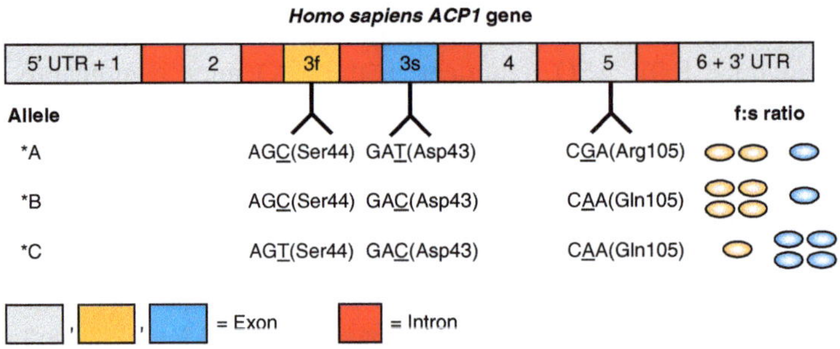

Fig. 11.2 Polymorphisms defining the *ACP1*A, ACP1*B,* and *ACP1*C* alleles. The fixed combinations of three exonic single nucleotide polymorphisms (SNPs) in the *ACP1* gene define the presence of the common codominant *ACP1**A, *ACP1**B, and *ACP1**C alleles. A SNP in exon 5 encodes for Arg at amino acid 105 in the *A allele, which is a Gln in the *B and *C alleles. Two synonymous SNPs in exons 3f and 3s do not change the encoded amino acids, but affect alternative splicing and the resulting f:s isoform ratios (2:1 in *A, 4:1 in *B, and 1:4 in *C)

Fig. 11.3 LMPTP enzymatic activities associated with *ACP1* genotypes. Carriers of the ACP1*B/*C and *C/*C genotypes exhibit the highest total LMPTP enzymatic activities, while carriers of the *A/*A and *A/*B genotypes exhibit the lowest. The rare *R, *GUA-1, and *TIC-1 alleles also lead to lower LMPTP enzymatic activity

High activity	f:s ratio
*B/*C	0.9
*C/*C	0.3
Medium activity	
*B/*B	4.2
*A/*C	0.6
Low activity	
*A/*A	2.4
*A/*B	3.6
*R	ND
*GUA-1	ND
*TIC-1	ND

ACP1 and Human Metabolic Syndrome

A compilation of genetic association studies strongly indicate that polymorphisms in the *ACP1* gene confer protection or predisposition to the dyslipidemia and insulin resistance often accompanying obesity (Table 11.1). As described below, *ACP1* genotypes encoding low LMPTP enzymatic activity (in carriers of the *A allele) appear to exert a protective influence against aspects of the metabolic syndrome.

Body Mass and Dyslipidemia

For decades, the *ACP1* gene has been demonstrated to associate with increases in body mass index (BMI). This was first shown in newborns, in which infants from both diabetic and normal pregnancies carrying the high enzymatic activity *C allele showed reduced risk of fetal macrosomia [33]. A subsequent series of studies in Italian subjects showed *ACP1* associates with the clinical variability of obesity, but not with the incidence of obesity itself. These reports showed significant positive association between the presence of the low enzymatic activity *A allele with severity of body mass deviations in obese, nondiabetic children and adults [5, 49, 59]. In these studies, the presence of the *A/*A and *A/*B genotypes was increased in severely obese subjects, however, did not differ between obese and non-obese subjects. The authors of these studies suggested that genetically determined variability in the levels of *ACP1* activity and/or A/f to B/s ratio influences the severity but not susceptibility to obesity.

Table 11.1 Genetic associations between *ACP1* and human metabolic syndrome

References	Population	Associated alleles	Major findings
[33]	Italian infants from diabetic and normal mothers	*C	Reduced risk of fetal macrosomia
[40]	Italian children ages 3–14 years	*A	Increased severity of obesity among obese children
[5]	Italian	*A	Increased severity of obesity among obese subjects
[59]	Italian	*A	Increased severity of obesity among obese subjects
[47]	Italian subjects with T2D	*A/*A	Positive association with BMI in subjects with T2D with normal lipid levels
[9]	US Caucasian postmenopausal females	*A	Decreased risk of total cholesterol, LDL cholesterol, and triglyceride levels in obese female subjects
[2]	Portuguese females	*A, *B and *C	Positive correlation between *ACP1* enzymatic activity genotype and LDL cholesterol; negative correlation between *ACP1* enzymatic activity genotype and GR activity
[21]	Italian females	*A/*A and *A/*B	*A/*A increased and *A/*B decreased in HFNB subjects; *A/*B increased in HFHB subjects; f:s isoform ratio decreased in HFNB subjects
[32]	Italian diabetic pregnant females	*A	Decreased severity of glycemia in diabetic pregnant subjects; positive correlation between *ACP1* activity genotype and glycemic levels
[46]	Italian subjects with T2D	*A	Decreased severity of glycemia in non-obese T2D subjects; positive association between concentration of A/f isoform and glycemic levels
[40]	Italian	*B/*B, *A/*C and *B/*C	Increased glycemic levels in healthy male subjects; positive correlation between *ACP1* enzymatic activity and glycemic levels
[66]	Mexican-American	T allele of rs3828329 SNP	Increased fasting insulin levels and decreased insulin sensitivity in male subjects
[41]	US children with acute lymphoblastic leukemia	A allele of rs12714403; T allele of rs10167992 SNPs	Increased risk of osteonecrosis; rs12714403 also associated with lower albumin and higher cholesterol levels
[7]	Italian subjects with T2D	High *ACP1*/low *ADA* and low *ACP1*/high *ADA* activity genotypes	High *ACP1*/low *ADA* activity genotypes positively associated with high glycemic levels; low *ACP1*/high *ADA* activity genotypes positively associated with dyslipidemia

Another large ($n=265$) study of Italian subjects reported that in T2D, *ACP1* associates with BMI in non-dyslipidemic subjects. In this report, the *A/*A genotype was positively associated with BMI, but only in subjects with normal blood lipid levels, proposing that at least in T2D, the *A allele may confer resistance to the dyslipidemia which typically occurs in overweight subjects [47].

Further investigation confirmed *ACP1* to be a predisposing factor for obesity-related dyslipidemia. In this report it was found that low-activity genotypes of *ACP1* protect against development of dyslipidemia in obese women [9]. This study of 277 postmenopausal Caucasian women in the United States revealed a significant decrease in total cholesterol, low-density lipoprotein (LDL) cholesterol, cholesterol/HDL ratio, and triglyceride levels in obese (BMI≥30) carriers of the low-activity *A allele vs. obese noncarriers. While no association between *ACP1* polymorphisms and BMI was shown in this cohort, the association of the *A allele with serum triglyceride levels increased with increasing BMI, and no significant association between *ACP1* and blood lipids was demonstrated in the non-obese. There was also a nonsignificant trend towards decreased waist-to-hip ratio in the *A carriers, a parameter associated with poorer medical prognosis in obesity. This study suggests that carriers of the *ACP1**A allele who become overweight may be partially protected against developing features of the metabolic syndrome—hypertriglyceridemia, hypercholesterolemia, and increased waist-to-hip ratio—that typically occur in obesity.

Another study of 318 Portuguese women showed significant association between *ACP1* genotype and blood LDL cholesterol levels. In this report, carriers of the lowest *ACP1* activity genotypes exhibited the lowest LDL cholesterol levels [2]. A similar trend was also shown for total cholesterol levels, although it did not reach statistical significance.

Another study showed that *ACP1* may regulate the deposition of adipose tissue. One hundred and thirty Caucasian women from Rome were studied for an effect of *ACP1* in women with high fat content but normal BMI (called HFNB) compared to women with high fat content and high BMI (called HFHB) [21]. In this cohort, low-activity genotypes (*A/*A and *A/*B) were significantly more prevalent in the whole sample of high fat women (HFNB and HFHB) compared to controls. Further analysis of the two groups of women showed that in HFNB subjects, there was a strong increase in *A/*A, but decrease in *A/*B genotypes, compared to controls. Conversely, in HFHB subjects, the *A/*B, but not the *A/*A, genotypes were strongly increased compared to controls. As the *A/*A and *A/*B genotypes differ in the A/f to B/s isoform ratio (2.4 for *A/*A and 3.6 for *B/*A), the authors tested the isoform ratio in these classes of subjects and indeed found it was significantly lower in HFNB subjects than in the HFHB or controls.

Glycemic Levels

The low-activity genotypes of *ACP1* (*A/*A and *A/*B) have also demonstrated a protective effect against severe hyperglycemia in diabetic disorders, suggesting that

ACP1 may influence susceptibility to insulin resistance. A study ($n = 214$) of pregnant women with type 1 diabetes (T1D), T2D, or gestational diabetes showed associations between both *ACP1* genotype and LMPTP activity levels and glucose levels [32]. The authors demonstrated a significant positive correlation between LMPTP enzymatic activity and glycemic level in all three classes of diabetes in the last trimester of pregnancy. There was also a significantly higher mean LMPTP activity in diabetic women with either high (6.67–8.9 mM) or very high (≥8.9 mM) glycemic levels compared to nondiabetic women. This study also showed a negative correlation between low-activity genotypes (*A/*A and *A/*B) and high or very high glycemic levels compared to low glycemic levels in all classes of diabetic subjects, and compared to nondiabetic subjects.

Another report showed that the *A allele occurs at a higher frequency in non-obese T2D diabetic subjects in a lower blood glucose range (<8.9 mM) than in the high blood glucose range (≥8.9 mM) [46]. No association occurred between *ACP1* and blood glucose level among the obese subjects. This study also found a positive association between glycemic level and concentration of the A/f, but not B/s, isoform. Interestingly, a more recent study on 137 employees of the University of Rome assessed the effect of *ACP1* polymorphisms on serum glucose concentrations in healthy adults [40]. The findings revealed a significant increase in glycemia in males with medium-high activity *ACP1* genotypes (*B/*B, *A/*C and *B/*C) compared to subjects with the low-activity genotypes (*A/*A and *A/*B), differences which were more marked with increasing age. An opposite, yet nonsignificant trend was shown in females. The authors also showed that in males, serum glucose levels positively correlated with total *ACP1* enzymatic activity, which was predominantly due to the A/f isoform. No correlation was found with the B/s isoform.

Association Studies with Additional SNPs

A comprehensive examination of the association of *ACP1* variants with T2D-related traits and adiposity was reported in 2009, in which a large study was conducted of 1,035 subjects from 339 Mexican-American families with probands with a previous diagnosis of gestational diabetes (GM) [66]. In this study, 15 SNPs in the coding and 5′ and 3′ untranslated regions of *ACP1* were assessed. The authors tested associations with a series of T2D-related traits, including insulin levels, glucose levels, insulin sensitivity, glucose sensitivity, BMI, % body fat, and blood lipid levels. The *ACP1* rs3828329 SNP was found to be significantly associated with fasting insulin levels and insulin sensitivity in males (these associations survived a Bonferroni correction for multiple testing). The T allele of this SNP increased fasting insulin levels by 24.7 % per allele and decreased insulin sensitivity in male subjects by 13 % per allele, associations that remained significant after adjustment for body fat. The SNP also displayed marginal association with 2-h insulin and percentage of body fat in males. Notably, no significant associations, for any of the variations, were found in females. There was an association in females of SNP rs11553742, which correlates with levels of the f and s isoforms, and triglycerides; however this did not remain significant after correction for multiple testing. This study suggests that *ACP1* variation affects insulin signaling in a sex-specific manner. The authors reasoned that the increase in fasting insulin could reflect compensation for an increase in insulin resistance, which was seen in the decreased insulin sensitivity conferred by T allele. The rs3828329 SNP is located in the 3′ untranslated region of the gene, and the functional effect of this variant is currently unclear. As the SNP is in strong linkage disequilibrium with several other *ACP1* SNPs in the region, further studies will need to be conducted in order to confirm an effect on insulin signaling. Another SNP in the analysis that had previously been reported to correlate with levels of the A/f and B/s isoforms [28], rs7576247, did not show any association with the T2D-related traits in this study.

A recent study of determinants associated with osteonecrosis, a glucocorticoid-induced complication of acute lymphoblastic leukemia, revealed that polymorphisms in the *ACP1* gene (rs12714403 and rs10167992) are associated with osteonecrosis. rs12714403 is also associated with lower albumin and higher cholesterol levels. In the same study, lower albumin and higher cholesterol were also shown to be significantly associated with symptomatic osteonecrosis [41]. Another recent genome-wide association study to identify *cis*-variants associated with plasma protein abundance levels identified *ACP1* SNP rs17713879 [45].

Interactions with ADA

Another interesting line of investigation into the genetics of *ACP1* has been the suggested interaction with the *ADA* gene. *ADA* encodes for adenosine deaminase (ADA), an enzyme that catalyzes the conversion of adenosine to inosine. *ADA* is also a polymorphic gene, with two codominant alleles, *ADA**1 and *ADA**2, which affect ADA enzymatic activity. *ADA**2 is associated with decreased activity [68]. The interaction between *ACP1* and *ADA* has been proposed to occur directly through an effect of ADA on LMPTP activity, and indirectly through a cooperative biological effect [7]. In an early study, *ADA* was shown to affect LMPTP activity [48]. As mentioned above, LMPTP activity is modulated by purine derivatives [26, 76]. In vitro, LMPTP activity in hemolysates was shown to be activated by adenosine, and to a much greater extent, by inosine [48]. This activation was dependent upon *ACP1* genotype, with the lowest increase in carriers of the *ACP1**B/*B and *A/*B genotypes and the highest increase in carriers of the *B/*C and *A/*C genotypes. The same study showed that *ADA* genotype affects LMPTP activity. The presence of the low-activity *ADA**2 allele lowers *ACP1* activity, as carriers of *ADA**2 generally showed lower LMPTP activity than noncarriers. The influence of *ADA* genotype on LMPTP activity is also dependent upon *ACP1* genotype, as the presence of the *ADA**2 allele contributed to the greatest reduction of LMPTP activity in *ACP1**A/*A carriers, less in *A/*B carriers, and no reduction in other *ACP1* genotypes [48]. The presence of *ADA**2 results in decreased conversion of adenosine to inosine, and thus it is likely that carriers of this allele contain lower intracellular concentrations of the LMPTP-activating inosine.

ADA may also interact with *ACP1* through a synergistic effect at the physiological level [7]. As adenosine has anti-insulin action in the liver through activation of adenosine A2B receptors [78], ADA exerts an indirect influence on glucose tolerance. A study of 280 adult subjects with T2D from Penne, Italy, was undertaken to investigate the effect of genetic interactions between *ACP1* and *ADA* on clinical manifestations of T2D [7]. This study showed that the high *ACP1*/low *ADA* activity genotypes (*ACP1**B/*B, *A/*C, *B/*C, *C/*C; ADA*1/*2 and *2/*2) were positively associated with high glycemic levels and high BMI. Low *ACP1*/high *ADA* activity genotypes (*ACP1**A/*A and *A/*B; *ADA**1/*1) were positively associated with dyslipidemia. These studies suggest that interactions with the *ADA* gene may influence the protective/predisposing effects of *ACP1* on aspects of the metabolic syndrome, either directly through a biochemical effect on LMPTP activity, or possibly indirectly through a cooperative effect on insulin signaling and subsequent glucose tolerance.

The Mechanism of Action of LMPTP in the Metabolic Syndrome

While much work is yet to be done to clearly elucidate the mechanism of action of LMPTP in the control of insulin sensitivity and lipidemia in the obese, multiple lines of evidence suggest that LMPTP is an in vivo negative regulator of signal transduction through the IR. According to this model, carriers of the low-activity *ACP1* genotypes would experience enhanced insulin sensitivity as a result of decreased tyrosine dephosphorylation of the IR by LMPTP. Lower dephosphorylation of the IR would increase insulin-stimulated signaling through the IR and activation of downstream pathways. Conversely, the presence of high activity *ACP1* genotypes would more easily favor insulin resistance through higher dephosphorylation of the IR by LMPTP, depressing insulin-induced activation of signaling pathways critical for control of metabolism. In support of this model are the strong in vivo data suggesting that inhibition of LMPTP in mouse models of obesity lowers insulin resistance and increases signaling through the IR, an action that is likely due to a direct effect on insulin-stimulated phosphorylation of the IR [60].

Additional Tyr phosphorylated substrates for LMPTP have also been proposed to affect its homeostatic maintenance of fat deposition and energy metabolism, such as the adipocyte lipid binding protein [65] and band III protein, a regulator of erythrocyte glycolysis [36, 69]. Alterations in the LMPTP-A to -B ratio may also influence predisposition to insulin resistance, as the two isoforms may display different affinities for the IR and/or other substrates.

Aside from activity as a tyrosine phosphatase, in vitro, both LMPTP-A and LMPTP-B can efficiently dephosphorylate flavin mononucleotide (FMN) to riboflavin [30]. Indeed it was shown that the majority of the intracellular phosphatase activity on FMN in extracts of Chinese hamster ovary cells is attributable to LMPTP [30]. It has been suggested that in vivo dephosphorylation of FMN by LMPTP would affect cellular respiration and antioxidant defense by regulating the activity of flavoenzymes, such as glutathione reductase (GR), by affecting the intracellular levels of the flavin cofactors FMN and flavin adenine dinucleotide [6, 35]. Several studies in human subjects support this notion. In two genetic association studies, carriers of low-activity *ACP1* alleles were shown to have increased GR activity [2, 55]. Additionally, cells subjected to over-expression of *ACP1* display reduced GR activity [51]. Flavoenzymes play essential roles in redox reactions in energy metabolism, and thus *ACP1* may exert its effects by influencing the activity of these enzymes. As this hypothesis was not assessed in the only study of in vivo inhibition of LMPTP conducted thus far [60], further examination of a role for LMPTP as a FMN phosphatase in vivo will be useful in elucidating the mechanism(s) of action of LMPTP in regulating metabolism.

Targeting LMPTP for Therapy of Metabolic Syndrome

In vivo and epidemiological evidence is accumulating in support of LMPTP as a target for therapeutic treatment of the complications accompanying obesity. Pharmacological agents capable of inhibiting the activity and/or expression of LMPTP could potentially prevent the lethal risk factors consequent to obesity. As the complications arising from obesity, namely insulin resistance and hyperlipidemia, rather than the weight gain itself are typically the risk factors that expose subjects to T2D and CAD, drugs capable of alleviating the metabolic syndrome in obese patients would greatly alleviate a current major health burden. Indeed, the compelling in vivo study by Pandey et al. provides the first validation that inhibition of LMPTP could be a beneficial approach for lowering obesity-associated insulin resistance. Furthermore, as mentioned above, multiple genetics studies in humans demonstrate that the lower enzymatic activity *ACP1* alleles favor protection against some of the metabolic complications often associated with extreme overweight.

Recent years have seen a rise in efforts aimed at developing small-molecule chemical inhibitors of human LMPTP, and for an excellent review regarding the progress in this field, the reader is referred to [50]. Although genetic association studies suggest that LMPTP-A may be the isoform most involved in metabolic pathologies, further in vivo validation is needed to confirm whether inhibition of both LMPTP-A and LMPTP-B, or of a single isoform, will provide the optimal pharmacological benefit. Among the most potent LMPTP inhibitors currently reported are a series of chromones [29] developed after previous findings that flavonoids display modulatory activity against LMPTP [53]. These compounds display low micromolar efficacy towards LMPTP-A, LMPTP-B, and PTP1B, another PTP well-validated as a target for therapy of T2D and obesity [82]. One such compound displayed submicromolar inhibition of LMPTP-A and PTP1B, with moderate selectivity over LMPTP-B. Several compounds within this series were efficacious at increasing phosphorylation of the IR in the HepG2 liver carcinoma cell line [29]. Structure-based optimization of benzoic acid derivatives has also yielded appreciable progress, providing low and submicromolar inhibitors of both LMPTP isoforms along with PTP1B [50, 58]. As the potential of LMPTP as a therapeutic target is approaching on the horizon, continued progress to provide chemical inhibitors of this PTP with desirable potency and selectivity will be essential to put the potential of LMPTP to the test.

References

1. Alonso A, Sasin J, Bottini N, Friedberg I, Osterman A, Godzik A et al (2004) Protein tyrosine phosphatases in the human genome. Cell 117:699–711
2. Apelt N, da Silva AP, Ferreira J, Alho I, Monteiro C, Marinho C et al (2009) ACP1 genotype, glutathione reductase activity, and riboflavin uptake affect cardiovascular risk in the obese. Metabolism 58:1415–1423
3. Blake NM, Kirk RL, Barnes KR, Thompson JM (1973) Expression of human red cell acid phosphatase activity in placenta and other tissues. Jinrui Idengaku Zasshi 18:10–23
4. Boivin B, Yang M, Tonks NK (2010) Targeting the reversibly oxidized protein tyrosine phosphatase superfamily. Sci Signal 3:pl2
5. Bottini E, Lucarini N, Gerlini G, Finocchi G, Scire G, Gloria-Bottini F (1990) Enzyme polymorphism and clinical variability of diseases: study of acid phosphatase locus 1 (ACP1) in obese subjects. Hum Biol 62:403–411
6. Bottini N, Bottini E, Gloria-Bottini F, Mustelin T (2002) Low-molecular-weight protein tyrosine phosphatase and human disease: in search of biochemical mechanisms. Arch Immunol Ther Exp (Warsz) 50:95–104
7. Bottini N, Gloria-Bottini F, Borgiani P, Antonacci E, Lucarelli P, Bottini E (2004) Type 2 diabetes and the genetics of signal transduction: a study of interaction between adenosine deaminase and acid phosphatase locus 1 polymorphisms. Metabolism 53:995–1001
8. Bottini N, Gloria-Bottini F, Lucarini N, Ronchetti PG, Fontana L (2000) Inflammatory bowel disease: are there gender differences in the genetics of signal transduction? A preliminary study of cytosolic low molecular weight protein tyrosine phosphatase. Dis Markers 16:163–166
9. Bottini N, MacMurray J, Peters W, Rostamkhani M, Comings DE (2002) Association of the acid phosphatase (ACP1) gene with triglyceride levels in obese women. Mol Genet Metab 77:226–229
10. Bottini N, Stefanini L, Williams S, Alonso A, Jascur T, Abraham RT et al (2002) Activation of ZAP-70 through specific dephosphorylation at the inhibitory Tyr-292 by the low molecular weight phosphotyrosine phosphatase (LMPTP). J Biol Chem 277:24220–24224
11. Bryson GL, Massa H, Trask BJ, Van Etten RL (1995) Gene structure, sequence, and chromosomal localization of the human red cell-type low-molecular-weight acid phosphotyrosyl phosphatase gene, ACP1. Genomics 30:133–140
12. Bucciantini M, Chiarugi P, Cirri P, Taddei L, Stefani M, Raugei G et al (1999) The low Mr phosphotyrosine protein phosphatase behaves differently when phosphorylated at Tyr131 or Tyr132 by Src kinase. FEBS Lett 456:73–78
13. Bugert P, Geider K (1997) Characterization of the amsI gene product as a low molecular weight acid phosphatase controlling exopolysaccharide synthesis of erwinia amylovora. FEBS Lett 400:252–256
14. Chernoff J, Li HC (1985) A major phosphotyrosyl-protein phosphatase from bovine heart is associated with a low-molecular-weight acid phosphatase. Arch Biochem Biophys 240:135–145

15. Chiarugi P, Cirri P, Marra F, Raugei G, Camici G, Manao G et al (1997) LMW-PTP is a negative regulator of insulin-mediated mitotic and metabolic signalling. Biochem Biophys Res Commun 238:676–682
16. Chiarugi P, Cirri P, Raugei G, Camici G, Dolfi F, Berti A et al (1995) PDGF receptor as a specific in vivo target for low M(r) phosphotyrosine protein phosphatase. FEBS Lett 372:49–53
17. Chiarugi P, Cirri P, Taddei ML, Giannoni E, Fiaschi T, Buricchi F et al (2002) Insight into the role of low molecular weight phosphotyrosine phosphatase (LMW-PTP) on platelet-derived growth factor receptor (PDGF-r) signaling. LMW-PTP controls PDGF-r kinase activity through TYR-857 dephosphorylation. J Biol Chem 277:37331–37338
18. Cirri P, Caselli A, Manao G, Camici G, Polidori R, Cappugi G et al (1995) Kinetic studies on rat liver low M(r) phosphotyrosine protein phosphatases. The activation mechanism of the isoenzyme AcP2 by cGMP. Biochim Biophys Acta 1243:129–135
19. Cirri P, Fiaschi T, Chiarugi P, Camici G, Manao G, Raugei G et al (1996) The molecular basis of the differing kinetic behavior of the two low molecular mass phosphotyrosine protein phosphatase isoforms. J Biol Chem 271:2604–2607
20. Corfield V, Jenkins T (1971) Red cell acid phosphatase: distribution in Southern Africa and first report of studies on families containing P r homozygotes. S Afr J Med Sci 36:51–56
21. De Lorenzo A, Di Renzo L, Puja A, Saccucci P, Gloria-Bottini F, Bottini E (2009) A study of acid phosphatase locus 1 in women with high fat content and normal body mass index. Metabolism 58:351–354
22. Dissing J (1987) Immunochemical characterization of human red cell acid phosphatase isozymes. Biochem Genet 25:901–918
23. Dissing J, Dahl O, Svensmark O (1979) Phosphonic and arsonic acids as inhibitors of human red cell acid phosphatase and their use in affinity chromatography. Biochim Biophys Acta 569:159–176
24. Dissing J, Johnsen AH (1992) Human red cell acid phosphatase (ACP1): the primary structure of the two pairs of isozymes encoded by the ACP1*A and ACP1*C alleles. Biochim Biophys Acta 1121:261–268
25. Dissing J, Johnsen AH, Sensabaugh GF (1991) Human red cell acid phosphatase (ACP1). The amino acid sequence of the two isozymes Bf and Bs encoded by the ACP1*B allele. J Biol Chem 266:20619–20625
26. Dissing J, Rangaard B, Christensen U (1993) Activity modulation of the fast and slow isozymes of human cytosolic low-molecular-weight acid phosphatase (ACP1) by purines. Biochim Biophys Acta 1162:275–282
27. Dissing J, Svensmark O (1990) Human red cell acid phosphatase: purification and properties of the A, B and C isozymes. Biochim Biophys Acta 1041:232–242
28. Faggioni G, Borgiani P, Bottini N, Gloria-Bottini F, Tontoli F, Contreas V et al (2002) Identification of two SNPs in the 5′ flanking region of the ACP1 gene and evaluation of disequilibrium among polymorphic sites. Ann Hum Genet 66:245–254
29. Forghieri M, Laggner C, Paoli P, Langer T, Manao G, Camici G et al (2009) Synthesis, activity and molecular modeling of a new series of chromones as low molecular weight protein tyrosine phosphatase inhibitors. Bioorg Med Chem 17:2658–2672
30. Fuchs KR, Shekels LL, Bernlohr DA (1992) Analysis of the ACP1 gene product: classification as an FMN phosphatase. Biochem Biophys Res Commun 189:1598–1605
31. Giannoni E, Raugei G, Chiarugi P, Ramponi G (2006) A novel redox-based switch: LMW-PTP oxidation enhances Grb2 binding and leads to ERK activation. Biochem Biophys Res Commun 348:367–373
32. Gloria-Bottini F, Gerlini G, Lucarini N, Borgiani P, Amante A, La Torre M et al (1996) Phosphotyrosine protein phosphatases and diabetic pregnancy: an association between low molecular weight acid phosphatase and degree of glycemic control. Experientia 52:340–343
33. Gloria-Bottini F, Gerlini G, Lucarini N, Borgiani P, Gori MC, Amante A et al (1988) Foetal macrosomia and erythrocyte acid phosphatase (ACP1) polymorphism in diabetic and normal pregnancy. Early Hum Dev 17:265–274

34. Grangeasse C, Doublet P, Vincent C, Vaganay E, Riberty M, Duclos B et al (1998) Functional characterization of the low-molecular-mass phosphotyrosine-protein phosphatase of Acinetobacter johnsonii. J Mol Biol 278:339–347
35. Greene LS, Bottini N, Borgiani P, Gloria-Bottini F (2000) Acid phosphatase locus 1 (ACP1): possible relationship of allelic variation to body size and human population adaptation to thermal stress—a theoretical perspective. Am J Hum Biol 12:688–701
36. Harrison ML, Rathinavelu P, Arese P, Geahlen RL, Low PS (1991) Role of band 3 tyrosine phosphorylation in the regulation of erythrocyte glycolysis. J Biol Chem 266: 4106–4111
37. Heinrikson RL (1969) Purification and characterization of a low molecular weight acid phosphatase from bovine liver. J Biol Chem 244:299–307
38. Hopkinson DA, Spencer N, Harris H (1963) Red cell acid phosphatase variants: a new human polymorphism. Nature 199:969–971
39. Huang L, Sankar S, Lin C, Kontos CD, Schroff AD, Cha EH et al (1999) HCPTPA, a protein tyrosine phosphatase that regulates vascular endothelial growth factor receptor-mediated signal transduction and biological activity. J Biol Chem 274:38183–38188
40. Iannaccone U, Bergamaschi A, Magrini A, Marino G, Bottini N, Lucarelli P et al (2005) Serum glucose concentration and ACP1 genotype in healthy adult subjects. Metabolism 54:891–894
41. Kawedia JD, Kaste SC, Pei D, Panetta JC, Cai X, Cheng C et al (2011) Pharmacokinetic, pharmacodynamic, and pharmacogenetic determinants of osteonecrosis in children with acute lymphoblastic leukemia. Blood 117:2340–2347; quiz 556
42. Kennelly PJ (2003) Archaeal protein kinases and protein phosphatases: insights from genomics and biochemistry. Biochem J 370:373–389
43. Kikawa KD, Vidale DR, Van Etten RL, Kinch MS (2002) Regulation of the EphA2 kinase by the low molecular weight tyrosine phosphatase induces transformation. J Biol Chem 277:39274–39279
44. Lazaruk KD, Dissing J, Sensabaugh GF (1993) Exon structure at the human ACP1 locus supports alternative splicing model for f and s isozyme generation. Biochem Biophys Res Commun 196:440–446
45. Lourdusamy A, Newhouse S, Lunnon K, Proitsi P, Powell J, Hodges A et al (2012) Identification of cis-regulatory variation influencing protein abundance levels in human plasma. Hum Mol Genet 21:3719–3726
46. Lucarini N, Antonacci E, Bottini N, Borgiani P, Faggioni G, Gloria-Bottini F (1998) Phosphotyrosine-protein-phosphatase and diabetic disorders. Further studies on the relationship between low molecular weight acid phosphatase genotype and degree of glycemic control. Dis Markers 14:121–125
47. Lucarini N, Antonacci E, Bottini N, Gloria Bottini F (1997) Low-molecular-weight acid phosphatase (ACP1), obesity, and blood lipid levels in subjects with non-insulin-dependent diabetes mellitus. Hum Biol 69:509–515
48. Lucarini N, Borgiani P, Ballarini P, Bottini E (1989) Erythrocyte acid phosphatase (ACP1) activity. In vitro modulation by adenosine and inosine and effects of adenosine deaminase (ADA) polymorphism. Hum Genet 81:185–187
49. Lucarini N, Finocchi G, Gloria-Bottini F, Macioce M, Borgiani P, Amante A et al (1990) A possible genetic component of obesity in childhood. Observations on acid phosphatase polymorphism. Experientia 46:90–91
50. Maccari R, Ottana R (2012) Low molecular weight phosphotyrosine protein phosphatases as emerging targets for the design of novel therapeutic agents. J Med Chem 55:2–22
51. Marraccini P, Iantomasi T, Rigacci S, Pacini S, Ruggiero M, Vincenzini MT et al (1994) Effect of phosphotyrosine phosphatase over-expression on glutathione metabolism in normal and oncogene-transformed cells. FEBS Lett 344:157–160
52. Marzocchini R, Bucciantini M, Stefani M, Taddei N, Thunnissen MG, Nordlund P et al (1998) Expression, purification and preliminary crystal analysis of the human low Mr phosphotyrosine protein phosphatase isoform 1. FEBS Lett 426:52–56

53. Miranda MA, Okamoto AK, Ferreira CV, Silva TL, Granjeiro JM, Aoyama H (2006) Differential effects of flavonoids on bovine kidney low molecular mass protein tyrosine phosphatase. J Enzyme Inhib Med Chem 21:419–425
54. Modesti A, Marzocchini R, Raugei G, Chiti F, Sereni A, Magherini F et al (1998) Cloning, expression and characterisation of a new human low Mr phosphotyrosine protein phosphatase originating by alternative splicing. FEBS Lett 431:111–115
55. Mohrenweiser HW, Novotny JE (1982) ACP1GUA-1: a low-activity variant of human erythrocyte acid phosphatase—association with increased glutathione reductase activity. Am J Hum Genet 34:425–433
56. Moller DE, Kaufman KD (2005) Metabolic syndrome: a clinical and molecular perspective. Annu Rev Med 56:45–62
57. Musumeci L, Bongiorni C, Tautz L, Edwards RA, Osterman A, Perego M et al (2005) Low-molecular-weight protein tyrosine phosphatases of bacillus subtilis. J Bacteriol 187:4945–4956
58. Ottana R, Maccari R, Ciurleo R, Paoli P, Jacomelli M, Manao G et al (2009) 5-arylidene-2-phenylimino-4-thiazolidinones as PTP1B and LMW-PTP inhibitors. Bioorg Med Chem 17:1928–1937
59. Paggi A, Borgiani P, Gloria-Bottini F, Russo S, Saponara I, Banci M et al (1991) Further studies on acid phosphatase in obese subjects. Dis Markers 9:1–7
60. Pandey SK, Yu XX, Watts LM, Michael MD, Sloop KW, Rivard AR et al (2007) Reduction of low molecular weight protein-tyrosine phosphatase expression improves hyperglycemia and insulin sensitivity in obese mice. J Biol Chem 282:14291–14299
61. Park EK, Warner N, Mood K, Pawson T, Daar IO (2002) Low-molecular-weight protein tyrosine phosphatase is a positive component of the fibroblast growth factor receptor signaling pathway. Mol Cell Biol 22:3404–3414
62. Ramponi G, Stefani M (1997) Structure and function of the low Mr phosphotyrosine protein phosphatases. Biochim Biophys Acta 1341:137–156
63. Rudbeck HB, Johnsen A, Dissing J (2003) Evolutionary aspects of the gene for the classical enzyme polymorphism, ACP1. Int Congr Ser 1239:733–736
64. Rudbeck L, Dissing J, Lazaruk KD, Sensabaugh G (2000) Human 18 kDa phosphotyrosine protein phosphatase (ACP1) polymorphism: studies of rare variants provide evidence that substitutions within or near alternatively spliced exons affect splicing result. Ann Hum Genet 64:107–116
65. Shekels LL, Smith AJ, Van Etten RL, Bernlohr DA (1992) Identification of the adipocyte acid phosphatase as a PAO-sensitive tyrosyl phosphatase. Protein Sci 1:710–721
66. Shu YH, Hartiala J, Xiang AH, Trigo E, Lawrence JM, Allayee H et al (2009) Evidence for sex-specific associations between variation in acid phosphatase locus 1 (ACP1) and insulin sensitivity in Mexican-Americans. J Clin Endocrinol Metab 94:4094–4102
67. Souza AC, Azoubel S, Queiroz KC, Peppelenbosch MP, Ferreira CV (2009) From immune response to cancer: a spot on the low molecular weight protein tyrosine phosphatase. Cell Mol Life Sci 66:1140–1153
68. Spencer N, Hopkinson DA, Harris H (1964) Quantitative differences and gene dosage in the human red cell acid phosphatase polymorphism. Nature 201:299–300
69. Stefani M, Caselli A, Bucciantini M, Pazzagli L, Dolfi F, Camici G et al (1993) Dephosphorylation of tyrosine phosphorylated synthetic peptides by rat liver phosphotyrosine protein phosphatase isoenzymes. FEBS Lett 326:131–134
70. Stefani M, Taddei N, Ramponi G (1997) Insights into acylphosphatase structure and catalytic mechanism. Cell Mol Life Sci 53:141–151
71. Swallow DM, Povey S, Harris H (1973) Activity of the "red cell" acid phosphatase locus in other tissues. Ann Hum Genet 37:31–38
72. Tabernero L, Aricescu AR, Jones EY, Szedlacsek SE (2008) Protein tyrosine phosphatases: structure-function relationships. FEBS J 275:867–882
73. Tailor P, Gilman J, Williams S, Couture C, Mustelin T (1997) Regulation of the low molecular weight phosphotyrosine phosphatase by phosphorylation at tyrosines 131 and 132. J Biol Chem 272:5371–5374

74. Tailor P, Gilman J, Williams S, Mustelin T (1999) A novel isoform of the low molecular weight phosphotyrosine phosphatase, LMPTP-C, arising from alternative mRNA splicing. Eur J Biochem 262:277–282
75. Vincent C, Doublet P, Grangeasse C, Vaganay E, Cozzone AJ, Duclos B (1999) Cells of Escherichia coli contain a protein-tyrosine kinase, Wzc, and a phosphotyrosine-protein phosphatase, Wzb. J Bacteriol 181:3472–3477
76. Wang S, Stauffacher CV, Van Etten RL (2000) Structural and mechanistic basis for the activation of a low-molecular weight protein tyrosine phosphatase by adenine. Biochemistry 39:1234–1242
77. Wo YY, McCormack AL, Shabanowitz J, Hunt DF, Davis JP, Mitchell GL et al (1992) Sequencing, cloning, and expression of human red cell-type acid phosphatase, a cytoplasmic phosphotyrosyl protein phosphatase. J Biol Chem 267:10856–10865
78. Yasuda N, Inoue T, Horizoe T, Nagata K, Minami H, Kawata T et al (2003) Functional characterization of the adenosine receptor contributing to glycogenolysis and gluconeogenesis in rat hepatocytes. Eur J Pharmacol 459:159–166
79. Yoshihara CM, Mohrenweiser HW (1980) Characterization of ACP1TIC-1, an electrophoretic variant of erythrocyte acid phosphatase restricted to the Ticuna Indians of central Amazonas. Am J Hum Genet 32:898–907
80. Zabell AP, Schroff AD Jr, Bain BE, Van Etten RL, Wiest O, Stauffacher CV (2006) Crystal structure of the human B-form low molecular weight phosphotyrosyl phosphatase at 1.6: a resolution. J Biol Chem 281:6520–6527
81. Zhang M, Stauffacher CV, Lin D, Van Etten RL (1998) Crystal structure of a human low molecular weight phosphotyrosyl phosphatase. Implications for substrate specificity. J Biol Chem 273:21714–21720
82. Zhang S, Zhang ZY (2007) PTP1B as a drug target: recent developments in PTP1B inhibitor discovery. Drug Discov Today 12:373–381
83. Zhang ZY, Van Etten RL (1990) Purification and characterization of a low-molecular-weight acid phosphatase–a phosphotyrosyl-protein phosphatase from bovine heart. Arch Biochem Biophys 282:39–49

Chapter 12
Mitogen-Activated Protein Kinase Phosphatases in Metabolism

Ahmed Lawan and Anton M. Bennett

Abstract Although we continue to learn much about how the mitogen-activated protein kinases (MAPKs) are involved in physiological and pathophysiological signaling in metabolism, a comparable level of understanding about the mechanisms of MAPK inactivation in the control of metabolic homeostasis is lacking. The family of enzymes known to specifically antagonize the MAPKs by direct dephosphorylation, the MAPK phosphatases (MKPs), are emerging as important players in the control of metabolic homeostasis. The MKPs regulate the MAPKs through a complex network of pathways within a spatio-temporal and tissue-specific manner. Here, we will review studies that have led to the realization that the MKPs play essential signaling roles in the control of metabolic homeostasis by counterbalancing the actions of the MAPKs. A growing body of evidence now demonstrates a critical role for the MKPs in the regulation of MAPK activity in metabolic signaling.

Abbreviations

DUSP	Dual-specificity protein phosphatase
ERK1/2	Extracellular signal-regulated kinases 1 and 2
FOXO	Forkhead box O protein
G6Pase	Glucose 6-phosphatase
JNK	c-Jun NH_2 kinases
KIM	Kinase interaction motif
MAPK	Mitogen-activated protein kinases
MKP	MAPK phosphatase
PEPCK	Phosphoenol pyruvate carboxykinase
PPAR	Peroxisome proliferator-activated receptor

A. Lawan Ph.D. (✉) • A.M. Bennett Ph.D.
Department of Pharmacology, Yale University School of Medicine,
New Haven, CT 06510, USA
e-mail: ahmed.lawan@yale.edu

K.K. Bence (ed.), *Protein Tyrosine Phosphatase Control of Metabolism*,
DOI 10.1007/978-1-4614-7855-3_12,

Introduction

Obesity is a growing epidemic that predisposes to conditions such as insulin resistance, cardiovascular disease, stroke, nonalcoholic fatty liver disease, and in some cases cancer [1–7]. Other risk factors include increasing age, genetic susceptibility, hormonal changes and, of course, diet. Because of the growing obesity problem and associated metabolic syndrome, there continues to be a tremendous effort towards identifying the molecules that participate in the regulation of metabolism. The mitogen-activated protein kinases (MAPKs) play a prominent role in processes that regulate cellular metabolism [8–15]. The stress-responsive MAPKs have figured prominently in the pathogenesis of metabolic disease. For example, the c-Jun NH_2 kinases (JNKs), specifically the JNK1 isoform, have been demonstrated to promote insulin resistance and obesity [10], the JNK2 isoform also appears to contribute, albeit to a lesser extent [16]. The other stress-responsive MAPK, p38α/β MAPK, has been shown to promote a variety of metabolic processes such as energy expenditure, glucose homeostasis, and lipid metabolism [14, 15, 17–19]. Whereas, the growth factor-responsive MAPK, the extracellular signal-regulated kinases 1 and 2 (ERK1/2), specifically ERK1 has been implicated in the maintenance of body mass and adipocyte differentiation [9, 20]. The wealth of knowledge on the MAPKs clearly establishes that these enzymes are critical for a wide variety of metabolic processes.

The MAPKs are activated by a variety of extracellular factors such as growth factors, cytokines, and hormones that engage a well-defined serine-threonine signaling cascade that includes the activation of the dual-specificity family of MAPK kinases [21]. The MAPK kinases phosphorylate the MAPKs on regulatory threonine and tyrosine residues resulting in MAPK activation [22, 23]. The MAPKs are inactivated by direct dephosphorylation of their regulatory threonine and tyrosine residues. Several protein phosphatases are capable of dephosphorylating one or both of the conserved threonine and/or tyrosine residues in the activation loop of the MAPKs [24, 25]. However, by far the most efficient enzymes that catalyze MAPK dephosphorylation are the subgroup of dual-specificity protein tyrosine phosphatases (DUSPs) called the MAPK phosphatases (MKPs) [26–28]. A large body of evidence now defines the MKPs as a major subfamily of DUSPs that specifically dephosphorylate the MAPKs on both regulatory threonine and tyrosine residues, rendering the MAPKs inactive [26]. Given the well-established role of the MAPKs in metabolic signaling, until recently it has been unclear whether the MKPs participate in the regulation of these processes. Based upon work from this laboratory and others, an emerging picture that the MKPs are major participants in metabolic signaling is beginning to take shape.

The Family of MAP Kinase Phosphatases

The MKPs constitute a family of ten catalytically active enzymes that share a common protein tyrosine phosphatase domain at their carboxyl terminus and a non-catalytic regulatory domain in the amino terminus [26, 28, 29]. The non-catalytic

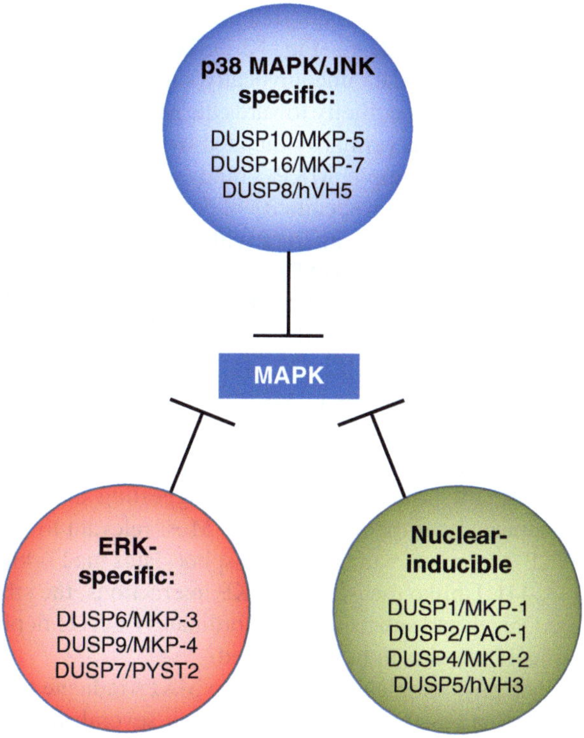

Fig. 12.1 Dual-specificity family of MAPK phosphatases. Schematic representation of the three groups of MKPs classified based upon their MAPK substrate specificity, and subcellular localization. Group 1, nuclear inducible Group 2, ERK-specific and Group 3, p38 MAPK/JNK-specific

region of the MKPs contains several motifs that are critical for MKP function. These include a kinase interaction motif (KIM) that directs binding to the MAPKs and in some cases targeting motifs that direct subcellular localization. The affinity of the KIM domain for its cognate MAPK is largely responsible for the selectivity of MKP-MAPK substrate dephosphorylation [29, 30]. However, MKP subcellular localization also contributes to its signaling specificity. As such MKPs can either be localized to the nucleus, cytosol, or may have the capacity to shuttle between both compartments. The non-catalytic region contains motifs that direct such subcellular localization as well as the ability by some MKPs to serve as an anchor for the MAPKs themselves [31–33]. Although highly similar, there appears to be a sufficient level of complexity and diversity within the non-catalytic regions of the MKPs to confer a very high level of functional diversity.

The ten catalytically active MKPs are grouped into three categories depending upon their MAPK substrate selectivity, subcellular localization, and sequence homology [25, 26]. The first group is represented by DUSP1/MKP-1, DUSP2/PAC-1, DUSP4/MKP-2, and DUSP5/hVH3; these MKPs are stress-responsive and dephosphorylate JNK and p38 MAPK primarily (Fig. 12.1). The second group comprises of DUSP6/MKP-3, DUSP7/MKP-X, and DUSP9/MKP-4 that specifically dephosphorylate ERK1/2 and localize to the cytoplasm (Fig. 12.1). The third group contains DUSP8/hVH5, DUSP10/MKP-5, and DUSP16/MKP-7; these MKPs predominately dephosphorylate JNK and p38 MAPK and can shuttle between the nucleus and cytoplasm (Fig. 12.1). To date a combination of structural, biochemical, and genetic data unequivocally supports the conclusion that the MKPs exhibit overwhelming specificity towards the MAPKs [34–40]. The specificity of the MKPs to the MAPKs is primarily dictated by the ability of the MKPs to interact with the MAPKs. For example, MKP-3 binds with high affinity to ERK1/2 and this interaction governs its high specificity of dephosphorylation towards ERK1/2 [30]. In contrast, MKP-3 fails to interact with either p38 MAPK or JNK, and hence fails to dephosphorylate p38 MAPK and JNK [30]. Whereas, MKP-1 which can dephosphorylate ERK1/2, p38 MAPK, and JNK is capable of binding all three MAPK family members and dephosphorylates them with a rank order of p38 MAPK=JNK>ERK1/2 [31, 41]. The resolution of the crystal structure of the MAPK binding domain of MKP-5 with its substrate p38α MAPK has revealed additional levels of MKP-MAPK substrate recognition whereby the KIM domain of MKP-5 engages in a unique manner with the CD domain of p38α MAPK [42]. Notably, there have been studies suggesting that the MKPs can dephosphorylate substrates other than the MAPKs. Most relevant to this discussion is the report that MKP-3 interacts with and dephosphorylates the Forkhead box O1 (FOXO1) protein [43, 44]. Unlike the MAPK-MKP interactions, the structural basis for how MKP-3 recognizes and subsequently coordinates FOXO1 dephosphorylation at the molecular level remains to be established.

The physiological effects of the MKPs indicate that these enzymes attain a high level of signaling complexity through several modes of operation with regard to MAPK dephosphorylation and cell signaling. Three general modes of operation can be considered. In the first, MKPs are induced in their expression levels in response to growth factors, hormones, cytokines, and other stresses. Typically, the induction of the MKPs is mediated by the activity of the MAPKs themselves. For example, MKP-1, MKP-2, and PAC-1 contain AP-1 sites within their proximal promoters that serve to promote the expression of these genes in response to MAPK activation. The induction of the MKPs in response to various stimuli plays a critical step in setting the kinetics of MAPK activity to evoke either a transient or sustained level of activation. Hence, the expression levels of the MKPs can set the temporal nature of MAPK activity and this in turn can influence gene expression. In the second mode of action, MKPs not only exert temporal control over MAPK activation, but also spatial regulation of the MAPKs. In this scenario, depending upon their subcellular localization, MKPs can dephosphorylate specific subcellular pools of MAPKs. For example, MKP-1 dephosphorylates the nuclear pool of MAPKs, independently of the pool of MAPKs that reside in the cytosol [25]. We have shown this to be an important aspect of MAPK regulation in the control of insulin resistance [45]. Hence, nuclear and cytosolic MAPK/MKP-mediated events occur in a spatio-temporal manner. In the final scenario, MKPs may also play key roles in facilitating pathway cross-talk. Here, activation of the MKP by a particular MAPK, which is not a substrate for the induced MKP, results in the dephosphorylation of a distinct MAPK module in another signaling pathway. Collectively, these modes of signaling illustrate the complexity of coordinating MKP-mediated MAPK activity.

It is also important to note that MKPs can inactivate several MAPKs simultaneously, giving rise to an additional level of signaling complexity whereby multiple MAPK signals are integrated. This aspect of MAPK regulation defines the MKPs as critical “signaling nodes” in the MAPK pathway. Information provided by multiple MAPK modules can be coordinated by a single MKP to control downstream phosphorylation of a number of distinct substrates and/or phosphorylation of distinct sites on a single substrate. As such, the ability of the MKPs to impact signaling through their influence on an integrated subset of MAPK activities in both a spatial and temporal manner can generate significant complexity [25, 46]. Therefore, the regulation of MAPK-dependent signaling by the MKPs could have profound and unanticipated effects on the control of metabolic homeostasis (Table 12.1).

MKPs in the Regulation of Body Mass and Glucose Homeostasis

The MAPKs have previously been implicated in the regulation of metabolic homeostasis through targeting the phosphorylation of transcription factors and cytosolic signaling proteins that control processes such as adipogenesis, insulin signaling, glucose uptake, fatty acid metabolism, lipogenesis, and energy expenditure [8–15]. As such, a role for the MKPs in the regulation of metabolic homeostasis could be anticipated. However, based upon both the overlapping effects of the MKPs on the MAPKs as well as the other issues of how the MKPs signal (see above), it was unclear as to whether the MKPs would play an essential physiological role in metabolism or a pathophysiological role in the progression of obesity and metabolic syndrome.

The initial observation suggesting that the MKPs are involved in the regulation of body mass came from studies showing that MKP-1-deficient mice (*mkp-1*$^{-/-}$) on a mixed genetic background [47] were resistant to weight gain when placed on a high-fat diet [45]. Even when bred to a pure C57BL/6 background *mkp-1*$^{-/-}$ mice were still resistant to weight gain when placed on a high-fat diet arguing against this metabolic phenotype occurring as a result of modifier alleles and/or strain-specific differences [45, 48]. It was determined that *mkp-1*$^{-/-}$ mice expended significantly more energy than wild-type mice despite exhibiting equivalent levels of activity [45]. To date, MKP-1 is the only MKP that has been reported to affect body mass. We have recently found that MKP-5-deficient mice gain weight at a similar rate to wild-type mice when placed on a high-fat diet despite the fact that MKP-5, like MKP-1, also dephosphorylates p38 MAPK and JNK (unpublished observations). These results suggest that the MKPs are likely to exert distinct effects on metabolism. Interestingly, one potential reason for such differences might lie in the subcellular localization of MKP-1 as compared with MKP-5 since MKP-5 is localized to both the nucleus and cytoplasm, whereas MKP-1 is resident only within the nucleus [31, 49]. It will be interesting to test the idea that these distinct phenotypes stem from compartment-specific differences in MAPK dephosphorylation.

Table 12.1 Metabolic effects of MKPs[a]

DUSP/MKP family member	MAPK substrate preference	Effect on metabolism	Site(s) of metabolic action(s)	Expression in obesity and HFD	References
DUSP1/MKP-1	p38 MAPK = JNK > ERK	MKP-1-deficient mice are resistant to diet-induced obesity	Liver	Increased	[55]
		Negatively regulates fatty acid oxidation, lipogenesis and mitochondrial oxidative phosphorylation	Skeletal muscle	Increased	[56]
DUSP2/PAC-1	ERK ≫ p38 MAPK ~ JNK	None reported	ND	ND	
DUSP4/MKP-2	ERK ~ JNK > p38 MAPK	Negative modulation of p38 MAPK-mediated gluconeogenic gene expression in hepatoma cell lines	Liver	ND	[6]
DUSP5/hVH3	ERK > JNK ~ p38 MAPK	None reported	ND	ND	
DUSP6/MKP-3	ERK ≫ p38 MAPK ~ JNK	Involved in promoting gluconeogenesis by dephosphorylating FOXO1	Liver	Increased	[74]
DUSP7/PYST2	ERK > p38 MAPK ≫ JNK	None reported	ND	ND	
DUSP8/hVh5	JNK ~ p38 MAPK ≫ ERK	None reported	ND	ND	
DUSP9/MKP-4	ERK ≫ p38 MAPK > JNK	Over-expression in adipocyte cell line impairs glucose uptake	Adipose	Increased	[20, 76]
DUSP10/MKP-5	p38 MAPK > JNK ≫ ERK	None reported	ND	ND	
DUSP16/MKP-7	JNK ~ p38 MAPK ≫ ERK	None reported	ND	ND	

[a]The table summarizes studies on the various effects the MKPs have on metabolic signaling. MKP-1 represents the only MKP studied in knockout mice that has been reported to exert a metabolic phenotype. The other reports on the MKPs were performed in various cultured cell systems or using over-expression approaches in mice. Many of the MKPs remain to be examined for their effects on metabolism

Given the initial observation that *mkp-1*$^{-/-}$ mice are resistant to diet-induced obesity, it was anticipated that these mice would be insulin hypersensitive. Surprisingly, hyperinsulinemic-euglycemic clamp experiments demonstrated little difference in whole body glucose infusion rates in mice fed a chow diet [45]. These results demonstrated that despite the fact that *mkp-1*$^{-/-}$ mice were resistant to diet-induced obesity, these mice were not insulin hypersensitive. This finding was further supported at the biochemical level since insulin-mediated activation of Akt in the liver, white adipose tissue, and skeletal muscle were unimpaired in *mkp-1*$^{-/-}$ mice [45]. Since *mkp-1*$^{-/-}$ mice exhibited enhanced JNK activation these results were at odds with previous reports suggesting that JNK phosphorylation of the insulin receptor substrate-1 resulted in impaired insulin-mediated Akt activation and subsequently, insulin resistance [50]. This conundrum was resolved by the finding that MKP-1 is localized to the nucleus and because of this it is restricted to the dephosphorylation of MAPKs that translocate to this location. As such, MKP-1-deficient mice exhibited enhanced nuclear but unaltered cytosolic JNK activity [45]. These results provided the first evidence for the importance of compartment-specific regulation of the MKPs, and hence MAPK signaling, in metabolism. Furthermore, the idea that MKP-1 regulates a discrete pool of MAPKs that are resident within the nucleus raised the possibility that aberrant MAPK dephosphorylation exclusively within the nucleus might represent a novel mechanism through which metabolic disease proceeds.

Role of MKPs in Liver Metabolism

The liver is the major site of action for processes such as gluconeogenesis, glycogen synthesis, and glycogenolysis. In addition, the liver plays an important role in the regulation of lipid metabolism. The MAPKs, in particular JNK1, have been implicated in the development of insulin resistance and type 2 diabetes [7]. p38 MAPK has also been suggested to promote hepatic gluconeogenesis and may contribute to the unrestrained hepatic gluconeogenesis observed in both type 1 and 2 diabetes [51]. With regard to ERK1/2, its role in glucose homeostasis in the liver is less clear [9]. How the MKPs control MAPK-dependent signaling in the liver has been explored and two MKPs, namely MKP-1 and MKP-3, have been demonstrated to play a role in the control of hepatic lipid metabolism and gluconeogenesis.

DUSP1/MKP-1

mkp-1$^{-/-}$ mice are resistant to the development of hepatic steatosis when placed on a high-fat diet [45]. Follow-up studies from this lab have utilized *mkp-1*$^{-/-}$ mice that were intercrossed with the leptin receptor-deficient (*db/db*) mice. The livers of *mkp-1*$^{-/-}$:*db/db* progeny were markedly less steatotic in comparison with wild-type mice [52]. These results demonstrate that MKP-1 plays an essential role in the management of

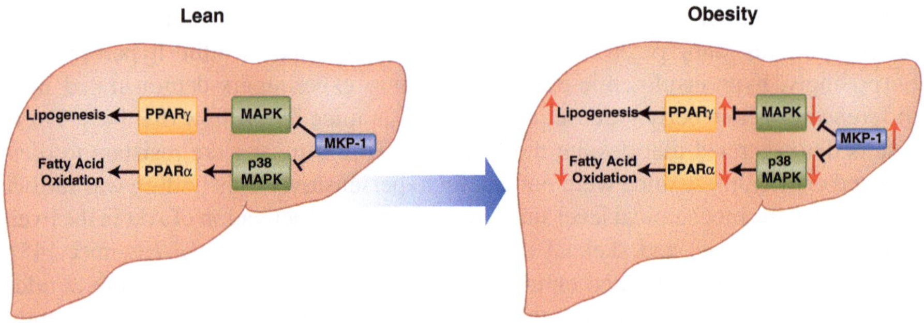

Fig. 12.2 MKP-1 in liver metabolism. Under lean conditions MKP-1 modulates MAPK-dependent signals that control PPARα and PPARγ activities. Whereas p38 MAPK promotes PPARα phosphorylation and subsequently fatty acid oxidation, MAPK phosphorylation of PPARγ inhibits its activity thereby reducing lipogenesis. During obesity MKP-1 expression levels increase resulting in decreased MAPK activities in the nucleus. Reduced MAPK-dependent phosphorylation of PPARα and PPARγ impairs fatty acid oxidation and enhances lipogenesis, respectively

hepatic lipid homeostasis. The resistance to the development of hepatic steatosis can be attributed, at least in part, to enhanced hepatic β-oxidation in the livers of *mkp-1*$^{-/-}$ mice suggesting that MKP-1 participates in a MAPK-dependent pathway that negatively regulates hepatic triglyceride metabolism [45, 52]. The molecular basis for this effect has been attributed to the observation that MKP-1 negatively regulates the peroxisome proliferator-activated receptor-alpha (PPARα). It has been shown that p38 MAPK phosphorylates PPARα, which promotes its sensitivity to both ligand-induced activation and co-activation by the PPARγ transcriptional co-activator (PGC-1α) [15]. Hepatocytes lacking MKP-1 exhibit increased PPARα-mediated transcriptional activation basally and in response to agonist, which can be restored by pharmacological inhibition of p38 MAPK [45]. In addition, a key rate-limiting enzyme in the fatty acid oxidation pathway, carnitine palmitoyl transferase I, is increased in the livers of *mkp-1*$^{-/-}$ mice [45]. Collectively, these results demonstrate that MKP-1 plays an essential role in the dephosphorylation of p38 MAPK to negatively regulate hepatic triglyceride metabolism (Fig. 12.2).

Since mice deficient in MKP-1 are resistant to the development of hepatic steatosis the question of whether MKP-1 is over-expressed in obesity and subsequently contributes to the development of hepatic steatosis has been examined. Indeed, MKP-1 expression levels are upregulated in the liver of mice fed either a high-fat or an atherogenic diet (unpublished observations and [53]). The increased expression levels of hepatic MKP-1 in obesity have been proposed to promote hepatic steatosis through a mechanism involving activation of PPARγ, which drives lipogenesis [54]. Under lean or non-obese conditions, hepatic PPARγ expression in mice is barely detectable [54]. However, in obesity PPARγ1 becomes over-expressed and promotes the development of hepatic steatosis [55–57]. PPARγ is negatively regulated by phosphorylation at Ser112 by either p38 MAPK or JNK [13, 58]. When over-expressed, MKP-1 enhances the dephosphorylation of PPARγ1 at Ser112, by inactivating either p38 MAPK and/or JNK, resulting in enhanced activation of PPARγ1 and its target genes that promote hepatic lipogenesis. A genome-wide microarray analysis of the livers from *db/db;mkp-1*$^{-/-}$ revealed an enrichment for a subset of highly repressed PPARγ1 target genes [52]. The two most highly repressed PPARγ1 target genes identified in *db/db:mkp-1*$^{-/-}$ mice were the cell death-inducing DFFA-like effector A (CIDEA) and CIDEC/fat-specific protein 27 (FSP27) [52]. Notably, FSP27 is required for the generation of lipid droplets, is over-expressed in obesity, and promotes hepatic steatosis [57, 59]. In support of this model, it was shown that PPARγ1 is hyperphosphorylated on Ser112 in hepatocytes-derived from MKP-1-deficient mice; furthermore, hepatocytes derived from *mkp-1*$^{-/-}$ mice are impaired in PPARγ1 activation and lipid droplet formation when treated with the PPARγ1 agonist rosiglitazone [52]. These data argue for a mechanism whereby MKP-1 over-expression promotes the activation of PPARγ1 leading to the upregulation of FSP27 to induce hepatic steatosis (Fig. 12.2). Consistent with this model, it has been shown that FSP27 over-expression is necessary and sufficient for the development of hepatic steatosis [57].

The other interesting observation is that it appears that MKP-1 is required to maintain PPARγ1 in a hypophosphorylated state on Ser112 through dephosphorylation of either p38 MAPK and/or JNK [52]. This implies that MKP-1 is required for optimal PPARγ1 activation since PPARγ1 Ser112 phosphorylation negatively regulates its activity [60]. Therefore, the role of MKP-1 on the actions of the anti-diabetic thiazolidinedione drugs that target PPARg will be an important avenue to investigate further. Uncovering new mechanisms by which PPARγ1 activation can be modulated may lead to novel therapeutic approaches in the treatment of diabetes.

DUSP6/MKP-3

Recently, it has been proposed that MKP-3 regulates hepatic gluconeogenesis, by dephosphorylating FOXO1 [43]. This result is particularly surprising given the highly specific regulatory mechanisms engaged by MKP-3 in the dephosphorylation of ERK1/2 [34]. MKP-3 expression levels increase dramatically in high fat-diet fed mice and its levels are upregulated by glucagon. Lean mice over-expressing MKP-3 in the liver have significantly increased blood glucose levels, whereas mice with reduced hepatic MKP-3 expression levels have reduced plasma blood glucose. These observations suggest that MKP-3 promotes hepatic gluconeogenesis. It was also shown that MKP-3 dephosphorylates FOXO1 on Ser256, and when dephosphorylated, FOXO1 translocates to the nucleus to activate the expression of the gluconeogenic gene program. Although it is clear that MKP-3 can interact with FOXO1 [44], whether MKP-3 directly dephosphorylates FOXO1 on Ser256 remains to be formally demonstrated [43, 44]. More recently, it has been shown that MKP-3 expression is extinguished by insulin. MKP-3 degradation is promoted by insulin through an ERK1/2 feedback pathway wherein ERK1/2 phosphorylates MKP-3 and targets it for degradation [61]. Over-expression of MKP-3 dephosphorylates both ERK1/2 and FOXO1, and ERK1/2 phosphorylates FOXO1 [62], raising the question as to whether MKP-3 indirectly regulates FOXO1 through ERK1/2 dephosphorylation in the control of hepatic gluconeogenesis. Studies utilizing MKP-3-deficient mice will be valuable in resolving some of these issues.

DUSP4/MKP-2

Although MKP-2 has a striking degree of structural similarity with MKP-1, it exhibits a distinct pattern of tissue distribution as compared with MKP-1 [63]. Interestingly, MKP-2 substrate specificity appears at least in vitro to favor ERK ~ JNK > p38 MAPK [63]. MKP-2 has been suggested to modulate the transcriptional machinery that regulates gluconeogenic gene expression [64]. The AMP-activated 5′-kinase (AMPK) suppresses gluconeogenesis when activated under conditions of low nutrient status. It was shown that AMPK activates the transcription factor Egr-1, which subsequently binds and promotes MKP-2 expression leading to the dephosphorylation of p38 MAPK [64]. A link between p38 MAPK and the activation of gluconeogenesis was demonstrated by showing that activating mutants of p38 MAPK promote the gluconeogenic program by stimulating the expression of glucose 6-phosphatase and PEPCK. Hence, it appears that MKP-2 dephosphorylates p38 MAPK in response to the activation of AMPK in order to suppress gluconeogenesis. Additional evidence testing this pathway in mice is still warranted. Nevertheless, these data raise an intriguing mechanism through which MKPs can modulate hepatic metabolic function.

Regulation of Metabolic Signaling by MKPs in Skeletal Muscle

Skeletal muscle tissue is a major site for the control of energy balance and a critical target tissue for insulin-mediated responses. The maintenance of the appropriate composition of skeletal myofiber composition is considered to be an important aspect of metabolic homeostasis. Dysfunctional energy metabolism in skeletal muscle has been attributed to the development of insulin resistance and obesity [65]. A clear role for skeletal muscle JNK1 has been shown to promote high-fat diet-induced insulin resistance [66]. In contrast, a role for p38 MAPK in skeletal muscle metabolism is somewhat controversial with some reporting a positive and negative role for p38 MAPK in glucose uptake and metabolism [67]. A role for skeletal muscle ERK1/2 is also less clear. Given the general level of uncertainty surrounding the role of the MAPKs in skeletal muscle metabolism, the influence of the MKPs in metabolic signaling in this tissue remained equally ambiguous.

DUSP1/MKP-1

MKP-1 is abundantly expressed in skeletal muscle and mice lacking MKP-1 show enhanced p38 MAPK, JNK, and ERK1/2 activities in this tissue [45]. MKP-1 negatively regulates oxidative phosphorylation as evidenced by increased mitochondrial oxygen consumption in skeletal muscle isolated from MKP-1-deficient mice [45].

In addition, the expression of the uncoupling protein-3 (UCP-3), which is an inner membrane mitochondrial protein involved in oxidative phosphorylation, was elevated in skeletal muscle isolated from MKP-1-deficient mice as compared with wild-type mice [45]. These observations are consistent with the increase in energy expenditure in MKP-1-deficient mice. Further, these results implied that MKP-1 regulates mitochondrial function and subsequently myofiber composition through a MAPK-dependent pathway. It has been suggested that the switch from oxidative to glycolytic myofibers contributes to obesity, presumably because glycolytic myofibers consume less energy [68]. In *mkp-1*$^{-/-}$ mice fed a high fat diet oxidative myofiber composition is maintained, whereas wild-type mice exhibit a significant loss of oxidative myofibers, hence shifting energy balance [48]. High fat diet-feeding increases the expression of MKP-1 suggesting that it promotes the loss of oxidative myofiber content, an event that could increase susceptibility to metabolic dysfunction. Mechanistically, over-expression of MKP-1 was found to inhibit the expression and activity of PGC-1α, a master regulator of mitochondrial biogenesis and energy expenditure, by impairing p38 MAPK-mediated PGC-1α phosphorylation, which reduced both its stability and co-activating activity [48]. Over-expression of MKP-1 impaired p38 MAPK-mediated PGC-1α phosphorylation at Ser265 and Thr298 [48], sites that are known to facilitate PGC-1α stability and co-transcriptional activity [11]. Hence, skeletal muscle MKP-1 coordinates energy sensing by controlling the ability of p38 MAPK to modulate mitochondrial function and myofiber composition through PGC-1α phosphorylation (Fig. 12.3).

Skeletal muscle is a target for adiponectin, which acts through the adiponectin receptor to promote skeletal muscle oxidative metabolism [69]. Recently, MKP-1 expression in skeletal muscle has been linked to the ability of adiponectin to stimulate skeletal muscle oxidative metabolism [70]. Qiao et al. showed that adiponectin induced p38 MAPK activation even in the absence of either MKK3 or MKK6; concomitantly, adiponectin repressed the expression of MKP-1 in skeletal muscle. Based upon previous work linking MKP-1 regulation to p38 MAPK phosphorylation of PGC-1α [48], Qiao et al. demonstrated that the ability of adiponectin to suppress MKP-1 was required to promote the activation of PGC-1α [70]. These results indicate that MKP-1 plays an important role in how adiponectin mediates oxidative metabolism in skeletal muscle. In obesity the levels of adiponectin are reduced which has led to the suggestion that this could promote insulin resistance. Interestingly, these data are consistent with the observation that in obesity the expression level of MKP-1 is elevated in skeletal muscle [48]. Hence, increased MKP-1 expression in skeletal muscle occurs not only through elevated levels of circulating free fatty acids [48], but also through a reduction in the expression levels of adiponectin. Together, these effects drive MKP-1 over-expression, which impairs p38 MAPK phosphorylation and PGC-1α activity. Given that skeletal muscle plays a major role in the development of insulin resistance in obesity, further studies employing conditional deletion of MKP-1 in skeletal muscle would be worthwhile in order to determine the importance of dysfunctional MKP-1 expression in this tissue in obesity. Nevertheless, these data provide a new level of understanding on how MKP-1 maintains energy balance through the control of p38 MAPK activity in skeletal muscle.

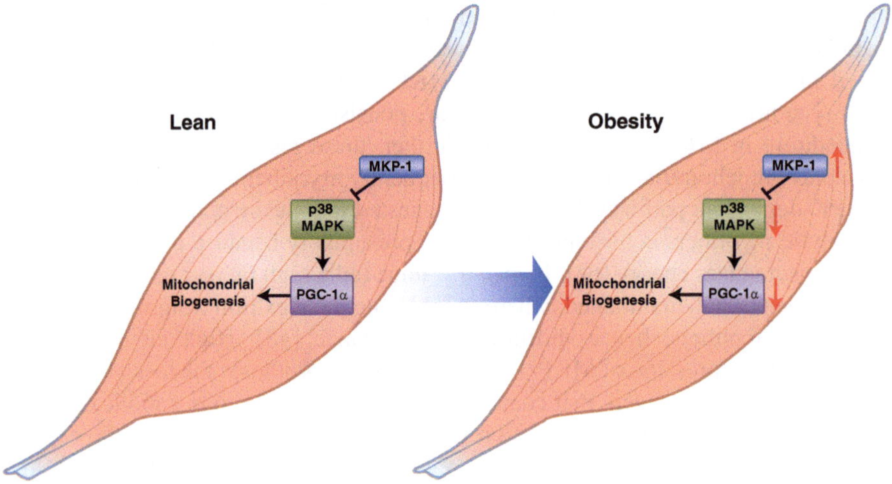

Fig. 12.3 MKP-1 in skeletal muscle metabolism. In skeletal muscle the transcriptional co-activator, PGC-1α, drives mitochondrial biogenesis. MKP-1 can regulate the ability of p38 MAPK to phosphorylate and subsequently activate PGC-1α. In obesity, it is proposed that increased MKP-1 expression levels impair PGC-1α activity leading to reduced mitochondrial function

Regulation of Adipose Metabolic Function by MKPs

Adipose tissue is a dynamic tissue that plays a key role in the maintenance of energy balance through its ability to secrete a variety of important hormones. An increase in fat mass as a result of excess nutrition, as seen in obesity, promotes adipocyte dysfunction leading to the development of insulin resistance. Therefore, understanding the mechanisms regulating adipocyte signaling in metabolism is critical. The MAPKs have been shown to play an important role in the regulation of adipocyte function [8, 9, 18, 51, 71]. However, it remains poorly understood what the effects of the MKPs are in adipocytes and whether MKPs in adipose tissue contribute to the maintenance of metabolic homeostasis.

DUSP1/MKP-1

In response to high-fat feeding, mice lacking the expression of MKP-1 exhibit a significant reduction in fat pad mass and enhanced p38 MAPK, JNK, and ERK1/2 activities [45]. Analysis of epididymal fat pads from high-fat diet fed MKP-1-deficient mice revealed that the reduced fat pad mass was due to smaller adipocytes [45]. These results tentatively suggested that MKP-1 may be required for the accumulation of lipid in adipocytes. Although these studies found no contribution for MKP-1 in adipocyte differentiation [45], another study reported a positive role for MKP-1 in adipocyte differentiation [72].

Obesity is considered to be a condition of chronic inflammation whereby adipose tissue becomes infiltrated by macrophages that subsequently promote metabolic dysfunction. MKP-1 plays a major role in the regulation of macrophage function and cytokine production [73]. Therefore, whether MKP-1 affects adipose tissue function in obesity by altering macrophage and/or adipocyte cytokine expression remains an important question. Insight into the relationship between MKP-1 and adipocyte function in obesity has been provided by Ito et al., who found that MKP-1 becomes downregulated in differentiated adipocytes undergoing hypertrophy concomitant with the upregulation of ERK1/2 and the cytokine monocyte chemoattractant protein-1 (MCP-1) [74]. MCP-1 expression levels also increased in adipose tissue in mice very early during the onset of high-fat feeding. The ability of MKP-1 to regulate MCP-1 expression in adipocytes through an ERK1/2-dependent pathway suggests that macrophage infiltration, which is driven at least in part, by local MCP-1 can be indirectly modulated by MKP-1. Therefore, MKP-1 may underlie the progression of the chronic inflammatory state in adipose tissue leading to the development of insulin resistance.

DUSP9/MKP-4

MKP-4 is a cytoplasmic resident MKP that is in the subclass of MKPs that specifically dephosphorylate ERK1/2 and it is expressed in white adipose tissue, kidney, and testis [75]. Interestingly, MKP-4 was identified as a candidate gene in a screen to identify genes involved in opposing the actions of insulin-mediated suppression of phosphoenol pyruvate carboxykinase (PEPCK) [76]. The expression levels of MKP-4 are increased in white adipose tissue of *ob/ob*, *db/db,* and in diet-induced obesity [76]. These findings suggest a potential role for MKP-4 in the development of insulin resistance and obesity. This notion was supported by the observation that over-expression of MKP-4 in adipocyte cell lines impaired insulin-mediated glucose uptake [77]. It remains unclear as to how MKP-4 mediates the actions of insulin. MKP-4 is localized to the cytoplasm and it appears to be quite specific to ERK1/2 suggesting that insulin-dependent targets are likely ERK1/2 substrates. Further work using MKP-4-deficient mice will be needed in order to gain the full picture of how MKP-4 regulates insulin-mediated signaling in adipocytes.

Conclusion

Collectively, the studies conducted in our lab [45, 48, 52] imply that inhibition of MKP-1 may provide a novel therapeutic strategy for the treatment of metabolic syndrome. Targeted inhibition of hepatic MKP-1 in obese patients using anti-sense approaches may have therapeutic value in the treatment of hepatic steatosis. Intuitively, MKP-1 inhibition, resulting in enhanced MAPK signaling, would be anticipated to be deleterious rather than beneficial in the treatment of metabolic disorders. However, it is important to note that MKP-1 inhibition results in the enhancement of the nuclear-restricted pool of MAPKs rather than the total cellular MAPK pool, an important distinction that is likely to be one reason that *mkp-1*$^{-/-}$ mice exhibit little apparent phenotype under unchallenged conditions [47]. It still needs to be determined whether MKP-1 is over-expressed in obesity and type II diabetes in humans. Provocatively, MKP-1 expression levels in fat are repressed in humans following bariatric surgery [78], suggesting a correlation between MKP-1 expression and fat mass. It is also unclear which tissue represents the major site through which MKP-1 influences its effect on metabolism. The involvement of MKP-1 as well as other MKPs in the hypothalamus as a mechanism for the central control of energy balance remains an open question. Future work will require metabolic analysis of MKP-1 tissue-specific knockout mice to decipher the contribution of peripheral and central effects of MKP-1 on metabolism.

With regard to other MKPs, MKP-3 has most recently emerged as a potential regulator of hepatic gluconeogenesis. The control of FOXO signaling by MKP-3 is an intriguing mechanism that suggests new modes of MKP regulation in the liver and possibly other tissues. With the profound selectivity of the MKPs to the MAPKs and its ability to confer signaling specificity at a variety of different levels, the complexity of the MKP/MAPK network in metabolic signaling is likely to be substantially more complex than imagined. Future work will continue to lay the foundation for what is likely to be a wealth of unexplored opportunities for the discovery of novel therapeutic strategies targeting the MKPs for the treatment of metabolic disease.

Acknowledgments A.M.B. was supported by NIH grants DK075776 and DK34989. We thank Florian Gatzke for the preparation of illustrations for this review.

References

1. Basen-Engquist K, Chang M (2011) Obesity and cancer risk: recent review and evidence. Curr Oncol Rep 13:71–76
2. Starley BQ, Calcagno CJ, Harrison SA (2010) Nonalcoholic fatty liver disease and hepatocellular carcinoma: a weighty connection. Hepatology 51:1820–1832
3. Siegel AB, Zhu AX (2009) Metabolic syndrome and hepatocellular carcinoma. Cancer 115:5651–5661
4. Farrell GC, Larter CZ (2006) Nonalcoholic fatty liver disease: from steatosis to cirrhosis. Hepatology 43:S99–S112
5. Flier JS (2004) Obesity wars: molecular progress confronts an expanding epidemic. Cell 116:337–350
6. Kopelman PG (2000) Obesity as a medical problem. Nature 404:635–643
7. Samuel VT, Shulman GI (2012) Mechanisms for insulin resistance: common threads and missing links. Cell 148:852–871
8. Bost F, Aouadi M, Caron L, Binetruy B (2005) The role of MAPKs in adipocyte differentiation and obesity. Biochimie 87:51–56
9. Bost F, Aouadi M, Caron L, Even P, Belmonte N, Prot M, Dani C, Hofman P, Pages G, Pouyssegur J, Le Marchand-Brustel Y, Binetruy B (2005) The extracellular signal-regulated kinase isoform ERK1 is specifically required for in vitro and in vivo adipogenesis. Diabetes 54:402–411

10. Hirosumi J, Tuncman G, Chang L, Gorgun CZ, Uysal KT, Maeda K, Karin M, Hotamisligil GS (2002) A central role for JNK in obesity and insulin resistance. Nature 420:333–336
11. Puigserver P, Rhee J, Lin J, Wu Z, Yoon JC, Zhang CY, Krauss S, Mootha VK, Lowell BB, Spiegelman BM (2001) Cytokine stimulation of energy expenditure through p38 MAP kinase activation of PPARgamma coactivator-1. Mol Cell 8:971–982
12. Xiong Y, Collins QF, An J, Lupo E Jr, Liu H-Y, Liu D, Robidoux J, Liu Z, Cao W (2007) p38 Mitogen-activated protein kinase plays an inhibitory role in hepatic lipogenesis. J Biol Chem 282:4975–4982
13. Burns KA, Vanden Heuvel JP (2007) Modulation of PPAR activity via phosphorylation. Biochim Biophys Acta 1771:952–960
14. Somwar R, Koterski S, Sweeney G, Sciotti R, Djuric S, Berg C, Trevillyan J, Scherer PE, Rondinone CM, Klip A (2002) A dominant-negative p38 MAPK mutant and novel selective inhibitors of p38 MAPK reduce insulin-stimulated glucose uptake in 3T3-L1 adipocytes without affecting GLUT4 translocation. J Biol Chem 277:50386–50395
15. Barger PM, Browning AC, Garner AN, Kelly DP (2001) p38 Mitogen-activated protein kinase activates peroxisome proliferator-activated receptor alpha: a potential role in the cardiac metabolic stress response. J Biol Chem 276:44495–44501
16. Tuncman G, Hirosumi J, Solinas G, Chang L, Karin M, Hotamisligil GS (2006) Functional in vivo interactions between JNK1 and JNK2 isoforms in obesity and insulin resistance. Proc Natl Acad Sci U S A 103:10741–10746
17. Somwar R, Perreault M, Kapur S, Taha C, Sweeney G, Ramlal T, Kim DY, Keen J, Cote CH, Klip A, Marette A (2000) Activation of p38 mitogen-activated protein kinase alpha and beta by insulin and contraction in rat skeletal muscle: potential role in the stimulation of glucose transport. Diabetes 49:1794–1800
18. Cao W, Daniel KW, Robidoux J, Puigserver P, Medvedev AV, Bai X, Floering LM, Spiegelman BM, Collins S (2004) p38 Mitogen-activated protein kinase is the central regulator of cyclic AMP-dependent transcription of the brown fat uncoupling protein 1 gene. Mol Cell Biol 24:3057–3067
19. Akimoto T, Pohnert SC, Li P, Zhang M, Gumbs C, Rosenberg PB, Williams RS, Yan Z (2005) Exercise stimulates Pgc-1{alpha} transcription in skeletal muscle through activation of the p38 MAPK pathway. J Biol Chem 280:19587–19593
20. Jager J, Corcelle V, Gremeaux T, Laurent K, Waget A, Pages G, Binetruy B, Le Marchand-Brustel Y, Burcelin R, Bost F, Tanti JF (2010) Deficiency in the extracellular signal-regulated kinase 1 (ERK1) protects leptin-deficient mice from insulin resistance without affecting obesity. Diabetologia 54(1):180–189
21. Raman M, Chen W, Cobb MH (2007) Differential regulation and properties of MAPKs. Oncogene 26:3100–3112
22. Derijard B, Raingeaud J, Barrett T, Wu I-H, Han J, Ulevitch RJ, Davis RJ (1995) Independent human MAP kinase signal transduction pathways defined by MEK and MKK isoforms. Science 267:682–685
23. Widmann C, Gibson S, Jarpe MB, Johnson GL (1999) Mitogen-activated protein kinase: conservation of a three-kinase module from yeast to human. Physiol Rev 79:143–180
24. Keyse SM (2000) Protein phosphatases and the regulation of mitogen-activated protein kinase signalling. Curr Opin Cell Biol 12:186–192
25. Karlsson M, Mandl M, Keyse SM (2006) Spatio-temporal regulation of mitogen-activated protein kinase (MAPK) signalling by protein phosphatases. Biochem Soc Trans 34:842–845
26. Boutros T, Chevet E, Metrakos P (2008) Mitogen-activated protein (MAP) kinase/MAP kinase phosphatase regulation: roles in cell growth, death, and cancer. Pharmacol Rev 60:261–310
27. Patterson KI, Brummer T, O'Brien PM, Daly RJ (2009) Dual-specificity phosphatases: critical regulators with diverse cellular targets. Biochem J 418:475–489
28. Dickinson RJ, Keyse SM (2006) Diverse physiological functions for dual-specificity MAP kinase phosphatases. J Cell Sci 119:4607–4615
29. Tanoue T, Adachi M, Moriguchi T, Nishida E (2000) A conserved docking motif in MAP kinases common to substrates, activators and regulators. Nat Cell Biol 2:110–116

30. Camps M, Nichols A, Gillieron C, Antonsson B, Muda M, Chabert C, Boschert U, Arkinstall S (1998) Catalytic activation of the phosphatase MKP-3 by ERK2 mitogen-activated protein kinase. Science 280:1262–1264
31. Wu JJ, Zhang L, Bennett AM (2005) The noncatalytic amino terminus of mitogen-activated protein kinase phosphatase 1 directs nuclear targeting and serum response element transcriptional regulation. Mol Cell Biol 25:4792–4803
32. Mandl M, Slack DN, Keyse SM (2005) Specific inactivation and nuclear anchoring of extracellular signal-regulated kinase 2 by the inducible dual-specificity protein phosphatase DUSP5. Mol Cell Biol 25:1830–1845
33. Karlsson M, Mathers J, Dickinson RJ, Mandl M, Keyse SM (2004) Both nuclear-cytoplasmic shuttling of the dual specificity phosphatase MKP-3 and its ability to anchor MAP kinase in the cytoplasm are mediated by a conserved nuclear export signal. J Biol Chem 279:41882–41891
34. Camps M, Nichols A, Arkinstall S (2000) Dual specificity phosphatases: a gene family for control of MAP kinase function. FASEB J 14:6–16
35. Jeong DG, Jung SK, Yoon TS, Woo EJ, Kim JH, Park BC, Ryu SE, Kim SJ (2009) Crystal structure of the catalytic domain of human MKP-2 reveals a 24-mer assembly. Proteins 76:763–767
36. Tao X, Tong L (2007) Crystal structure of the MAP kinase binding domain and the catalytic domain of human MKP5. Protein Sci 16:880–886
37. Jeong DG, Yoon TS, Kim JH, Shim MY, Jung SK, Son JH, Ryu SE, Kim SJ (2006) Crystal structure of the catalytic domain of human MAP kinase phosphatase 5: structural insight into constitutively active phosphatase. J Mol Biol 360:946–955
38. Farooq A, Zhou MM (2004) Structure and regulation of MAPK phosphatases. Cell Signal 16:769–779
39. Farooq A, Plotnikova O, Chaturvedi G, Yan S, Zeng L, Zhang Q, Zhou MM (2003) Solution structure of the MAPK phosphatase PAC-1 catalytic domain. Insights into substrate-induced enzymatic activation of MKP. Structure 11:155–164
40. Farooq A, Chaturvedi G, Mujtaba S, Plotnikova O, Zeng L, Dhalluin C, Ashton R, Zhou MM (2001) Solution structure of ERK2 binding domain of MAPK phosphatase MKP-3: structural insights into MKP-3 activation by ERK2. Mol Cell 7:387–399
41. Franklin CC, Kraft AS (1997) Conditional expression of the mitogen-activated protein kinase (MAPK) phosphatase MKP-1 preferentially inhibits p38 MAPK and stress-activated protein kinase in U397 cells. J Biol Chem 272:16917–16923
42. Zhang YY, Wu JW, Wang ZX (2011) A distinct interaction mode revealed by the crystal structure of the kinase p38alpha with the MAPK binding domain of the phosphatase MKP5. Sci Signal 4:ra88
43. Wu Z, Jiao P, Huang X, Feng B, Feng Y, Yang S, Hwang P, Du J, Nie Y, Xiao G, Xu H (2010) MAPK phosphatase-3 promotes hepatic gluconeogenesis through dephosphorylation of forkhead box O1 in mice. J Clin Invest 120:3901–3911
44. Jiao P, Feng B, Xu H (2012) Mapping MKP-3/FOXO1 interaction and evaluating the effect on gluconeogenesis. PLoS One 7:e41168
45. Wu JJ, Roth RJ, Anderson EJ, Hong E-G, Lee M-K, Choi CS, Neufer PD, Shulman GI, Kim JK, Bennett AM (2006) Mice lacking MAP kinase phosphatase-1 have enhanced MAP kinase activity and resistance to diet-induced obesity. Cell Metab 4:61–73
46. Soulsby M, Bennett AM (2009) Physiological signaling specificity by protein tyrosine phosphatases. Physiology (Bethesda) 24:281–289
47. Dorfman K, Carrasco D, Gruda M, Ryan C, Lira SA, Bravo R (1996) Disruption of the *erp/mkp-1* gene does not affect mouse development: normal MAP kinase activity in ERP/MKP-1-deficient fibroblasts. Oncogene 13:925–931
48. Roth RJ, Le AM, Zhang L, Kahn M, Samuel VT, Shulman GI, Bennett AM (2009) MAPK phosphatase-1 facilitates the loss of oxidative myofibers associated with obesity in mice. J Clin Invest 119:3817–3829
49. Tanoue T, Moriguchi T, Nishida E (1999) Molecular cloning and characterization of a novel dual specificity phosphatase, MKP-5. J Biol Chem 274:19949–19956

50. Aguirre V, Uchida T, Yenush L, Davis R, White MF (2000) The c-Jun NH2-terminal kinase promotes insulin resistance during association with insulin receptor substrate-1 and phosphorylation of Ser307. J Biol Chem 275:9047–9054
51. Liu Z, Cao W (2009) p38 mitogen-activated protein kinase: a critical node linking insulin resistance and cardiovascular diseases in type 2 diabetes mellitus. Endocr Metab Immune Disord Drug Targets 9:38–46
52. Flach RJ, Qin H, Zhang L, Bennett AM (2011) Loss of mitogen-activated protein kinase phosphatase-1 protects from hepatic steatosis by repression of cell death-inducing DNA fragmentation factor A (DFFA)-like effector C (CIDEC)/fat-specific protein 27. J Biol Chem 286: 22195–22202
53. Reddy ST, Nguyen JT, Grijalva V, Hough G, Hama S, Navab M, Fogelman AM (2004) Potential role for mitogen-activated protein kinase phosphatase-1 in the development of atherosclerotic lesions in mouse models. Arterioscler Thromb Vasc Biol 24:1676–1681
54. Lehrke M, Lazar MA (2005) The many faces of PPARgamma. Cell 123:993–999
55. Gavrilova O, Haluzik M, Matsusue K, Cutson JJ, Johnson L, Dietz KR, Nicol CJ, Vinson C, Gonzalez FJ, Reitman ML (2003) Liver peroxisome proliferator-activated receptor gamma contributes to hepatic steatosis, triglyceride clearance, and regulation of body fat mass. J Biol Chem 278:34268–34276
56. Matsusue K, Haluzik M, Lambert G, Yim SH, Gavrilova O, Ward JM, Brewer B Jr, Reitman ML, Gonzalez FJ (2003) Liver-specific disruption of PPARgamma in leptin-deficient mice improves fatty liver but aggravates diabetic phenotypes. J Clin Invest 111:737–747
57. Matsusue K, Kusakabe T, Noguchi T, Takiguchi S, Suzuki T, Yamano S, Gonzalez FJ (2008) Hepatic steatosis in leptin-deficient mice is promoted by the PPARgamma target gene Fsp27. Cell Metab 7:302–311
58. Yin R, Dong YG, Li HL (2006) PPARgamma phosphorylation mediated by JNK MAPK: a potential role in macrophage-derived foam cell formation. Acta Pharmacol Sin 27:1146–1152
59. Traini M, Jessup W (2009) Lipid droplets and adipose metabolism: a novel role for FSP27/CIDEC. Curr Opin Lipidol 20:147–149
60. Hu E, Kim J, Sarraf P, Spiegelman B (1996) Inhibition of adipogenesis through MAP kinase-mediated phosphorylation of PPARγ. Science 274:2100–2103
61. Feng B, Jiao P, Yang Z, Xu H (2012) MEK/ERK pathway mediates insulin-promoted degradation of MKP-3 protein in liver cells. Mol Cell Endocrinol 361:116–123
62. Asada S, Daitoku H, Matsuzaki H, Saito T, Sudo T, Mukai H, Iwashita S, Kako K, Kishi T, Kasuya Y, Fukamizu A (2007) Mitogen-activated protein kinases, Erk and p38, phosphorylate and regulate Foxo1. Cell Signal 19:519–527
63. Misra-Press A, Rim CS, Yao H, Roberson MS, Stork PJS (1995) A novel mitogen-activated protein kinase phosphatase. J Biol Chem 270:14587–14596
64. Berasi SP, Huard C, Li D, Shih HH, Sun Y, Zhong W, Paulsen JE, Brown EL, Gimeno RE, Martinez RV (2006) Inhibition of gluconeogenesis through transcriptional activation of EGR1 and DUSP4 by AMP-activated kinase. J Biol Chem 281:27167–27177
65. Cahova M, Vavrinkova H, Kazdova L (2007) Glucose-fatty acid interaction in skeletal muscle and adipose tissue in insulin resistance. Physiol Res 56:1–15
66. Sabio G, Davis RJ (2010) cJUN NH2-terminal kinase 1 (JNK1): roles in metabolic regulation of insulin resistance. Trends Biochem Sci 35:490–496
67. Cao W, Collins QF, Becker TC, Robidoux J, Lupo EG Jr, Xiong Y, Daniel KW, Floering L, Collins S (2005) p38 mitogen-activated protein kinase plays a stimulatory role in hepatic gluconeogenesis. J Biol Chem 280:42731–42737
68. Zierath JR, Hawley JA (2004) Skeletal muscle fiber type: influence on contractile and metabolic properties. PLoS Biol 2:e348
69. Kadowaki T, Yamauchi T (2005) Adiponectin and adiponectin receptors. Endocr Rev 26:439–451
70. Qiao L, Kinney B, Yoo HS, Lee B, Schaack J, Shao J (2012) Adiponectin increases skeletal muscle mitochondrial biogenesis by suppressing mitogen-activated protein kinase phosphatase-1. Diabetes 61:1463–1470

71. Sabio G, Das M, Mora A, Zhang Z, Jun JY, Ko HJ, Barrett T, Kim JK, Davis RJ (2008) A stress signaling pathway in adipose tissue regulates hepatic insulin resistance. Science 322:1539–1543
72. Sakaue H, Ogawa W, Nakamura T, Mori T, Nakamura K, Kasuga M (2004) Role of MAPK phosphatase-1 (MKP-1) in adipocyte differentiation. J Biol Chem 279:39951–39957
73. Lawan A, Shi H, Gatzke F, Bennett AM (2012) Diversity and specificity of the mitogen-activated protein kinase phosphatase-1 functions. Cell Mol Life Sci 70(2):223–237
74. Ito A, Suganami T, Miyamoto Y, Yoshimasa Y, Takeya M, Kamei Y, Ogawa Y (2007) Role of MAPK phosphatase-1 in the induction of monocyte chemoattractant protein-1 during the course of adipocyte hypertrophy. J Biol Chem 282:25445–25452
75. Hong SB, Lubben TH, Dolliver CM, Petrolonis AJ, Roy RA, Li Z, Parsons TF, Li P, Xu H, Reilly RM, Trevillyan JM, Nichols AJ, Tummino PJ, Gant TG (2005) Expression, purification, and enzymatic characterization of the dual specificity mitogen-activated protein kinase phosphatase, MKP-4. Bioorg Chem 33:34–44
76. Xu H, Dembski M, Yang Q, Yang D, Moriarty A, Tayber O, Chen H, Kapeller R, Tartaglia LA (2003) Dual specificity mitogen-activated protein (MAP) kinase phosphatase-4 plays a potential role in insulin resistance. J Biol Chem 278:30187–30192
77. Emanuelli B, Eberle D, Suzuki R, Kahn CR (2008) Overexpression of the dual-specificity phosphatase MKP-4/DUSP-9 protects against stress-induced insulin resistance. Proc Natl Acad Sci U S A 105:3545–3550
78. Dankel SN, Fadnes DJ, Stavrum AK, Stansberg C, Holdhus R, Hoang T, Veum VL, Christensen BJ, Vage V, Sagen JV, Steen VM, Mellgren G (2010) Switch from stress response to homeobox transcription factors in adipose tissue after profound fat loss. PLoS One 5:e11033

Chapter 13
Glycogen Metabolism and Lafora Disease

Peter J. Roach and Anna A. DePaoli-Roach

Abstract Lafora disease is a juvenile-onset, fatal epilepsy that is characterized by the formation of Lafora bodies in many tissues, including skeletal muscle, heart, and neurons. Lafora bodies are insoluble deposits that contain polyglucosan, a poorly branched form of glycogen, and associated proteins. Evidence is mounting that Lafora bodies either cause or contribute to the pathology of the disease. It is a genetic disease caused by mutation in one of two genes, *EPM2A* and *EPM2B*, which encode, respectively, a phosphatase called laforin and an E3 ubiquitin ligase called malin. Laforin is a phosphatase of the atypical dual specificity phosphatase subfamily that, in vitro and in vivo, removes phosphate monoesters from glycogen. Normal glycogen contains trace amounts of phosphate introduced as a minor side reaction by the synthetic enzyme, glycogen synthase. In laforin knockout mice, glycogen becomes hyperphosphorylated and, as the mice age, acquires structural abnormalities and the tendency to come out of solution, consistent with Lafora body formation. Mutation of malin or laforin results in similar symptoms in patients and malin and laforin knockout mice exhibit phenotypic similarity in terms of neurological defects and abnormal glycogen metabolism, including hyperphosphorylation and Lafora body formation.

Abbreviations

AGL	Glycogen debranching enzyme and gene
AMPK	AMP-activated protein kinase
CBM	Carbohydrate-binding module

P.J. Roach, Ph.D. (✉) • A.A. DePaoli-Roach, Ph.D.
Department of Biochemistry and Molecular Biology, Indiana University School of Medicine, Indianapolis, IN 46202, USA
e-mail: proach@iu.edu

K.K. Bence (ed.), *Protein Tyrosine Phosphatase Control of Metabolism*,
DOI 10.1007/978-1-4614-7855-3_13,

GAA	Lysosomal α-glucosidase and gene
GABARAP	Gamma aminobutyric acid receptor-associated protein
GABARAPL1	GABARAP-like 1
GL	Product of *PPP1R3B* gene
GSD	Glycogen storage disease
GSK-3	Glycogen synthase kinase 3
mTOR	Mammalian target of rapamycin
PAS	Periodic acid-Schiff
PTEN	Phosphatase and tensin homolog
PTG	Protein targeting-to-glycogen product of the *PPP1R3C* gene
R6	Product of the *PPP1R3D* gene
SEX4	Starch excess 4
Stbd1	Starch-binding domain-containing protein 1 also called genethonin 1

Introduction

Tyrosine phosphorylation of proteins is intimately linked to cell signaling pathways and thus it is not surprising that protein tyrosine phosphatases (PTPs) have an important impact on metabolism, as described in other chapters of this monograph. Many receptors and downstream signaling proteins are dephosphorylated by PTPs. Even a phosphatase like PTEN, which has small molecule substrates, appears to function by modulating signaling [1]. In contrast, the phosphatase featured in this chapter, laforin, appears to have a more direct linkage to metabolism: the highly specialized metabolism of the phosphate in the carbohydrate storage reserve, glycogen.

Lafora disease, first described a century ago by Dr. Gonzago Lafora [2], is a rare, autosomal recessive, genetic form of epilepsy that is usually diagnosed in the second decade of life following an acute epileptic incident. It is a progressive myoclonus epilepsy for which there is no cure and no effective treatment; afflicted individuals undergo continued neurological deterioration, normally succumbing within 10 years of onset [3–7]. It is a tragic disease claiming the lives of young people just as they should start the trajectory to adulthood. A consistent feature of Lafora disease is the presence of Lafora bodies, described initially by Lafora himself. These are deposits identified microscopically in many tissues, including neurons, muscle, heart, and skin [2, 8]. The initial assumption was that the Lafora bodies, presumably in neurons in the brain, caused cell death and hence the disease. Lafora bodies characteristically contain polyglucosan, a term used to describe a glycogen-like polymer that is less branched and less soluble than glycogen, resembling more the amylopectin of plant starch [9, 10]. Recent reviews have addressed the function of laforin [11] and malin, the other main Lafora disease protein [12]. The focus of this chapter will be on the role of glycogen and abnormal glycogen metabolism in Lafora disease.

Glycogen Metabolism

Glycogen, like amylopectin, is a branched polymer of glucose that serves as a storage form of glucose in many species and in various tissues of higher eukaryotes (see [13] for a recent review). The main deposits of mammalian glycogen, in absolute amount, are in liver and skeletal muscle but glycogen accumulates also in heart, kidney, and brain, and many cell types have the machinery to synthesize glycogen. De novo glycogen synthesis begins with a specialized initiation step mediated by the self-glucosylating protein glycogenin [14–16]. Glycogen synthase then catalyzes formation of the main polymerizing α-1,4-glycosidic linkages and the branchpoints are introduced as α-1,6-glycosidic linkages created by the branching enzyme ([13]; see Fig. 13.1). Two processes lead to glycogen breakdown. In one pathway, glycogen is degraded in the cytosol by the actions of glycogen phosphorylase, which catalyzes the phosphorolysis of the nonreducing ends of the outer chains to release glucose-1-P, and the debranching enzyme (AGL) which is ultimately responsible for the hydrolysis of the α-1,6-glycosidic linkages to yield glucose. In the second pathway, glycogen is hydrolyzed to glucose in lysosomes by an α-glucosidase

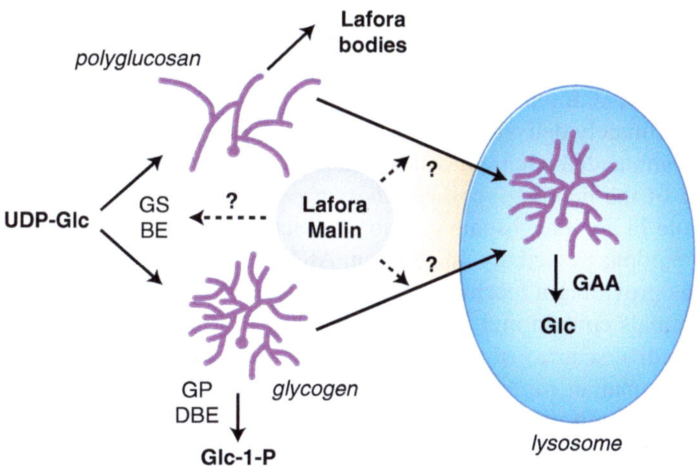

Fig. 13.1 Glycogen metabolism. The *lower portion* of the figure shows normal glycogen metabolism. The bulk synthesis of glycogen is mediated by glycogen synthase (GS) and the branching enzyme (BE). Cytosolic degradation of glycogen is catalyzed by the actions of glycogen phosphorylase (GP) and the debranching enzyme (DBE). Glycogen can also be degraded by transfer to the lysosome where an α-glucosidase (GAA) breaks it down to glucose. The *upper portion* of the figure depicts the anomalous glycogen metabolism associated with the formation of polyglucosan, poorly branched glycogen, and ultimately insoluble Lafora bodies. Malin and laforin, whose defects are implicated in Lafora disease, affect glycogen metabolism by various potential mechanisms discussed in this chapter. Possible links to glycogen metabolism include involvement in cytosolic biosynthesis, in cytosolic degradation, and in the lysosomal disposal pathway. Modified from Arch Biochem Biophys, 457, Wang et al., Glycogen metabolism in tissues from a mouse model of Lafora disease, pp. 264–9, Copyright (2007), with permission from Elsevier

(GAA, acid maltase). The importance of the lysosomal pathway is emphasized by the severity of the most extreme forms of Pompe disease (GSD II) in which the GAA gene is mutated [17]. In the classic form of infantile onset Pompe disease, death usually occurs within the first year primarily due to cardiac dysfunction.

The mechanism by which glycogen is delivered to the lysosome is not understood in detail, but it is likely that autophagy or an autophagy-like process is involved. Autophagy describes a family of pathways by which intracellular material is engulfed by membranes to form vesicles or autophagosomes that are transferred to the lysosome, in response to a variety of signals, including nutritional deprivation [18–22]. Evidence supporting involvement of an autophagic process in vesicular trafficking of glycogen comes from the work of Raben and Plotz [23] on a mouse model of Pompe disease in which, besides the expected over-accumulation of glycogen in lysosomes of skeletal muscle, there was a massive increase in glycogen-laden autophagosomes termed "autophagic build-up." In its early conception, autophagy was considered to be mainly a mechanism for the random recycling of cellular components, allowing for retrieval of building block molecules like amino acids and providing additional sources for metabolic energy production. Furthermore, autophagy would be a means to dispose of damaged molecules or organelles. The current, and still evolving, view of autophagy is more complex and, while not discounting a role for the random, housekeeping disposal of cellular content, also embraces the existence of more specific processes directed at particular classes of molecules or organelles, processes that can be subject to individual regulation [24, 25]. Formation of autophagosomes, the vesicles surrounded by a double membrane that will ultimately deliver their contents to the lysosome, is a complicated process that has commanded intense investigation [18–22]. Many advances were made from the study of the budding yeast *Saccharomyces cerevisiae*. In yeast, Atg8 plays a central role in autophagosome creation and its lipidation is often monitored as an index of autophagic activity. In addition, Atg8 interacts with cargo-specifying proteins, thus providing a mechanism for more specialized autophagic pathways [24, 25]. Mammals contain multiple Atg8-related genes [26, 27], potentially allowing for a more diverse array of autophagic pathways as compared with yeast. The best-studied mammalian Atg8-family member is LC3.

We recently reported on a protein, starch-binding domain-containing protein 1 (Stbd1 or genethonin 1) [28] that consists of a highly conserved N-terminal hydrophobic segment and a C-terminal carbohydrate-binding module (CBM) of the CBM20 family [29], which is shared by laforin and the β-subunit of AMP-activated protein kinase (AMPK). Stbd1 binds glycogen and amylopectin in vitro and, when over-expressed in cultured cells, co-localizes with glycogen in large perinuclear structures. Deletion of the N-terminus eliminates the perinuclear localization and disabling the CBM20 eliminates the appearance of glycogen in the perinuclear structures [28]. The hypothesis is that Stbd1 may be involved in anchoring glycogen to membranes and may in some way participate in the trafficking of glycogen to the lysosome. Furthermore, Stbd1 interacts with GABARAP and GABARAPL1, two members of the ATG8/LC3 family of autophagy proteins that might also contribute to the process. A specific Atg8 interacting motif (AIM) was identified in Stbd1 [30]. Stbd1 binds more tightly to amylopectin or glycogen purified from laforin knockout mice, both poorly branched and phosphorylated (see below), than it binds to normal glycogen. One possibility, therefore, is that Stbd1 serves to preferentially aid disposal of abnormally structured glycogen as a kind of quality control. Whether this potentially glycogen-specific pathway, dubbed glycophagy [30], operates in vivo will require further exploration. Even if this mechanism does occur, it does not preclude the existence of other autophagic pathways by which glycogen is transferred to lysosomes in response to other stimuli, such as starvation for example.

Diseases of Glycogen Metabolism

Several inheritable glycogen storage diseases (GSDs) or glycogenoses have been identified [31–33]. Some arise from direct genetic defects in the enzymes of glycogen metabolism whereas others are indirect, due to perturbations of intermediary metabolism in which abnormal glycogen metabolism is one of the sequelae. In a genome-wide survey of the *S. cerevisiae* deletion series, 12.4 % of 4,600 strains analyzed had altered glycogen accumulation [34] underscoring how responsive glycogen stores can be to functionally distal processes. Though coverage of the mouse genome is much more limited, a surprising number of genetically modified mice have altered glycogen accumulation even when none of the glycogen metabolizing enzymes are targeted. Some human GSDs, like Lafora disease, result in the formation of poorly branched glycogen. For example, GSD IV, which includes Andersen disease and Adult Polyglucosan Body Disease, is caused by mutations in the branching enzyme and, although the disease presents with a largely unexplained diversity in terms of severity and tissue involvement, the predictable formation of poorly branched glycogen or polyglucosan is a common feature [35, 36]. In GSD VII, Tarui disease, a genetic defect in phosphofructokinase [32, 37] blocks glycolysis and leads to a back-up of glycolytic intermediates, including glucose-6-P, a potent activator of glycogen synthase. Activation of glycogen synthase together with the mass action effect of increased levels of precursors could then account for both over-accumulation of glycogen as well as the formation of polyglucosan. In both of these human disorders, the altered branching pattern of the abnormal glycogen can be explained by an imbalance between glycogen elongation via glycogen synthase and branching mediated by the branching enzyme. A similar scenario is observed in polysaccharide storage myopathy (PSSM), an often serious genetic condition in horses, which is also linked to the formation of polyglucosan [38]. A gain-of-function mutation in glycogen synthase was identified as a frequent cause of the disorder, providing yet another mechanism for disturbing the balance between elongation and branching during glycogen synthesis. A related example comes from the study of transgenic mice engineered to over-express a hyperactive mutated form of glycogen synthase in skeletal muscle [39]. One line of these mice, GSL30, has massive over-accumulation of muscle glycogen that is less branched [40] and which histochemically can resemble Lafora bodies [41]. The mice displayed no obvious muscle dysfunction and even had normal capacity for exhaustive exercise [42]. From these studies of polyglucosan formation, two important points emerge: (1) the pathology associated with polyglucosan accumulation is quite variable, suggesting that not all polyglucosan is equal, and (2) it is clear that an imbalance between elongation and branching can result in poorly branched glycogen.

Mechanistic Basis of Lafora Disease

The modern era of Lafora research began in the late 1990s when the first causative gene, *EPM2A*, was identified [43, 44]. Not long thereafter, a second gene *EPM2B (also called NHLRC1)* was found [45] and most Lafora cases can be attributed to mutations in these two genes [3–7]. The genes encode, respectively, laforin and malin. Laforin, based on primary structure, belongs to the dual specificity protein phosphatase (DSP) family and also contains a predicted CBM20 CBM [43]. Malin, based on its sequence, contains a ring finger domain characteristic of E3 ubiquitin ligases [45]. In Lafora patients, the presentation and symptoms are very similar whether mutation occurs in *EPM2A* or *EPM2B*. A recent neurological and behavioral study indicated very similar, though not identical, results for *Epm2a*$^{-/-}$ and *Epm2b*$^{-/-}$ mice, which in general mimicked what is seen in patients [46]. However, there has been a suggestion that malin mutations may lead to a later onset and slower progression of the disease [47].

Not surprisingly, a primary goal of Lafora research since identification of the two main causative genes has been to define the functions of laforin and malin with the hope of understanding the molecular basis for the disease and how *EPM2A* or *EPM2B* mutations lead to such similar phenotypes. Knowledge of function might also suggest possible sites of therapeutic intervention. Because of the association of Lafora bodies with the disease, one obvious line of investigation has been to assess whether laforin and malin are involved in glycogen metabolism. Possible sites of action, as laid out in a 2007 paper [48], were (1) cytosolic glycogen metabolism, notably the biosynthetic pathway and (2) the transfer of glycogen and/or polyglucosan to the lysosome (Fig. 13.1). Does laforin and/or malin affect the ratio of elongating to branching activity during glycogen synthesis, the "imbalance hypothesis" as discussed above? Or do these proteins influence how glycogen is trafficked to the lysosome by autophagy or some autophagy-related pathway? Or is some other aspect of glycogen metabolism affected?

Laforin

Laforin is unique among mammalian phosphatases in harboring a CBM of the CBM20 subfamily in the same polypeptide as the phosphatase domain; all other protein phosphatases that associate with glycogen, such as the glycogen-associated type 1 phosphatases (PP1G), appear to achieve polysaccharide binding via a separate glycogen-binding subunit ([49]; Fig. 13.2). There is evidence from in vitro [50, 51]

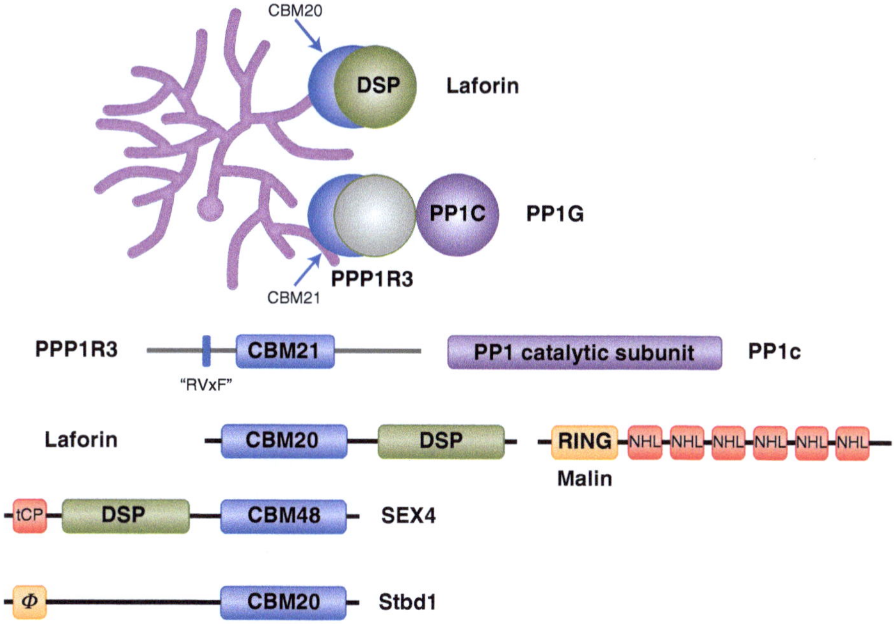

Fig. 13.2 Recruitment of phosphatase activity to glycogen. The figure illustrates different molecular strategies for the recruitment of phosphatase activity to a glycogen particle. The glycogen-associated type 1 phosphatases (PP1G) utilize a glycogen targeting subunit, PPP1R3, of which seven are known to bind to glycogen via a carbohydrate-binding module (CBM) of the CBM21 subfamily and to bind to the phosphatase catalytic subunit, PP1c, via interactions that include an "RVxF" motif. In laforin, the CBM, in this case of the CBM20 subfamily, is housed in the same polypeptide as the dual specificity phosphatase domain (DSP). The plant phosphatase SEX4 uses a similar strategy for its association with amylopectin. Shown also is Stbd1, a glycogen-binding protein that has a similar CBM as laforin that may locate glycogen close to membranes via its N-terminal hydrophobic region (ϕ). Malin, the other Lafora disease protein, may also be recruited to glycogen by interaction with malin. *tCP* chloroplast targeting peptide

and cellular [50] experiments that laforin does bind to glycogen and interacts more strongly with poorly branched polysaccharides such as amylopectin [51–53]. In subcellular fractionation of glycogen, laforin tracks with the polysaccharide. Disease-related mutations of laforin include many that are clearly loss-of-function mutations that lead to truncations or loss of phosphatase activity [44, 54]. However, some are in the CBM20 domain and one, W32G, has been shown biochemically to disable glycogen binding [50, 51, 53].

The other conserved domain in laforin is predicted to be a dual specificity protein phosphatase, more specifically one in the atypical DSP subfamily [55]. Indeed, initial study of laforin established that in vitro it could dephosphorylate Ser, Thr, and Tyr phosphates in model protein substrates [56]. Searches for physiological protein substrates of laforin, however, have not been successful. A potential protein substrate, GSK-3, has emerged [57, 58], but this has proven controversial. The hypothesis is that dephosphorylation of the inhibitory N-terminal site in GSK-3 would activate GSK-3, thereby leading to increased phosphorylation and decreased activity of glycogen synthase. Lack of laforin could thus inappropriately activate glycogen synthase and, via the imbalance hypothesis, result in poorly branched glycogen. Several studies, however, found no evidence for dephosphorylation of GSK-3 (or glycogen synthase) by laforin, whether in vitro, in cultured cell systems, or in *Epm2a*$^{-/-}$ mice [48, 59, 60].

Laforin as a Glycogen Phosphatase

A rather different role for laforin was suggested by the study of Worby et al. [59] who showed that laforin, in vitro, could remove phosphate esterified to amylopectin. Both amylopectin [61] and glycogen [60, 62, 63] have been reported to contain small amounts of phosphate esters. In amylopectin, the phosphate is present as C3 and C6 phosphomonoesters of glucose [61] and in mammalian glycogen as C2 and C3 phosphomonoesters of glucose [64]. Tagliabracci et al. [60] went on to demonstrate that laforin could also dephosphorylate glycogen. The W32G disease mutation in the CBM20 domain, noted above, eliminated the ability of laforin to dephosphorylate glycogen without abolishing activity towards *p-nitrophenylphosphate*, PNPP, a frequently used generic phosphatase substrate. Since the W32G mutation causes Lafora disease, one could postulate that glycogen dephosphorylation is important in vivo. Further evidence that the glycogen phosphatase activity of laforin is physiologically relevant came from the observation that glycogen isolated from *Epm2a* knockout mice had elevated levels of covalent phosphate [60, 65]. Thus, the hypothesis emerged that the function of laforin might be to dephosphorylate glycogen. Interestingly, other atypical DSPs, such as PTEN and myotubularins, also have non-proteinaceous substrates [1]. In the plant kingdom, the SEX4 phosphatase, another atypical DSP family member, is responsible for the dephosphorylation of amylopectin [66, 67]. Like laforin, it contains a CBM domain, in this case of the CBM48 family and located at the C-terminus (Fig. 13.2).

Analysis of glycogen purified from young, 3-month-old *Epm2a* knockout mice revealed that, while covalent phosphate increased ~4-fold, it was associated with only modest changes in glycogen properties, including solubility, degree of branching, and appearance when analyzed by electron microscopy [65]. Mice of this age are beginning to show the appearance of Lafora bodies in various tissues, including muscle. When similar analyses were applied to glycogen from 9- to 12-month-old mice, major differences were noted. First, glycogen phosphorylation was further increased to ~6-fold and total glycogen levels were about threefold greater than in wild-type littermates. The abundance of Lafora bodies was also greatly increased and their polyglucosan likely contributed much more to the bulk properties of the purified glycogen. The typical rosette-like appearance of glycogen particles was not evident and larger, amorphous conglomerates were visible. The polysaccharide was much less soluble in water and its branching frequency was significantly reduced. This chemical progression parallels the development of Lafora bodies and the disease symptoms.

A question that follows is how the small amount of phosphate in glycogen, one phosphate per 1,500 glucose residues in mouse skeletal muscle, could lead to such a major disruption in overall structure. The structure of glycogen is not amenable to traditional three-dimensional determination because of its polydisperse nature. However, there is considerable evidence that polyglucose forms helices in both amylopectin and glycogen. The structure of a 26-residue cycloamylose has been solved by X-ray crystallography [68], and it shows two anti-parallel polyglucose helices. Although this structure is unnaturally constrained by virtue of being cyclic, it likely has many features of polyglucose helices in polysaccharides and in particular indicates a rich network of intra-helix hydrogen bonds. It is easy to see how the introduction of phosphate at C2 or C3 on glucose would destabilize and disrupt the helix. In this way, one can rationalize how a relatively small number of extra phosphates could have wider impact on the structure and properties of a glycogen molecule, and how increased phosphate content could underlie the pathology of Lafora disease.

Less clear is the reason for the decrease in branching frequency associated with the polyglucosan of Lafora bodies. In the laforin knockout mice, a significant proportion of glycogen and glycogen synthase is associated with the pellet obtained by low speed centrifugation and the overall level of glycogen synthase protein is significantly increased in the older mice [65]. Presumably malin is present in these mice (see discussion of malin function below). One might therefore invoke the imbalance hypothesis discussed above. Inconsistent with this idea, though, is the fact that the glycogen synthase associated with glycogen in the low speed pellet from knockout animals had low enzymatic activity, even when measured in the presence of its activator glucose-6-P. The branching enzyme levels were not substantially altered by the absence of laforin so that there was no shift towards increased elongating vs. branching activity. We therefore do not believe that formation of polyglucosan associated with the lack of laforin activity can be attributed to the imbalance hypothesis. The reason for decreased branching remains unresolved. One possibility is that the disruption of glycogen structure elicited by the covalent attachment of phosphate, whether at a local or a global level, has an impact on the degradation mediated by phosphorylase and/or debranching enzyme, resulting in longer average chain lengths. Another possibility is that lysosomal disposal of poorly branched glycogen is impaired, either by overwhelming the capacity of the pathway or because of a positive role for laforin in that process.

Origin and Function of Glycogen Phosphate: Is Laforin a "Damage Control" Enzyme?

In plants, phosphate in amylopectin is introduced through the action of two enzymes, glucan water dikinase (GWD) and phosphoglucan water dikinase (PWD), and has its origin as the β-phosphate of the ATP substrate [69]. The role of amylopectin phosphorylation seems to be disruption of the semi-crystalline, helical regions of the polysaccharide so that degradative enzymes can access cleavage sites. The plant enzyme SEX4 [66, 67], which like laforin contains a DSP domain and a CBM, would then dephosphorylate the resulting phospho-oligosaccharides [10, 70]. Bioinformatic analyses of mammalian genomes do not identify any enzymes comparable to the plant GWD and PWD nor do simple biochemical assays reveal such activities. The best evidence to date for the origin of the phosphate in mammalian glycogen comes from the work of Tagliabracci et al. [64] who demonstrated that the biosynthetic enzyme glycogen synthase can infrequently, once in every ~10,000 catalytic cycles, transfer the β-phosphate of its substrate UDP-glucose into glycogen. In other words, rather than attaching a glucose moiety to the nonreducing end of a growing chain, glycogen synthase occasionally adds a glucose phosphate. A mechanism was proposed whereby the reaction could progress through the formation of a C1-C2 or a C1-C3 cyclic phosphodiester of glucose that results in the addition of a C2- or C3-glucose phosphate. Whereas a reasonable evolved purpose can be assigned to starch phosphorylation, the biological rationale for glycogen phosphorylation is not obvious. Glycogen phosphorylation occurs in nature and in mouse muscle remains at a constant basal level between 3 and 12 months of age [65]. One might argue that glycogen phosphate might have some as yet unappreciated biological function. However, if the origin of the phosphate is indeed from the action of the normal biosynthetic enzyme glycogen synthase, glycogen phosphorylation may be classifiable as a catalytic error [71]. It is not uncommon, even among enzymes that have evolved to mediate very specific reactions, for there to be minor side reactions. In many instances, the generation of trace amounts of a soluble side product will be innocuous. In the case of longer-lived molecules, like polymers, the results of such side reactions can be more persistent. For example, glycogen molecules can undergo multiple cycles of expansion and contraction in response to metabolic conditions, and any blemishes in structure can be perpetuated in the absence of a repair mechanism. Perhaps the best example of the significance of errors in polymers is that of DNA synthesis where erroneous copying of a DNA template could potentially have enduring and devastating consequences. Therefore, elaborate and evolutionarily expensive mechanisms have evolved to improve the fidelity of DNA synthesis and to enable its repair. Viewed in this way, laforin might be considered a "repair" or "damage control" enzyme, whose function is to curtail glycogen phosphorylation and maintain it within tolerable levels that are not excessively disruptive of glycogen structure.

Is Lafora Disease a Glycogenosis?

Lafora disease is characterized by the accumulation of Lafora bodies in multiple tissues, including the brain where their disruption of nerve function could provide a tangible connection with the neurological aspects of the disease. Still, the question can be asked whether abnormal glycogen and/or Lafora bodies actually cause Lafora disease or are simply a non-pathological consequence of the genetic defects [4, 72]. Some arguments center around the timing and location of the development of Lafora bodies as compared to the onset of symptoms. For example, in the initial characterization of *Epm2a*$^{-/-}$ mice [73], it was reported that neuron degeneration was detected around 2 months of age when Lafora bodies just began to appear. Basically, one idea is that genetic defects in laforin or malin could trigger some primary neural dysfunction and that the aberrant glycogen metabolism is secondary and not causative of the disease. Several considerations, however, link the disease to glycogen.

First, as for any genetic disease, it is necessary to view the molecular defect(s) in relation to the onset of the disease. In traditional glycogenoses, including Andersen and Pompe diseases, gross impairments in basic glycogen metabolism exist from an early developmental stage and yet, even when there is early lethality, the over-accumulation of glycogen can long precede organ damage and malfunction. In Lafora disease, the metabolic deficiency is more subtle and in mice the time frame for detection of any abnormal glycogen is measured in months. Substantially elevated glycogen phosphorylation in mouse muscle is apparent at 3 months, well before clear changes in glycogen total level or branching have developed, even though Lafora bodies have begun to appear. The relationship between the abundance of Lafora bodies in tissues and biochemically detectable differences in glycogen in *Epm2a*$^{-/-}$ mice is complex. Recall that a Lafora body, as defined histochemically, even when small, still represents a conglomerate of individual glycogen molecules and hence a relatively late stage in the derangement of glycogen metabolism (Fig. 13.3). Another point is that the formation of abnormal glycogen molecules almost certainly has a stochastic element so that the bulk properties of tissue glycogen need not reflect the anomalies possible in individual molecules resulting in Lafora bodies. The presence of excess phosphate in a glycogen molecule might favor further perturbations in metabolism and structure, putting a subset of molecules on the "fast-track" for becoming part of Lafora bodies. Only in the later stages, as in the 12-month-old *Epm2a*$^{-/-}$ mice where some 75 % of muscle glycogen can be harvested by low speed centrifugation, do the bulk properties of the glycogen transition to those characteristic of Lafora bodies.

Glycogen over-accumulation in the brain has been noted as contributing to neural defects in glycogenoses such as Adult Polyglucosan Body Disease. Recently, Duran et al. [74] reported an elegant study in which an activated form of muscle glycogen synthase, achieved by mutating all nine phosphorylated Ser residues to Ala, was expressed in neurons of flies or mice. The resulting over-accumulation of glycogen in neurons led to neuronal loss and neurological defects. Two main lines of reasoning support the idea that glycogen contributes to the pathology of Lafora disease. First, studies of laforin function and properties, both in vitro and in mice, argue strongly that laforin has evolved to act physiologically as a glycogen phosphatase. It is hard to overlook laforin's uniqueness as the only phosphatase known to bind directly to glycogen and the fact that defects in glycogen phosphorylation and structure accrue in vivo from its absence. Secondly, genetic reduction of glycogen levels in mice lacking a functional laforin gene alleviates both the formation of Lafora bodies and the onset of neurological symptoms. The type 1 phosphatase glycogen targeting subunit PTG is widely expressed and its disruption in mice causes decreased glycogen accumulation in several tissues including brain [75].

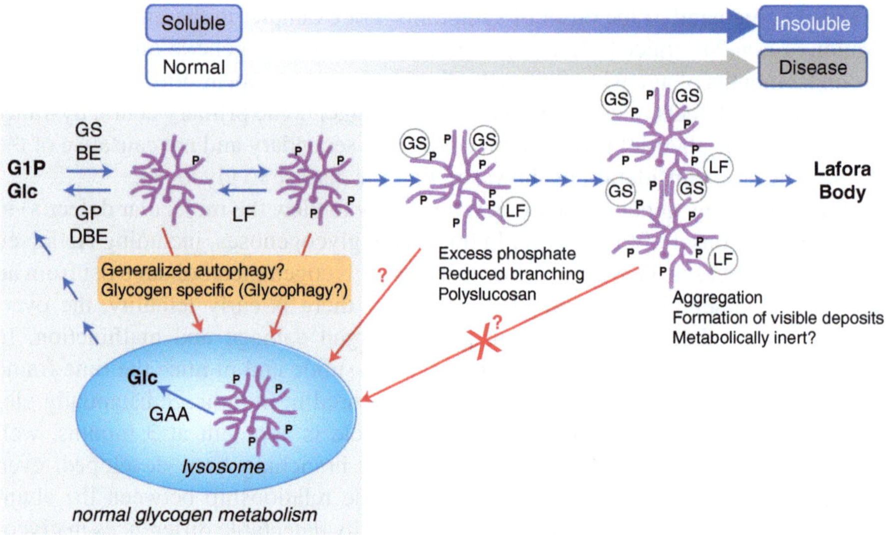

Fig. 13.3 Formation of polyglucosan and Lafora bodies. The intent of the figure is to emphasis the transition from the normal metabolism of soluble glycogen to the gradual and irreversible formation of the poorly structured glycogen characteristic of Lafora disease. Some of the general ideas would of course be relevant to other glycogenoses in which insoluble glycogen accumulates. In Lafora disease, the defects in glycogen structure seem to be related to glycogen phosphorylation which is opposed by the actions of laforin (LF) and malin. Poorly branched, hyperphosphorylated glycogen, which is less soluble, ultimately can no longer undergo degradation by the usual pathways, resulting in the sequestration of metabolic proteins like glycogen synthase (GS), phosphorylase (GP), and laforin, in metabolically inert deposits. In sensitive cells like neurons, dysfunction and cell death lead to the onset of disease. *Glc-1-P* glucose-1-P, *Glc* glucose, *BE* branching enzyme, *DBE* debranching enzyme or AGL, *GAA* lysosomal α-glucosidase

Turnbull et al. [76] described how deletion of the PTG gene in an *Epm2a*$^{-/-}$ background returned the elevated muscle glycogen levels of 12-month-old *Epm2a*$^{-/-}$ mice to wild-type values while substantially suppressing the formation of Lafora bodies. Concomitantly, neurological deficits observed in the laforin knockout mice were eliminated in the double knockout animals. This result makes a powerful argument that the pathology associated with the lack of laforin function is indeed linked to abnormal glycogen if not the formation of Lafora bodies. The only other explanation would require that laforin and PTG, both proteins with evolved glycogen-binding domains, collaborate in some function unrelated to glycogen metabolism. Interestingly, a Lafora patient with an *EPM2B* mutation was found who also carried a mutation in the PTG gene that in tissue culture led to decreased glycogen levels [77]. This patient exhibited a milder form of the disease, consistent with the results obtained with the *Epm2a*$^{-/-}$ *PTG*$^{-/-}$ double knockout mice.

The arguments outlined above lead us to the conclusion that laforin functions physiologically as a glycogen phosphatase, that the absence of laforin activity leads to aberrant glycogen whose presence correlates with neurological dysfunction, and that Lafora disease is, in essence, a form of glycogenosis.

Does Laforin Have Additional Functions?

Even if laforin has evolved to serve as a glycogen phosphatase, could it have other roles? The answer is, of course, yes. There is a growing list of proteins with true secondary or "moonlighting" activities [78]. More common, however, is the existence of truly multifunctional proteins, whose activity or other function can be applied in multiple contexts. Prime examples are protein kinases, protein phosphatases, and ubiquitin ligases that can exist in multiple complexes with distinct functions. Substrate selectivity is achieved in part through the catalytic site itself, but with additional specificity frequently imparted by the other associated proteins. In this way, a single protein can be involved in diverse cellular processes via its participation in distinct protein complexes that can have their own individual regulatory inputs. The type 1 protein phosphatase family, already discussed, is an excellent example in which one of three very similar catalytic subunits partners with over 200 ancillary subunits to serve a myriad of distinguishable functions [79]. Proteins can also serve in multiple functionally distinct complexes by virtue of their protein-protein interactions, their ability to act as "scaffolds," and by this mechanism also participate in diverse cellular processes without contributing enzymatic activity. Could laforin's contribution to Lafora disease, therefore, arise from some such function separate from its role as a glycogen phosphatase?

One recent suggestion for an additional laforin function, based mainly on analysis of *Epm2a*$^{-/-}$ mice, was that laforin can down-regulate insulin signaling; thus, in its absence, glycogen synthesis might be enhanced due to increased insulin signaling [80]. A second, independent study of these mice and genetically matched wild-type controls, however, found no such alterations in insulin sensitivity [81] and a physiologically relevant role in insulin signaling is, in our opinion, unlikely. Several potential laforin-interacting proteins have been described, as is summarized by Gentry et al. [11]. Of these candidates (malin, PTG, GL, R6, glycogen synthase, GSK-3, and AMPK β-subunit), most are known or possible glycogen-binding proteins in their own right. Almost all were identified by yeast two-hybrid screening and/or immunoprecipitation from cultured cells over-expressing the various proteins. An ongoing and nagging worry in assessing interactions among glycogen-associating proteins is possible interference by glycogen itself acting as a molecular bridge. Typical yeast two-hybrid reporter strains produce glycogen and the presence of trace amounts of glycogen in the nucleus is difficult to exclude with absolute certainty. Similarly, while cultured cells do not normally accumulate much glycogen, it can undoubtedly be present. Therefore, more than the usual caution in validating potential interacting proteins is probably warranted in this area.

One laforin-interacting protein described in multiple studies is malin [12], a protein not known to be able to bind to glycogen directly. After identification by yeast two-hybrid screening, this interaction was confirmed by co-immunoprecipitation and direct protein binding [82]. The idea has evolved that laforin and malin form a functional complex (reviewed in [12]). This is a compelling hypothesis since it would provide a molecular basis for the clear genetic relationship between the malin and laforin genes as causes of Lafora disease. It would explain why defects in either malin or laforin can lead to very similar phenotypes and pathologies. It could explain how defective laforin might affect processes unrelated to glycogen metabolism. A teleological rationale for complex formation is provided by the suggestion that laforin binding to glycogen recruits malin and thereby allows it access to a number of potential ubiquitylation targets associated with the polysaccharide. It has been proposed that the malin-laforin association is a regulated process. AMPK has been reported to phosphorylate laforin [83, 84] and enhance malin binding, thus promoting modification of potential malin targets such as laforin, PTG, and glycogen synthase. However, activation of muscle AMPK by exercise did not cause changes in laforin, PTG, or glycogen synthase protein levels [85].

It is important to the field to establish whether or not laforin and malin indeed act together as a complex, either exclusively or partially, in vivo. Note that if, as some propose, laforin is a substrate for malin, then a malin-laforin complex might be akin to a transient enzyme-substrate complex rather than a stable entity, and correspondingly may prove difficult to identify. Though good evidence for malin-laforin association has been obtained from cell over-expression experiments with epitope-tagged malin, it has been challenging to analyze endogenous malin. To our knowledge, no one has succeeded in obtaining or producing antibodies able to detect endogenous malin whether in cells or tissues. Therefore, it has not been possible to validate or disprove the existence of endogenous laforin-malin complexes in vivo, particularly in tissues, by co-localization and/or co-immunoprecipitation experiments.

Malin

One of the least understood areas of Lafora research is the physiological function of malin and, based on its potential role as a ubiquitin ligase, its protein targets. Malin contains an N-terminal RING finger domain of the C3HC4 family, characteristic of E3 ubiquitin ligases, flanked by six NHL domains that are believed to mediate protein-protein interactions ([45]; Fig. 13.2). It has been shown to be active biochemically as a ubiquitylating enzyme [82, 86, 87] and several proteins involved in glycogen metabolism have been proposed as targets of malin action. These include laforin [82], glycogen synthase [88], PTG [84, 88, 89], the debranching enzyme (AGL) [90], AMPK [86], and neuronatin [91]. All would be modified by K48-linked ubiquitin and targeted for proteosomal degradation except the AMPK β-subunit which was reported to be modified by K63-linked ubiquitin [86]. These studies, for the most part, have been performed using cultured cell systems and so it was of interest to examine the levels of these putative malin substrates in vivo in mice with the *Epm2b* gene disrupted. If malin acts in vivo to target these proteins for degradation, one might expect, even from early stages of development, that the protein levels would be increased in the absence of malin.

To date, four independent *Epm2b*$^{-/-}$ mouse models have been developed and, although different published studies have emphasized different aspects of the phenotype, many of the findings are quite consistent [81, 92–95]. All of the mouse lines develop Lafora bodies in multiple tissues and, as the mice age, glycogen over-accumulates in an insoluble fraction that can be harvested by low speed centrifugation. Minassian and colleagues [92, 95] reported that the phosphate content of glycogen from *Epm2b*$^{-/-}$ mice was elevated, though perhaps to a slightly lesser degree than in *Epm2a*$^{-/-}$ mice. We have confirmed this observation (Irimia, DePaoli-Roach and Roach, unpublished results). This result is important in linking an increase in glycogen phosphorylation to loss of function of both malin and laforin.

The first report on *Epm2b*$^{-/-}$ mice, by DePaoli-Roach et al. [81], analyzed 3-month-old animals and found no differences in the skeletal muscle protein levels of PTG, glycogen synthase, and AGL as compared to wild-type animals. These results suggested that in vivo, PTG, glycogen synthase, and AGL are not targeted for degradation by malin. The total laforin level however increased, especially in brain. There was also a clear redistribution of laforin from a low speed supernatant to the insoluble low speed pellet. Since then, the increase in laforin and its association with insoluble glycogen was observed also in the other three *Epm2b*$^{-/-}$ mouse models [92–95], some differences likely depending on the age of the mice analyzed and perhaps procedural variations from laboratory to laboratory.

The major point of discussion is the reason for the increased laforin. One school of thought is that malin ubiquitylates laforin and targets it for proteolytic degradation, consistent with the in vitro and cell-based analyses noted above. Therefore, in the absence of malin, laforin degradation would be reduced and its level correspondingly increased. Evidence in favor of this hypothesis includes the observation by Tiberia et al. [95] of elevated laforin in 1-month-old knockout mice in which classic Lafora bodies are not yet seen histochemically. Similarly, Criado et al. [94] reported increased laforin in brain tissue from 16-day-old malin knockout mice. One might note that glycogen structure can be defective well before its deposition as a multimolecular conglomerate that is a Lafora body.

A second interpretation, which we proposed [85] based on our studies of the 3-month-old *Epm2b*$^{-/-}$ mice and more recently of 6–7-month-old animals (Segvich, Roach and DePaoli-Roach, unpublished data), is that laforin is simply sequestered by binding to abnormal glycogen and thereby protected from degradation. This would be sufficient to account for the increased laforin levels in malin knockout mice. In addition, laforin was depleted from the soluble fraction, where it would normally function to remove phosphate from soluble glycogen, thus providing an explanation for how loss of malin could correlate with loss of function of laforin. There is good precedent for the level of glycogen-associated proteins in general, and laforin in particular, tracking with the amount of accumulated glycogen. When muscle glycogen levels are manipulated genetically by mutations of glycogen synthase or the glycogen-associated PP1 targeting subunit R_{GL} [96], the laforin concentration follows the glycogen level [48], increased glycogen correlating with increased laforin. These changes in laforin level presumably occur in the presence of malin, since the *Epm2b* gene is intact. A similar scenario holds for the glycogen-binding protein Stbd1: when muscle or liver glycogen is eliminated by disruption of the corresponding glycogen synthase gene, the liver or muscle Stbd1 levels are reduced [28].

In the first analyses of *Epm2b*$^{-/-}$ mice, at 3 [81] or 6 [92] months of age, there was no change in glycogen synthase protein levels but it appears that, as the mice age, the glycogen synthase levels do increase ([93–95]; Segvich, Roach and DePaoli-Roach, unpublished data). Similarly, the levels of glycogen phosphorylase are also increased in both malin [85, 93] and laforin knockout mice in brain and skeletal muscle (Segvich, Roach and DePaoli-Roach, unpublished data) even though there is no evidence that phosphorylase is a substrate for malin. The same debate posed for laforin can be applied to glycogen synthase and phosphorylase. Is the increase because they are malin substrates or because they are segregated with the over-accumulated, aberrant glycogen and removed from their normal degradation pathway? An argument for the latter explanation is that the same pattern of increased glycogen synthase and phosphorylase, associated with insoluble glycogen, is seen in laforin knockout mice which have an intact malin gene ([65]; Segvich, Roach and DePaoli-Roach, unpublished data). In both laforin and malin knockout animals, the hyperaccumulated glycogen synthase, based on the amount of protein present, has much lower than expected enzyme activity as measured by the standard assay [65, 93]—in other words, it has a low specific activity. Whether this is a technical issue with the assay or some stable modification of the enzyme is unclear.

Since the initial suggestion that malin might ubiquitylate laforin and target its degradation [82], it has been hard to rationalize how increased laforin levels could produce a similar phenotype to loss-of-function mutations in the *Epm2a* gene. Yet, as discussed above, there is a good consensus that laforin protein is increased in the *Epm2b*$^{-/-}$ mouse models of Lafora disease. Tiberia et al. [95] have recently proposed a model in which excess laforin, resulting from lack of malin-directed proteolysis, is actively involved in the generation of polyglucosan and abnormal glycogen deposits. Laforin, in this model, would bind to glycogen and disrupt its structure somewhat analogously to phosphorylation. The key experimental evidence for this proposal comes from transient over-expression of laforin in HEK293 cells. Although the total glycogen measured biochemically was little changed, over-expression of wild-type laforin or the catalytically inactive C265S mutant correlated with the appearance of PAS-stained structures in the cells. The PAS-stained structures were amylase resistant and had reduced branching according to iodine spectra. These structures were not observed after expression of laforin mutants disabled for glycogen binding. These are interesting data that merit further exploration. It will be important to see whether over-expression of laforin at levels comparable to what is seen in *Epm2b*$^{-/-}$ mice also generates these punctate glycogen deposits. The model of Tiberia et al. [95] also raises an issue that has not been addressed in previous work. Their scheme suggests that malin functions to cause laforin degradation once it has catalyzed glycogen dephosphorylation, following on from some of the original suggestions of Gentry et al. [82]. In the extreme, this would imply that laforin as an enzyme has limited catalytic turnover of substrates, being degraded after it has acted. If so, one might expect laforin to have a relatively short half-life. In any event, it would be valuable to have information on the half-life of laforin in wild-type and *Epm2b*$^{-/-}$ mice.

Autophagy

From traditional genetic screens, Wang et al. [97] linked autophagy with glycogen accumulation in *S. cerevisiae* in 2001. In subsequent work, some 60 genes that affected yeast glycogen levels were identified as being involved in vesicular trafficking and vacuolar function [34]. The importance of lysosomal glycogen disposal in humans, as noted above, is underscored by the severity of Pompe disease [17]. Insofar as Lafora disease is a glycogenosis, then impaired lysosomal glycogen disposal could contribute to the pathology ([48]; Fig. 10.1). Aguado et al. [98] linked an impairment of autophagy with laforin by analyzing LC-3-II in human fibroblasts from Lafora patients with mutated laforin as well as embryonic fibroblasts (MEFs) from *Epm2a*$^{-/-}$ mice. In this study, they implicated the mTOR pathway. Puri et al. [99] proposed involvement of endosomal-lysosomal and autophagy pathways in the pathology of *Epm2a*$^{-/-}$ mice, but did not report alterations in autophagosome formation as monitored by LC3-II production. A similar suppression of autophagy was observed in the brains of *Epm2b*$^{-/-}$ mice and in MEFs from these animals [94]. However, in this case, the effects seemed independent of the mTOR pathway. Clearly, some details of the mechanisms still need to be worked out, but there is evidence to link mutations associated with Lafora disease with alterations in autophagic processes. As detailed above, autophagy, or better macroautophagy, is really a collection of processes involved in the lysosomal disposal of a variety of cellular constituents. de Cordoba and colleagues [100], who have been strong proponents of the concept that laforin and malin are positive regulators of autophagy, would certainly view its defects in Lafora disease as potentially having more wide ranging impact than only modulating glycogen disposal. They also argue that defects in autophagy precede the detection of classically defined Lafora bodies and could have independent links to neurodegeneration. Similarly, Ganesh and colleagues would argue for a broader role for the lysosomal pathway, including the disposal of proteins, as well as an active, intermediate role for Lafora bodies in instigating the pathology of the disease [101]. These views, however, could certainly embrace a role for autophagy, or some subset of autophagy, in the clearance of abnormal glycogen.

Polyglucosan and Lafora Body Formation

The correlation between Lafora bodies and Lafora disease has been known for as long as the disease itself. The model in Fig. 13.3 attempts to provide a view of the progression of normal glycogen to the insoluble, structurally abnormal polymer associated with the disease. In a soluble compartment, glycogen molecules undergo repeated cycles of breakdown and resynthesis and maintain their branching structure within some acceptable range. There is also a lysosomal pathway for glycogen disposal whose importance relative to phosphorylase-mediated degradation is not really known and may vary with cell type and prevailing metabolic status. The function of this pathway may be for the generation of glucose, under certain conditions, or for the recycling of glycogen molecules whose structure deviates too far from normal. Glycogen contains trace amounts of esterified phosphate, likely resulting from the action of glycogen synthase. Laforin counteracts this phosphorylation which, in mouse muscle, is maintained at a constant low level from 3 to 12 months of age. In Lafora disease, glycogen phosphorylation is unchecked, because of the absence of laforin due to *EPM2A* mutation or by some other mechanism when malin is defective, as discussed in this chapter. Excessive phosphorylation of glycogen disrupts the hydrogen-bonding structure of the polyglucose helices of the polysaccharide, rendering it less water soluble and correlating with decreased branching frequency. We envision a gradual transition from glycogen molecules that fall within the normal range to what is typically defined as polyglucosan. At some point during this transition, the glycogen molecule will no longer be amenable to its usual degradative pathways, by phosphorylase or via lysosomal disposal, and it would become insoluble and develop into a metabolically inert state. Glycogen metabolizing enzymes, such as glycogen synthase, phosphorylase, and laforin, bound to the abnormal glycogen are effectively sequestered and unable to perform their normal functions. The insoluble polyglucosan aggregates to form larger structures that ultimately are visible by light microscopy. At some stage, the aberrant glycogen plus associated proteins would progress to generate the classic Lafora body. Tragic as Lafora disease is, an immediate question is why the onset of symptoms does not occur even earlier since the genetic defects are present during embryogenesis and immediately after birth. If Lafora disease is indeed a glycogenosis, then a model as in Fig. 13.3 can partly explain the later onset of the disease. Only after multiple cycles of synthesis and degradation do the structural defects in glycogen become severe. Because glycogen phosphorylation, even in the absence of laforin, is a relatively rare event, mechanisms for glycogen disposal—one might speculate particularly lysosomal disposal—provide a buffer to the over-accumulation of pathological glycogen deposits. Ultimately, however, the abnormal glycogen accumulation prevails and, in neurons, leads to cell death and the initiation of symptoms.

Conclusion

Lafora disease research continues to be a remarkably active field and progress has been strong the last few years. While different research groups will surely not agree on every experimental outcome and interpretation, there is definitely convergence on some key points. Most groups now consider that Lafora disease is, at least in part, a glycogenosis and that Lafora bodies contribute to the pathology. There is good consistency in the experimental finding that elevated laforin bound to increased amounts of abnormal glycogen accompanies the loss of malin in a mouse model of the disease. Most seem to accept a possible link between laforin and/or malin and autophagy. There seems a general acceptance that laforin is a glycogen phosphatase and that glycogen hyperphosphorylation is observed when malin or laforin is defective. The idea that polyglucosan formation can be explained solely by the "imbalance" hypothesis, increased elongation by glycogen synthase compared to branching activity, is no longer a leading theory.

In our opinion, the following are critical outstanding questions/controversies in the field. (1) What is the function of malin and what are its physiological substrates? Is the elevation of laforin levels in malin knockout mice due to its being a malin substrate or due to its sequestration in metabolically inactive polyglucosan deposits? (2) The genetic linkage between the *Epm2a* and *Epm2b* genes is well established, first by the symptoms of the disease but also by a number of other assessments of a common phenotype in the laforin and malin knockout mouse models—these include aberrant glycogen metabolism and phosphorylation as well as neurological defects. Does physical interaction between malin and laforin proteins, as enzyme and substrate or as a malin-laforin complex, underlie the genetic linkage? (3) There is solid evidence that laforin is a glycogen phosphatase. Does it have other roles, perhaps by virtue of its interaction with malin? Can excess laforin of itself disrupt glycogen structure? (4) How do defects in autophagy, or in some specific autophagic pathway, contribute to Lafora disease? Does autophagy or an autophagy-like pathway impact polyglucosan accumulation and/or other relevant cellular processes? All of these questions warrant additional investigation in the future.

Acknowledgments Research from the authors' laboratories was supported by NIH grants R37 DK27221, R01 NS056454, and R21 HL108301.

References

1. Maehama T, Taylor GS, Dixon JE (2001) PTEN and myotubularin: novel phosphoinositide phosphatases. Annu Rev Biochem 70:247–279
2. Lafora GR, Glueck B (1911) Beitrag zur histopathologie der myoklonischen epilepsie. Z Gesamte Neurol Psychiatr 6:1–14
3. Gentry MS, Dixon JE, Worby CA (2009) Lafora disease: insights into neurodegeneration from plant metabolism. Trends Biochem Sci 34(12):628–639
4. Ganesh S, Puri R, Singh S, Mittal S, Dubey D (2006) Recent advances in the molecular basis of Lafora's progressive myoclonus epilepsy. J Hum Genet 51(1):1–8
5. Delgado-Escueta AV (2007) Advances in Lafora progressive myoclonus epilepsy. Curr Neurol Neurosci Rep 7(5):428–433
6. Andrade DM, Turnbull J, Minassian BA (2007) Lafora disease, seizures and sugars. Acta Myol 26(1):83–86
7. Ramachandran N, Girard JM, Turnbull J, Minassian BA (2009) The autosomal recessively inherited progressive myoclonus epilepsies and their genes. Epilepsia 50(suppl 5):29–36
8. Collins GH, Cowden RR, Nevis AH (1968) Myoclonus epilepsy with Lafora bodies. An ultrastructural and cytochemical study. Arch Pathol 86(3):239–254
9. Ball S, Guan HP, James M, Myers A, Keeling P, Mouille G et al (1996) From glycogen to amylopectin: a model for the biogenesis of the plant starch granule. Cell 86(3):349–352
10. Zeeman SC, Smith SM, Smith AM (2007) The diurnal metabolism of leaf starch. Biochem J 401(1):13–28

11. Gentry MS, Roma-Mateo C, Sanz P (2012) Laforin, a protein with many faces: glucan phosphatase, adapter protein, and others. FEBS J 280(2):525–537
12. Roma-Mateo C, Sanz P, Gentry MS (2012) Deciphering the role of malin in the Lafora progressive myoclonus epilepsy. IUBMB Life 64(10):801–808
13. Roach PJ, Depaoli-Roach AA, Hurley TD, Tagliabracci VS (2012) Glycogen and its metabolism: some new developments and old themes. Biochem J 441(3):763–787
14. Roach PJ, Skurat AV (1997) Self-glucosylating initiator proteins and their role in glycogen biosynthesis. Prog Nucleic Acid Res Mol Biol 57:289–316
15. Smythe C, Cohen P (1991) The discovery of glycogenin and the priming mechanism for glycogen biogenesis. Eur J Biochem 200(3):625–631
16. Lomako J, Lomako WM, Whelan WJ (2004) Glycogenin: the primer for mammalian and yeast glycogen synthesis. Biochim Biophys Acta 1673(1–2):45–55
17. Hirschhorn R, Reuser AJ (2000) Glycogen storage disease type II: acid alpha-glucosidase (acid maltase) deficiency. In: Scriver CR, Beaudet AL, Sly WS, Valle D (eds) He metabolic and molecular basis of inherited disease. McGraw-Hill, New York, pp 3389–3420
18. Yang Z, Klionsky DJ (2009) An overview of the molecular mechanism of autophagy. Curr Top Microbiol Immunol 335:1–32
19. Yang Z, Klionsky DJ (2010) Mammalian autophagy: core molecular machinery and signaling regulation. Curr Opin Cell Biol 22(2):124–131
20. Nakatogawa H, Suzuki K, Kamada Y, Ohsumi Y (2009) Dynamics and diversity in autophagy mechanisms: lessons from yeast. Nat Rev Mol Cell Biol 10(7):458–467
21. Rabinowitz JD, White E (2010) Autophagy and metabolism. Science 330(6009):1344–1348
22. Mizushima N, Yoshimori T, Ohsumi Y (2011) The role of Atg proteins in autophagosome formation. Annu Rev Cell Dev Biol 27:107–132
23. Raben N, Roberts A, Plotz PH (2007) Role of autophagy in the pathogenesis of Pompe disease. Acta Myol 26(1):45–48
24. Johansen T, Lamark T (2011) Selective autophagy mediated by autophagic adapter proteins. Autophagy 7(3):279–296
25. Komatsu M, Ichimura Y (2010) Selective autophagy regulates various cellular functions. Genes Cells 15(9):923–933
26. Geng J, Klionsky DJ (2008) The Atg8 and Atg12 ubiquitin-like conjugation systems in macroautophagy. 'Protein modifications: beyond the usual suspects' review series. EMBO Rep 9(9):859–864
27. Noda NN, Ohsumi Y, Inagaki F (2010) Atg8-family interacting motif crucial for selective autophagy. FEBS Lett 584(7):1379–1385
28. Jiang S, Heller B, Tagliabracci VS, Zhai L, Irimia JM, DePaoli-Roach AA et al (2010) Starch binding domain-containing protein 1/genethonin 1 is a novel participant in glycogen metabolism. J Biol Chem 285(45):34960–34971
29. Christiansen C, Abou Hachem M, Janecek S, Vikso-Nielsen A, Blennow A, Svensson B (2009) The carbohydrate-binding module family 20—diversity, structure, and function. FEBS J 276(18):5006–5029
30. Jiang S, Wells CD, Roach PJ (2011) Starch-binding domain-containing protein 1 (Stbd1) and glycogen metabolism: identification of the Atg8 family interacting motif (AIM) in Stbd1 required for interaction with GABARAPL1. Biochem Biophys Res Commun 413(3):420–425
31. Chen Y-T, Burchell A (1995) Glycogen storage diseases. In: Scriver CR, Beaudet AL, Sly WS, Valle D (eds) The metabolic and molecular bases of inherited disease, 7th edn. McGraw-Hill, New York, pp 935–965
32. DiMauro S, Lamperti C (2001) Muscle glycogenoses. Muscle Nerve 24(8):984–999
33. DiMauro S, Spiegel R (2011) Progress and problems in muscle glycogenoses. Acta Myol 30(2):96–102
34. Wilson WA, Wang Z, Roach PJ (2002) Systematic identification of the genes affecting glycogen storage in the yeast Saccharomyces cerevisiae: implication of the vacuole as a determinant of glycogen level. Mol Cell Proteomics 1(3):232–242

35. Andersen DH (1956) Familial cirrhosis of the liver with storage of abnormal glycogen. Lab Invest 5(1):11–20
36. Moses SW, Parvari R (2002) The variable presentations of glycogen storage disease type IV: a review of clinical, enzymatic and molecular studies. Curr Mol Med 2(2):177–188
37. Nakajima H, Raben N, Hamaguchi T, Yamasaki T (2002) Phosphofructokinase deficiency; past, present and future. Curr Mol Med 2(2):197–212
38. McCue ME, Valberg SJ, Lucio M, Mickelson JR (2008) Glycogen synthase 1 (GYS1) mutation in diverse breeds with polysaccharide storage myopathy. J Vet Intern Med 22(5):1228–1233
39. Manchester J, Skurat AV, Roach P, Hauschka SD, Lawrence JC Jr (1996) Increased glycogen accumulation in transgenic mice overexpressing glycogen synthase in skeletal muscle. Proc Natl Acad Sci U S A 93(20):10707–10711
40. Pederson BA, Csitkovits AG, Simon R, Schroeder JM, Wang W, Skurat AV et al (2003) Overexpression of glycogen synthase in mouse muscle results in less branched glycogen. Biochem Biophys Res Commun 305(4):826–830
41. Raben N, Danon M, Lu N, Lee E, Shliselfeld L, Skurat AV et al (2001) Surprises of genetic engineering: a possible model of polyglucosan body disease. Neurology 56(12):1739–1745
42. Pederson BA, Cope CR, Irimia JM, Schroeder JM, Thurberg BL, Depaoli-Roach AA et al (2005) Mice with elevated muscle glycogen stores do not have improved exercise performance. Biochem Biophys Res Commun 331(2):491–496
43. Minassian BA, Lee JR, Herbrick JA, Huizenga J, Soder S, Mungall AJ et al (1998) Mutations in a gene encoding a novel protein tyrosine phosphatase cause progressive myoclonus epilepsy. Nat Genet 20(2):171–174
44. Serratosa JM, Gomez-Garre P, Gallardo ME, Anta B, de Bernabe DB, Lindhout D et al (1999) A novel protein tyrosine phosphatase gene is mutated in progressive myoclonus epilepsy of the Lafora type (EPM2). Hum Mol Genet 8(2):345–352
45. Chan EM, Young EJ, Ianzano L, Munteanu I, Zhao X, Christopoulos CC et al (2003) Mutations in NHLRC1 cause progressive myoclonus epilepsy. Nat Genet 35(2):125–127
46. Garcia-Cabrero AM, Marinas A, Guerrero R, de Cordoba SR, Serratosa JM, Sanchez MP (2012) Laforin and malin deletions in mice produce similar neurologic impairments. J Neuropathol Exp Neurol 71(5):413–421
47. Gomez-Abad C, Gomez-Garre P, Gutierrez-Delicado E, Saygi S, Michelucci R, Tassinari CA et al (2005) Lafora disease due to EPM2B mutations: a clinical and genetic study. Neurology 64(6):982–986
48. Wang W, Lohi H, Skurat AV, DePaoli-Roach AA, Minassian BA, Roach PJ (2007) Glycogen metabolism in tissues from a mouse model of Lafora disease. Arch Biochem Biophys 457(2):264–269
49. Ceulemans H, Bollen M (2004) Functional diversity of protein phosphatase-1, a cellular economizer and reset button. Physiol Rev 84(1):1–39
50. Wang J, Stuckey JA, Wishart MJ, Dixon JE (2002) A unique carbohydrate binding domain targets the Lafora disease phosphatase to glycogen. J Biol Chem 277(4):2377–2380
51. Wang W, Roach PJ (2004) Glycogen and related polysaccharides inhibit the laforin dual-specificity protein phosphatase. Biochem Biophys Res Commun 325(3):726–730
52. Chan EM, Ackerley CA, Lohi H, Ianzano L, Cortez MA, Shannon P et al (2004) Laforin preferentially binds the neurotoxic starch-like polyglucosans, which form in its absence in progressive myoclonus epilepsy. Hum Mol Genet 13(11):1117–1129
53. Ganesh S, Tsurutani N, Suzuki T, Hoshii Y, Ishihara T, Delgado-Escueta AV et al (2004) The carbohydrate-binding domain of Lafora disease protein targets Lafora polyglucosan bodies. Biochem Biophys Res Commun 313(4):1101–1109
54. Minassian BA, Ianzano L, Meloche M, Andermann E, Rouleau GA, Delgado-Escueta AV et al (2000) Mutation spectrum and predicted function of laforin in Lafora's progressive myoclonus epilepsy. Neurology 55(3):341–346
55. Alonso A, Sasin J, Bottini N, Friedberg I, Friedberg I, Osterman A et al (2004) Protein tyrosine phosphatases in the human genome. Cell 117(6):699–711

56. Ganesh S, Agarwala KL, Ueda K, Akagi T, Shoda K, Usui T et al (2000) Laforin, defective in the progressive myoclonus epilepsy of Lafora type, is a dual-specificity phosphatase associated with polyribosomes. Hum Mol Genet 9(15):2251–2261
57. Lohi H, Ianzano L, Zhao XC, Chan EM, Turnbull J, Scherer SW et al (2005) Novel glycogen synthase kinase 3 and ubiquitination pathways in progressive myoclonus epilepsy. Hum Mol Genet 14(18):2727–2736
58. Wang Y, Liu Y, Wu C, Zhang H, Zheng X, Zheng Z et al (2006) Epm2a suppresses tumor growth in an immunocompromised host by inhibiting Wnt signaling. Cancer Cell 10(3):179–190
59. Worby CA, Gentry MS, Dixon JE (2006) Laforin: a dual specificity phosphatase that dephosphorylates complex carbohydrates. J Biol Chem 281:30412–30418
60. Tagliabracci VS, Turnbull J, Wang W, Girard JM, Zhao X, Skurat AV et al (2007) Laforin is a glycogen phosphatase, deficiency of which leads to elevated phosphorylation of glycogen in vivo. Proc Natl Acad Sci U S A 104(49):19262–19266
61. Blennow A, Nielsen TH, Baunsgaard L, Mikkelsen R, Engelsen SB (2002) Starch phosphorylation: a new front line in starch research. Trends Plant Sci 7(10):445–450
62. Fontana JD (1980) The presence of phosphate in glycogen. FEBS Lett 109(1):85–92
63. Lomako J, Lomako WM, Kirkman BR, Whelan WJ (1994) The role of phosphate in muscle glycogen. Biofactors 4(3–4):167–171
64. Tagliabracci VS, Heiss C, Karthik C, Contreras CJ, Glushka J, Ishihara M et al (2011) Phosphate incorporation during glycogen synthesis and Lafora disease. Cell Metab 13(3):274–282
65. Tagliabracci VS, Girard JM, Segvich D, Meyer C, Turnbull J, Zhao X et al (2008) Abnormal metabolism of glycogen phosphate as a cause for Lafora disease. J Biol Chem 283(49):33816–33825
66. Hejazi M, Fettke J, Kotting O, Zeeman SC, Steup M (2010) The laforin-like dual-specificity phosphatase SEX4 from arabidopsis hydrolyzes both C6- and C3-phosphate esters introduced by starch-related dikinases and thereby affects phase transition of alpha-glucans. Plant Physiol 152(2):711–722
67. Niittyla T, Comparot-Moss S, Lue WL, Messerli G, Trevisan M, Seymour MD et al (2006) Similar protein phosphatases control starch metabolism in plants and glycogen metabolism in mammals. J Biol Chem 281(17):11815–11818
68. Gessler K, Uson I, Takaha T, Krauss N, Smith SM, Okada S et al (1999) V-amylose at atomic resolution: X-ray structure of a cycloamylose with 26 glucose residues (cyclomaltohexaicosaose). Proc Natl Acad Sci U S A 96(8):4246–4251
69. Ritte G, Heydenreich M, Mahlow S, Haebel S, Kotting O, Steup M (2006) Phosphorylation of C6- and C3-positions of glucosyl residues in starch is catalysed by distinct dikinases. FEBS Lett 580(20):4872–4876
70. Fettke J, Hejazi M, Smirnova J, Hochel E, Stage M, Steup M (2009) Eukaryotic starch degradation: integration of plastidial and cytosolic pathways. J Exp Bot 60(10):2907–2922
71. Roach PJ (2011) Are there errors in glycogen biosynthesis and is laforin a repair enzyme? FEBS Lett 585(20):3216–3218
72. Machado-Salas J, Avila-Costa MR, Guevara P, Guevara J, Duron RM, Bai D et al (2012) Ontogeny of Lafora bodies and neurocytoskeleton changes in laforin-deficient mice. Exp Neurol 236(1):131–140
73. Ganesh S, Delgado-Escueta AV, Sakamoto T, Avila MR, Machado-Salas J, Hoshii Y et al (2002) Targeted disruption of the Epm2a gene causes formation of Lafora inclusion bodies, neurodegeneration, ataxia, myoclonus epilepsy and impaired behavioral response in mice. Hum Mol Genet 11(11):1251–1262
74. Duran J, Tevy MF, Garcia-Rocha M, Calbo J, Milan M, Guinovart JJ (2012) Deleterious effects of neuronal accumulation of glycogen in flies and mice. EMBO Mol Med 4(8): 719–729
75. Zhai L, Choi CS, Irimia-Dominguez J, McGuire AC, Kim S, Bock CB et al (2007) Enhanced insulin sensitivity and energy expenditure in PPP1R3C (PTG) deleted mice. Diabetes 56(1):A62

76. Turnbull J, DePaoli-Roach AA, Zhao X, Cortez MA, Pencea N, Tiberia E et al (2011) PTG depletion removes Lafora bodies and rescues the fatal epilepsy of Lafora disease. PLoS Genet 7(4):e1002037
77. Guerrero R, Vernia S, Sanz R, Abreu-Rodriguez I, Almaraz C, Garcia-Hoyos M et al (2011) A PTG variant contributes to a milder phenotype in Lafora disease. PLoS One 6(6):e21294
78. Khersonsky O, Tawfik DS (2010) Enzyme promiscuity: a mechanistic and evolutionary perspective. Annu Rev Biochem 79:471–505
79. Bollen M, Peti W, Ragusa MJ, Beullens M (2010) The extended PP1 toolkit: designed to create specificity. Trends Biochem Sci 35(8):450–458
80. Vernia S, Heredia M, Criado O, Rodriguez de Cordoba S, Garcia-Roves PM, Cansell C et al (2011) Laforin, a dual specificity phosphatase involved in Lafora disease, regulates insulin response and whole-body energy balance in mice. Hum Mol Genet 20:2571–2584
81. DePaoli-Roach AA, Segvich DM, Meyer CM, Rahimi Y, Worby CA, Gentry MS et al (2012) Laforin and malin knockout mice have normal glucose disposal and insulin sensitivity. Hum Mol Genet 21(7):1604–1610
82. Gentry MS, Worby CA, Dixon JE (2005) Insights into Lafora disease: malin is an E3 ubiquitin ligase that ubiquitinates and promotes the degradation of laforin. Proc Natl Acad Sci U S A 102(24):8501–8506
83. Roma-Mateo C, Solaz-Fuster Mdel C, Gimeno-Alcaniz JV, Dukhande VV, Donderis J, Worby CA et al (2011) Laforin, a dual-specificity phosphatase involved in Lafora disease, is phosphorylated at Ser25 by AMP-activated protein kinase. Biochem J 439(2):265–275
84. Solaz-Fuster MC, Gimeno-Alcaniz JV, Ros S, Fernandez-Sanchez ME, Garcia-Fojeda B, Criado Garcia O et al (2008) Regulation of glycogen synthesis by the laforin-malin complex is modulated by the AMP-activated protein kinase pathway. Hum Mol Genet 17(5):667–678
85. DePaoli-Roach AA, Tagliabracci VS, Segvich DM, Meyer CM, Irimia JM, Roach PJ (2010) Genetic depletion of the malin E3 ubiquitin ligase in mice leads to Lafora bodies and the accumulation of insoluble laforin. J Biol Chem 285(33):25372–25381
86. Moreno D, Towler MC, Hardie DG, Knecht E, Sanz P (2010) The laforin-malin complex, involved in Lafora disease, promotes the incorporation of K63-linked ubiquitin chains into AMP-activated protein kinase beta subunits. Mol Biol Cell 21(15):2578–2588
87. Sharma J, Mulherkar S, Mukherjee D, Jana NR (2012) Malin regulates Wnt signaling pathway through degradation of dishevelled2. J Biol Chem 287(9):6830–6839
88. Vilchez D, Ros S, Cifuentes D, Pujadas L, Valles J, Garcia-Fojeda B et al (2007) Mechanism suppressing glycogen synthesis in neurons and its demise in progressive myoclonus epilepsy. Nat Neurosci 10(11):1407–1413
89. Worby CA, Gentry MS, Dixon JE (2008) Malin decreases glycogen accumulation by promoting the degradation of protein targeting to glycogen (PTG). J Biol Chem 283(7):4069–4076
90. Cheng A, Zhang M, Gentry MS, Worby CA, Dixon JE, Saltiel AR (2007) A role for AGL ubiquitination in the glycogen storage disorders of Lafora and Cori's disease. Genes Dev 21(19):2399–2409
91. Sharma J, Rao SN, Shankar SK, Satishchandra P, Jana NR (2011) Lafora disease ubiquitin ligase malin promotes proteasomal degradation of neuronatin and regulates glycogen synthesis. Neurobiol Dis 44(1):133–141
92. Turnbull J, Wang P, Girard JM, Ruggieri A, Wang TJ, Draginov AG et al (2010) Glycogen hyperphosphorylation underlies Lafora body formation. Ann Neurol 68(6):925–933
93. Valles-Ortega J, Duran J, Garcia-Rocha M, Bosch C, Saez I, Pujadas L et al (2011) Neurodegeneration and functional impairments associated with glycogen synthase accumulation in a mouse model of Lafora disease. EMBO Mol Med 3(11):667–681
94. Criado O, Aguado C, Gayarre J, Duran-Trio L, Garcia-Cabrero AM, Vernia S et al (2012) Lafora bodies and neurological defects in malin-deficient mice correlate with impaired autophagy. Hum Mol Genet 21(7):1521–1533
95. Tiberia E, Turnbull J, Wang T, Ruggieri A, Zhao XC, Pencea N et al (2012) Increased laforin and laforin binding to glycogen underlie Lafora body formation in malin-deficient Lafora disease. J Biol Chem 287(30):25650–25659

96. Suzuki Y, Lanner C, Kim JH, Vilardo PG, Zhang H, Yang J et al (2001) Insulin control of glycogen metabolism in knockout mice lacking the muscle-specific protein phosphatase PP1G/RGL. Mol Cell Biol 21(8):2683–2694
97. Wang Z, Wilson WA, Fujino MA, Roach PJ (2001) Antagonistic controls of autophagy and glycogen accumulation by Snf1p, the yeast homolog of AMP-activated protein kinase, and the cyclin-dependent kinase Pho85p. Mol Cell Biol 21(17):5742–5752
98. Aguado C, Sarkar S, Korolchuk VI, Criado O, Vernia S, Boya P et al (2010) Laforin, the most common protein mutated in Lafora disease, regulates autophagy. Hum Mol Genet 19(14):2867–2876
99. Puri R, Suzuki T, Yamakawa K, Ganesh S (2012) Dysfunctions in endosomal-lysosomal and autophagy pathways underlie neuropathology in a mouse model for Lafora disease. Hum Mol Genet 21(1):175–184
100. Knecht E, Criado-Garcia O, Aguado C, Gayarre J, Duran-Trio L, Garcia-Cabrero AM et al (2012) Malin knockout mice support a primary role of autophagy in the pathogenesis of Lafora disease. Autophagy 8(4):701–703
101. Puri R, Ganesh S (2012) Autophagy defects in Lafora disease: cause or consequence? Autophagy 8(2):289–290

Index

K.K. Bence (ed.), *Protein Tyrosine Phosphatase Control of Metabolism*,
DOI 10.1007/978-1-4614-7855-3, © Springer Science+Business Media New York 2013

MIX
Papier aus verantwortungsvollen Quellen
Paper from responsible sources
FSC® C105338

If you have any concerns about our products,
you can contact us on
ProductSafety@springernature.com

In case Publisher is established outside the EU,
the EU authorized representative is:
Springer Nature Customer Service Center GmbH
Europaplatz 3, 69115 Heidelberg, Germany

Printed by Libri Plureos GmbH
in Hamburg, Germany